FOUNDATION OF COGNITIVE RADIO SYSTEMS

Edited by **Samuel Cheng**

Foundation of Cognitive Radio Systems
Edited by Samuel Cheng

Published by InTech

Janeza Trdine 9, 51000 Rijeka, Croatia

Publishing Process Manager Bojan Rafaj

Technical Editor Teodora Smiljanic

Cover Designer InTech Design Team

Additional hard copies can be obtained from orders@intechopen.com

Foundation of Cognitive Radio Systems, Edited by Samuel Cheng
p. cm.
978-953-51-0268-7

For EU product safety concerns:
IN TECH d.o.o., Prolaz Marije Krucifikse Kozulić 3, 51000 Rijeka, Croatia,
info@intechopen.com or visit our website at intechopen.com.

Meet the editor

Samuel Cheng received the B.S. degree in Electrical and Electronic Engineering from the University of Hong Kong, and the M.Phil. degree in Physics and the M.S. degree in Electrical Engineering from Hong Kong University of Science and Technology and the University of Hawaii, Honolulu, respectively. He received the Ph.D. degree in Electrical Engineering from Texas A&M University in 2004. He worked in Microsoft Asia, China, and Panasonic Technologies Company, New Jersey, during the summers of 2000 and 2001. In 2004, he joined Advanced Digital Imaging Research, a research company based near Houston, Texas, as a Research Engineer and was promoted to Senior Research Engineer the next year. Since 2006, he joined the Telecommunications Program of the School of Electrical and Computer Engineering at the University of Oklahoma and is currently an assistant professor. He has been awarded five US patents in miscellaneous areas of signal processing and also coauthored a paper that received the 2006 IEEE Best Signal Processing Magazine Paper Award. He is a member of IEEE and Sigma Xi. His research interests include information theory, image/signal processing, and pattern recognition.

Contents

Preface

The wide proliferation of wireless communications unavoidably leads to the scarcity of frequency spectra. On one hand, wireless users become increasingly difficult to find available frequency bands for communications. On the other hand, many preallocated frequency bands are ironically under-utilized and thus the resources there are simply wasted. This situation leads to the introduction of cognitive radio, which was proposed in the last decade to address this dilemma.

Under the cognitive radio model, there are mainly two tiers of users. Primary users (PU), who typically require stable frequency spectra for communications, are licensed users and have the rights of priority in using certain frequency spectra. Secondary users (SU), who typically only need to access frequency spectra momentarily, are allowed to use frequency spectra only if they do not interfere with the PU. As a consequence, one can easily see that the ability of sensing an idle spectrum (i.e., spectrum sensing) and the ability to momentarily utilize a spectrum without interfering with PU (i.e., spectrum management) are two essential elements required for the success of cognitive radios.

In this book we have comprised 13 chapters from experts of this field to address these two essential elements. We have organized these chapters into four different parts. The first part, spectrum sensing foundation, collects four chapters on fundamental methods on spectrum sensing. The first chapter explains eigenvalue based detectors and also compares that with eigenvalue ratio based detectors. The second chapter provides a practical demonstration of spectrum sensing using cyclostationarity-based detector. It also includes experimental analysis under a realistic WiMAX environment. Chapter 3 explains the use of modulation classification techniques for cognitive radio. Hardware implementations are also described in detail. In Chapter 4, the authors review the most prominent link quality prediction techniques which have been published over the last fifteen years and point out the limitations of existing approaches.

In the second part, we move from sensing methods for individual SU to cooperative sensing methods in which multiple SU collaborate to attain better sensing performance. In the first chapter of this part, the author give an overview of cooperative sensing. They also identify research challenges and unsolved issues. In

Chapter 6, the second chapter of this part, the authors look into the possibility of improving spectrum sensing performance by exploiting multiuser diversity. They also describe a practical MAC protocol to coordinate transmissions between SU and a fusion center that handles fusion of sensing information.

The remaining two parts deal with the aspects of spectrum management. Part 3 includes chapters that layout the foundation of interference management. Chapter 7, the first chapter of Part 3, introduces partial response signaling as a tool for spectrum shaping in cognitive radio. It addresses the potential weakness of having large out-of-band spectrum components in conventional OFDM based systems. In Chapter 8, the second chapter of Part 3, the authors discuss the structure, reconfiguration, and the parameter selection when adopting the dynamic frequency-band allocation (DFBA) and and dynamic frequency-band reallocation (DFBR) for cognitive radios. They also demonstrate how the reconfigurability can be achieved either by a channel switch or by variable multipliers/commutators. In Chapter 9, the third chapter of Part 3, the authors present detail theoretical analysis of opportunistic spectrum access using continuous-time Markov chains. In particular, they explain how SU may be forced to terminate under different traffic scenarios.

The last part of this book includes chapters that focus more on the interaction among multiple SU, which naturally leads to the problem of resource allocation. Chapter 10, the first chapter of Part 4, considers different system models in which SUs compete for a chance to transmit simultaneously or orthogonally with the PU. On the basis of these models, the chapter also offers insights into how to design such scenario in a cognitive radio network environments. Chapter 11, the second chapter of Part 4, focuses on the issue of how to implement interference mitigation by power control techniques amongst multiple cognitive radios in the light of game theory. In Chapter 12, the third chapter of Part 4, the authors design a joint PHY layer cooperative spectrum sensing and MAC layer resource scheduling scheme. They also formulate the joint design as a non-convex optimization problem and derive the asymptotic optimum solution based on recent advances in duality optimization theory. The last chapter of this book is rather different from the rest but is equally important. The author presents a summary of standards activities related to cognitive radio systems and dynamic spectrum access systems.

Samuel Cheng, Ph.D.
Assistant Professor, School of Electrical and Computer Engineering,
University of Oklahoma, Tulsa
USA

Part 1

Spectrum Sensing Foundation

1

Exact and Asymptotic Analysis of Largest Eigenvalue Based Spectrum Sensing

Olav Tirkkonen and Lu Wei
Aalto University
Finland

1. Introduction

Cognitive radio (CR) is a promising technique for future wireless communication systems. In CR networks, dynamic spectrum access (DSA) of frequency is implemented to mitigate spectrum scarcity. Specifically, a secondary (unlicensed) user may be allowed to access the temporarily unused frequency bands granted to a primary (licensed) user. DSA has to be implemented so that the quality of service (QoS) promised to the primary user must be satisfied. The key point for this is the secondary user's ability to detect the presence of the primary user correctly. Therefore a quick and reliable spectrum occupancy decision based on spectrum sensing becomes a critical issue irrespective of the architecture of the CR networks.

Several spectrum sensing methods exist in the literature. Energy detection has been considered in (Digham et al., 2003; Sahai & Cabric, 2005; Tandra & Sahai, 2005), matched filter detection in (Kay, 1993), cyclostationary feature detection in (Gardner, 1991). Recently, eigenvalue based detection has been proposed in (Penna et al., 2009A;B; Penna & Garello, 2010; Wei & Tirkkonen, 2009; Zeng et al., 2008; Zeng & Liang, 2008). Each of these techniques has its strengths and weaknesses. For example, matched filter detection and cyclostationary feature detection require knowledge on the waveform of the primary user, which is impractical for certain applications. Energy detection and eigenvalue based detection are so-called blind detection methods which do not need any a priori information of the signal. Eigenvalue based detection can be further divided into eigenvalue ratio based (ER) detectors and largest eigenvalue based (LE) detectors. The ER detection circumvents the need to know the noise power, since asymptotically its test statistics does not depend on the noise power. Noise uncertainty (Tandra & Sahai, 2005) may has important consequences for detector performance. For example, ER outperforms energy detector, when there is uncertainty of the noise level. In the literature, performance analysis of the ER detector relies on the limiting laws of the largest and the smallest eigenvalue distributions. These limiting laws are valid for large numbers of sensors and samples and are not able to characterize detection performance when the number of sensors and the sample size are small. On the other hand, exact characterization of the ER detection requires knowledge of the condition number distribution of finite dimensional covariance matrices, which is generically mathematically intractable. A semi-analytical expression of the condition number distribution is presented in (Penna et al., 2009A). This result becomes rather complicated to implement when the number of sensors and the sample sizes are large. For the LE detector, asymptotical performance analysis based on the Tracy-Widom distribution is proposed in (Zeng et al., 2008). There, the limiting law of the largest eigenvalue distribution is utilized to set a decision threshold, considering only the false alarm probability. This result characterizes the LE detector performance in the asymptotical region where the sample sizes and the number of cooperating sensors are huge. In (Kritchman & Nadler, 2009) a more general problem of estimating the number of signals using the largest eigenvalue is studied, where the estimation probability is obtained using the Tracy-Widom distribution as well. Finally we note that the LE detector is similar to the energy detector in that the test statistics are functions of the noise variance. Therefore the LE detector is pestered by the noise uncertainty problem as well.

In this chapter, the analysis of eigenvalue detector is carried out in a setting where there is only one primary user transmitting. The detection problem is a hypothesis test between two possible hypotheses; either there is a primary user, or there is none. The covariance matrices under these hypotheses can be formulated as central and non-central Wishart matrices, respectively. Empirically we found that the largest eigenvalue calculated from the received covariance matrix is an efficient quantity to discriminate between the two hypotheses, which motivates the investigation of the LE detection.

The contribution of this chapter is two-fold. Firstly we derive the exact largest eigenvalue distributions for central and non-central Wishart matrices. We modify the results on the largest eigenvalue distributions from (Dighe et al., 2003; Kang & Alouini, 2003; Khatri, 1964) in order to derive distribution functions suitable for performance analysis. As a result we obtain exact characterizations for both the false alarm probability and the probability of missed detection. Secondly, we investigate the detection performance in the asymptotical region where both the number of sensors and the sample size are large. Specifically we derive closed-form asymptotic largest eigenvalue distributions for central and non-central Wishart matrices. These results are possible due to recent breakthrough in random matrix theory. Moreover a simple closed-form formula for the receiver operating characteristics (ROC) can also be derived. Besides gaining more insights into the detection performance, the low complexity asymptotic results can be used for the implementation of the LE detector.

The accuracy of the asymptotic approximations is investigated by comparing to the exact distributions through various realistic spectrum sensing scenarios. The results confirm the usefulness of the asymptotic distributions in analyzing the detection performance in practice. We also compare the detection performance of the LE detection with other well-known detection schemes. It turns out that in the case of perfectly estimated noise power the LE detector performs best among the detectors considered. In order to see the whole picture, we extend the analysis to the case where the noise power is not perfectly known. With worst case noise uncertainty, the LE detector performs worse than the ER detector, but is by far more robust against noise uncertainty the energy detector.

The rest of this chapter is organized as follows. In Section 2, we formulate the primary user detection problem in a multi-antenna spectrum sensing setting. We then motivate the choice of largest eigenvalue as the test statistics. Section 3 is devoted to deriving the exact as well as the asymptotical largest eigenvalue distributions. In Section 4, we first study the impact of approximation accuracy of the asymptotic distributions on the detection performance. We then compare the detection performance of the LE detector with that of other detection methods. Lastly, we investigate the impact of the noise uncertainty on the detection performance. Finally in Section 5 we conclude the main results of this chapter.

2. Problem formulation

2.1 Signal model

Consider a primary signal detection problem with K collaborating sensors. These sensors may be, for example, K receive antennas in one secondary terminal or K collaborating secondary devices each with a single antenna, or any combination of these. We assume periodical sensing, where each sensor periodically collects N samples during a sensing time. This collaborative sensing scenario is more relevant if the K sensors are in one device, i.e. for multi-antenna assisted spectrum sensing. For multiple collaborating devices, communication to the fusion center by sensors of different locations becomes a problem even for a small sample size N.

The received $K \times N$ data matrix $\mathbf{Y}$ is represented as

$$\mathbf{Y} = \begin{pmatrix} y_{1,1} & y_{1,2} & \cdots & y_{1,N} \\ y_{2,1} & y_{2,2} & \cdots & y_{2,N} \\ \vdots & \vdots & \ddots & \vdots \\ y_{K,1} & y_{K,2} & \cdots & y_{K,N} \end{pmatrix}. \tag{1}$$

Mathematically, the primary user detection problem is a hypothesis test between two hypotheses. Hypothesis 0 ($\mathbf{H_0}$) denotes the absence of the primary user and hypothesis 1 ($\mathbf{H_1}$) denotes the presence of the primary user. If we assume no fading in the temporal domain, i.e. the channel stays constant during the sensing time, the two hypotheses can be represented as:

$$\mathbf{H_0} : y_{k,n} = n_{k,n} \tag{2}$$

$$\mathbf{H_1} : y_{k,n} = \sum_{i=1}^{P} h_k^{(i)} s_n^{(i)} + n_{k,n}, \tag{3}$$

where $k = 1, \ldots, K$ and $n = 1, \ldots, N$. Here $n_{k,n}$ is the complex Gaussian noise with zero mean and variance σ_{cn}^2, P denotes the number of simultaneously transmitting primary users. The receive covariance matrix $\mathbf{R}$ is defined as $\mathbf{R} = \mathbf{Y}\mathbf{Y}^H$, where H denotes the Hermitian conjugate operator. Throughout this chapter, we make the following assumption.

Assumption: There is at most one primary user transmitting ($P = 1$), and its signal amplitude is drawn independently from a Gaussian process for every sample.

Under hypothesis $\mathbf{H_0}$, the receive covariance matrix $\mathbf{R}$ follows the complex central Wishart distribution, denoted as (Gupta, 2000) $\mathbf{R} \sim \mathcal{W}_K\left(N, \sigma_{\text{cn}}^2 \mathbf{I}_K\right)$, where $\mathbf{I}_K$ denotes the identity matrix of dimension K. Under hypothesis $\mathbf{H_1}$, by our assumption the covariance $\mathbf{R}$ follows the complex non-central Wishart distribution, which is denoted as (Gupta, 2000) $\mathbf{R} \sim \mathcal{W}_K\left(N, \sigma_{\text{cn}}^2 \mathbf{I}_K, \mathbf{M}\mathbf{M}^H\right)$. Here $\mathbf{M}\mathbf{M}^H$ is the non-centrality parameter matrix with $\mathbf{M} = \mathbf{h}_k^{(1)}(\mathbf{s}_n^{(1)})^T$. The complex vectors $\mathbf{h}_k^{(1)} = [h_1^{(1)}, h_2^{(1)}, \ldots, h_K^{(1)}]'$ and $\mathbf{s}_n^{(1)} = [s_1^{(1)}, s_2^{(1)}, \ldots, s_N^{(1)}]'$ are column vectors. From the assumption that there is one primary user, it follows that the matrix $\mathbf{M}$ is rank one, because $\text{Rank}\left(\mathbf{M}\right) = \text{Rank}\left(\mathbf{M}\mathbf{M}^H\right) = \text{Rank}\left(||\mathbf{s}^{(1)}||^2 \mathbf{h}_k^{(1)} (\mathbf{h}_k^{(1)})^H\right) = 1$. In other words, the Hermitian matrix $\mathbf{M}\mathbf{M}^H$ has only one non-zero eigenvalue, which we denote by ϕ_1.

Strictly speaking the non-centrality parameter $\mathbf{M}\mathbf{M}^H$ is not a constant matrix, since the norm of the transmit signal $||\mathbf{s}^{(1)}||^2$ is still a random variable. However, the randomness in $||\mathbf{s}^{(1)}||^2$ is diminishing very fast as the sample size N increases and $||\mathbf{s}^{(1)}||^2/N$ can be well approximated by the signal variance σ_s^2 for sufficiently large N. On the other hand, if we assume the primary user's signal $s_n^{(i)}$ is of constant modulus, for example, MPSK modulation, the non-centrality parameter matrix is a strictly constant matrix. We note that under $\mathbf{H_1}$ it is also possible to model $\mathbf{R}$ as the so-called spike correlation model (Penna & Garello, 2010). Whilst the spike correlation model is mathematically more tractable, it is formally valid only for Gaussian signals. Finally, the average SNR under hypothesis $\mathbf{H_1}$ is defined as SNR $= \frac{\sigma_s^2\sigma_h^2}{\sigma_{\text{cn}}^2}$. In practice, σ_s^2 can be estimated by $||\mathbf{s}^{(1)}||^2/N$.

2.2 Test statistics

We want to discriminate between the two hypotheses based on the eigenvalues $\lambda_1 \geq \lambda_2 \geq \ldots \geq \lambda_K$ of the observed covariance matrix $\mathbf{R}$. The fact that $\mathbf{M}$ is of rank one leads to a major difference on the numerical value of the largest eigenvalue λ_1, but the impact on other eigenvalues is much smaller. This fact is firstly explored and studied in the statistics literature, where it is known as Roy's largest root test (Roy, 1953). No explicit expression for its distribution is given in (Roy, 1953). To motivate this approach, in Table 1 we empirically calculated the sample mean of the ordered eigenvalues of the covariance matrix $\mathbf{R}$ under both hypotheses, where we set the parameters $K = 4$, $N = 30$, SNR $= -5$ dB and $\sigma_{\text{cn}}^2 = 1$. From Table 1 we can see that the largest eigenvalue λ_1 provides a most prominent candidate to discriminate $\mathbf{H_0}$ from $\mathbf{H_1}$. Specifically in this case, the difference between the mean values of the largest eigenvalues can be as large as 28.447, whereas the difference between the mean values of the smallest eigenvalues is only 1.587.

$K = 4, N = 30$	$\bar{\lambda}_1$	$\bar{\lambda}_2$	$\bar{\lambda}_3$	$\bar{\lambda}_4$
$\mathbf{H_1}$(SNR = -5 dB)	73.215	38.385	27.523	18.820
$\mathbf{H_0}$	44.768	33.265	24.747	17.233

Table 1. Sample mean of the ordered eigenvalues under both hypotheses.

Using the received data matrix (1), several other sensing algorithms can be proposed. For example, the test statistics T_{ED} of the energy detector relies on the norm of the data matrix, i.e. $||\mathbf{X}||_F^2$ (Digham et al., 2003). The test statistics of the eigenvalue ratio based detector (Penna et al., 2009A;B; Penna & Garello, 2010; Zeng & Liang, 2008) is defined as $T_{\text{ER}} = \lambda_1/\lambda_K$, which is the condition number of Wishart matrices.

For detection, a test variable is calculated, which is compared with its corresponding precalculated decision threshold γ to decide the presence or absence of a primary user. If $T < \gamma$ the detector chooses $\mathbf{H_0}$, otherwise $\mathbf{H_1}$ is chosen. In order to calculate the decision thresholds we need to know the distributions of the respective test statistics. For the energy detector, the test statistics under $\mathbf{H_0}$ follows a central Chi-square distribution and under $\mathbf{H_1}$ it follows a non-central Chi-square distribution (Digham et al., 2003). For the ER detector, asymptotical condition number distributions under both hypotheses are derived in (Penna et al., 2009B; Penna & Garello, 2010; Zeng & Liang, 2008). Moreover under $\mathbf{H_0}$, the exact condition number distribution can be calculated (Penna et al., 2009A). For the LE detector, asymptotical method for computing the test statistics distribution under $\mathbf{H_0}$ is presented in (Zeng et al., 2008). However, the resulting distribution can only be evaluated numerically. In the next section, we will derive exact largest eigenvalue distributions as well as closed-form asymptotical largest eigenvalue distributions for both hypotheses.

3. The Test statistics distributions under both hypotheses

In order to analyze the detection performance of the LE detector we need to know the distributions of the largest eigenvalue under both hypotheses. In this section, we first derive the exact distributions of the largest eigenvalue by making use of finite dimensional results on Wishart matrices. The exact distributions can be utilized to calculate detection performance metrics, such as the false alarm probability, the missed detection probability or the decision threshold. On the other hand, due to the complexity of the exact results, they are most useful when the number of sensors and sample size are small. In order to characterize the detection performance in the asymptotical region where both the number of sensors and sample sizes are large, we derive asymptotic largest eigenvalue distributions under both hypotheses. Specifically by exploring recent results in random matrix theory, we derive analytical Gaussian approximations to the largest eigenvalue distributions of central and non-central Wishart matrices. The derived closed-form asymptotical distributions provide accurate approximations in realistic spectrum sensing scenarios. Due to the simplicity of the asymptotic results, computation of various performance metrics of the LE detector can be easily performed on-line.

3.1 Exact characterizations

For the signal model in the last section, computable closed-form expressions for the largest eigenvalue distributions of central and non-central (non-central matrix $\mathbf{M}$ being rank one) Wishart matrices can be derived from the results in (Kang & Alouini, 2003; Khatri, 1964). Specifically, assuming independent and identically distributed (i.i.d) entries in the received data matrices $\mathbf{Y}$ for both hypotheses, the matrix variate distribution of $\mathbf{Y}$ under $\mathbf{H_0}$ is

$$\mathbf{Y} \sim \frac{1}{\pi^{KN}\left(\sigma_{\mathrm{cn}}^2\right)^{KN}}\, \mathrm{e}^{-\mathrm{tr}\{\mathbf{Y}\mathbf{Y}^H\}/\sigma_{\mathrm{cn}}^2}. \tag{4}$$

The corresponding Wishart distribution $\mathbf{R} = \mathbf{Y}\mathbf{Y}^H$ can be trivially derived from Theorem 3.2.2 in (Gupta, 2000) by placing σ_{cn}^2 in appropriate equations. Then the joint eigenvalue distribution of the covariance matrix $\mathbf{R}$ is derived by using the result in (James, 1964). Finally, following steps in (Khatri, 1964) the cumulative distribution function (CDF) of the largest eigenvalue of the covariance matrix $\mathbf{R}$, denoted by $F_{\mathrm{c}}(x\,|\sigma_{\mathrm{cn}}^2)$, is derived as

$$F_{\mathrm{c}}\left(x\,|\sigma_{\mathrm{cn}}^2\right) = \frac{\det \mathbf{A}}{\left(\sigma_{\mathrm{cn}}^2\right)^{KN}\prod_{k=1}^{K}\Gamma(N-k+1)\Gamma(K-k+1)} \tag{5}$$

where $\det(\cdot)$ denotes the matrix determinant operator and $\Gamma(\cdot)$ is the Gamma function. The i,j th entry of matrix $\mathbf{A}$ $(K\times K)$ is defined through the regularized incomplete Gamma function $\gamma_R(\cdot,\cdot)$ as $\mathbf{A}_{i,j} = (\sigma_{\mathrm{cn}}^2)^{N-K+i+j-1}\Gamma(N-k+i+j-1)\gamma_R\left(N-K+i+j-1, \frac{x}{\sigma_{\mathrm{cn}}^2}\right)$. When $\sigma_{\mathrm{cn}}^2 = 1$, the above result reduces to the result in(Khatri, 1964). By elementary manipulations, $F_{\mathrm{c}}\left(x\,|\sigma_{\mathrm{cn}}^2\right)$ can be simplified to

$$F_{\mathrm{c}}\left(x\,|\sigma_{\mathrm{cn}}^2\right) = |\det \widehat{\mathbf{A}}|, \tag{6}$$

with $\widehat{\mathbf{A}}_{i,j} = \binom{N-j+i-1}{i-1}\gamma_R(N+i-j, \frac{x}{\sigma_{\text{cn}}^2})$.

Under $\mathbf{H_1}$, the matrix variate distribution of $\mathbf{Y}$ is

$$\mathbf{Y} \sim \frac{1}{\pi^{KN}\left(\sigma_{\text{cn}}^2\right)^{KN}}\, \mathrm{e}^{-\text{tr}\{(\mathbf{Y}-\mathbf{M})(\mathbf{Y}-\mathbf{M})^H\}/\sigma_{\text{cn}}^2}. \tag{7}$$

The corresponding Wishart form distribution $\mathbf{R} = \mathbf{Y}\mathbf{Y}^H$ can directly be obtained from Theorem 3.5.1 in (Gupta, 2000) by placing σ_{cn}^2 in appropriate lines. The corresponding joint eigenvalue distribution of the covariance matrix $\mathbf{R}$ is derived from (James, 1964). When matrix $\mathbf{M}$ is rank one, the CDF of the largest eigenvalue of the covariance matrix $\mathbf{R}$, denoted $F_{\text{nc}}(x\,|\sigma_{\text{cn}}^2)$, can be calculated by following the derivations in (Kang & Alouini, 2003) as

$$F_{\text{nc}}\left(x\,|\sigma_{\text{cn}}^2\right) = \frac{\left(\prod_{k=1}^{K-1}\Gamma(N-k)\Gamma(K-k)\right)^{-1}\det\mathbf{B}}{\mathrm{e}^{\frac{\phi_1}{\sigma_{\text{cn}}^2}}\left(\sigma_{\text{cn}}^2\right)^{(KN-2K+2)}\phi_1^{K-1}\Gamma(N-K+1)} \tag{8}$$

where elements in the first column of $\mathbf{B}$ are $\mathbf{B}_{i,1} = \int_0^x \beta^{N-i}\,\mathrm{e}^{-\frac{\beta}{\sigma_{\text{cn}}^2}}\,{}_0F_1\left(N-K+1; \frac{\phi_1\beta}{\sigma_{\text{cn}}^4}\right)d\beta$ and in the second to Kth column
$\mathbf{B}_{i,j} = (\sigma_{\text{cn}}^2)^{N+K-i-j+1}\Gamma(N+K-i-j+1)\gamma_R\left(N+K-i-j+1, \frac{x}{\sigma_{\text{cn}}^2}\right)$, $j = 2, \ldots, N$. Here, ${}_0F_1(\cdot;\cdot)$ is the hypergeometric function of Bessel type. Recall that ϕ_1 is the only non-zero eigenvalue of the Hermitian matrix $\mathbf{M}\mathbf{M}^H$. When $\sigma_{\text{cn}}^2 = 1$, the result above reduces to the result in (Kang & Alouini, 2003). After some manipulations, $F_{\text{nc}}\left(x\,|\sigma_{\text{cn}}^2\right)$ can be simplified to

$$F_{\text{nc}}\left(x\,|\sigma_{\text{cn}}^2\right) = |\det\widehat{\mathbf{B}}|, \tag{9}$$

with elements in the first column $\widehat{\mathbf{B}}_{i,1} = \frac{\Gamma(K)}{(N-K)\Gamma(N-i)}\int_0^{x/\sigma_{\text{cn}}^2}\beta^{N-i}\,\mathrm{e}^{-\beta}\,{}_0F_1\left(N-K+1; \frac{\phi_1\beta}{\sigma_{\text{cn}}^2}\right)d\beta$ and in the other $K-1$ columns $\widehat{\mathbf{B}}_{i,j} = \binom{N-i+j-2}{j-1}\gamma_R(N-i+j-1, \frac{x}{\sigma_{\text{cn}}^2})$, $j = 2, \ldots, N$.

Based on distribution of the test statistics under $\mathbf{H_0}$, for a given threshold γ, the false alarm probability can be calculated as

$$P_{\text{fa}}(\gamma) = 1 - F_{\text{c}}\left(\gamma\,|\sigma_{\text{cn}}^2\right). \tag{10}$$

Similarly, for a given SNR and threshold, the missed detection probability can be obtained by using the distribution under $\mathbf{H_1}$ as

$$P_{\text{m}}(\gamma) = F_{\text{nc}}\left(\gamma\,|\sigma_{\text{cn}}^2\right). \tag{11}$$

Equivalently, for a target false alarm probability or missed detection probability the resulting threshold can be calculated by numerically inverting (10) or (11).

Assuming a worst case SNR, a performance bound can be obtained by the analysis above. This worst case SNR value can, for example, be set based on the receiver sensitivity of the primary system of interest (Ruttik et al., 2009). Notice that the actual detection performance could be better or worse depending on the validity of the assumed worst case SNR value. When taking into account the test statistics distributions under both hypotheses, the decision threshold can be obtained by the Neyman-Pearson criterion (Kay, 1993) or by minimizing a weighted sum of the false alarm probability and the missed detection probability (Wei & Tirkkonen, 2009). Once we obtain the decision threshold the detection procedure is as follows. Firstly, the K cooperative sensors form the $K \times N$ received data matrix $\mathbf{Y}$ as in (1). Secondly, the largest eigenvalue λ_1 of the covariance matrix $\mathbf{R} = \mathbf{Y}\mathbf{Y}^H$ is calculated. Finally, we make a decision; if λ_1 is larger than the threshold γ it is decided that the primary user is present and if λ_1 is smaller than γ it is decided that the primary user is absent.

3.2 Asymptotical characterizations

Although the previous derived results could capture the detection performance exactly, the computation complexity increases rapidly as the number of sensors and sample sizes increase. To circumvent this numerical burden and gain more insights into this detection problem, we derive asymptotic Gaussian approximations to the largest eigenvalue distributions under both hypotheses. As a consequence, simple closed-form results for the false alarm and missed detection probabilities are possible. In addition, a closed-form expression for the receiver operating characteristics is derived as well.

For the largest eigenvalue λ_1 of covariance matrix $\mathbf{R}$ under $\mathbf{H_0}$, it is known that (Johansson, 2000) there exists proper centering sequence, $a_1(K,N) = (\sqrt{K} + \sqrt{N})^2$, and scaling sequence $b_1(K,N) = (\sqrt{K} + \sqrt{N})(\sqrt{\frac{1}{K}} + \sqrt{\frac{1}{N}})^{1/3}$, such that the distribution of the random variable

$$\Lambda_1 = \frac{\lambda_1 - \sigma_{\text{cn}}^2 a_1(K,N)}{\sigma_{\text{cn}}^2 b_1(K,N)}, \tag{12}$$

converges to the Tracy-Widom distribution of order two (Tracy & Widom, 1996), denoted as F_{TW2}. The convergence occurs when $K \to \infty$, $N \to \infty$ and $\frac{K}{N} \to c \in (0,1)$. This asymptotical result provides us an approximation to the largest eigenvalue distribution for a given matrix size K and N. Namely, the CDF for the largest eigenvalue of a covariance matrix with N degrees of freedom can be approximated by a linear transform of the Tracy-Widom variable as

$$F_{\text{c}}\left(x \left|\sigma_{\text{cn}}^2\right.\right) \approx F_{\text{TW2}}\left(\frac{x - \sigma_{\text{cn}}^2 a_1(K,N)}{\sigma_{\text{cn}}^2 b_1(K,N)}\right). \tag{13}$$

The distribution function of the Tracy-Widom distribution of order two can be represented as

$$F_{\text{TW2}}(x) = \exp\left\{-\int_x^{\infty} (s-x) q^2(s) ds\right\}, \tag{14}$$

where $q(s)$ is the solution to the Painlevé II differential equation $q''(s) = sq(s) + 2q^3(s)$ with boundary condition $q(s) \sim \text{Ai}(s)$ $(s \to \infty)$, where $\text{Ai}(s)$ is the Airy function. Numerically it is possible to compute the value of $F_{\text{TW2}}(x)$ for a given x by using software packages such as (Dieng, 2006; Perry et al., 2009). This facilitates efficient calculations of the approximative CDF in (13). However, equation (13) is not a closed-form approximation, since it depends on the numerical solution of (14). On the other hand, it is shown in (Anderson, 1963) that the largest eigenvalue distribution converges to a Gaussian distribution when N goes to infinity for any fixed K. Although this asymptotic result gives only a loose bound for finite-dimensional expressions, it motivates us to adopt the Gaussian approximation to the largest eigenvalue distribution. In order to obtain a closed-form Gaussian approximation we need to calculate the first two moments of λ_1, which is a non-trivial problem from the exact distribution (6). However the asymptotic moments of λ_1 via the Tracy-Widom distribution are readily obtained. From (12), the first moment of λ_1 is

$$E[\lambda_1] = \sigma_{\rm cn}^2(a_1(K,N) + b_1(K,N)E[\Lambda_1]), \tag{15}$$

where $E[\cdot]$ donates the expected value. The second moment of λ_1 is

$$V[\lambda_1] = \left(\sigma_{\rm cn}^2 b_1(K,N)\right)^2 V[\Lambda_1], \tag{16}$$

where $V[\cdot]$ denotes the variance. Since the distribution of Λ_1 converge to $F_{\rm TW2}$ as $K \to \infty$, $N \to \infty$ and $\frac{K}{N} \to c \in (0,1)$, the mean and variance of Λ_1 also converges to the ones of the Tracy-Widom variable; $E[\Lambda_1] \to E[x_{\rm TW2}] = -1.7711$, $V[\Lambda_1] \to V[x_{\rm TW2}] = 0.8132$. These numerical values are obtained by using (Dieng, 2006). A closed-form Gaussian approximation is obtained by fitting these two moments to the corresponding Gaussian moments. Note that by matching higher moments of the Tracy-Widow distribution to higher moments of other distributions, for example, the generalized lambda distribution (Karian & Dudewicz, 2000), we expect to achieve more accurate approximations. Finally, the largest eigenvalue distribution is approximated by a Gaussian distribution $\mathcal{N}(\mu_1, \sigma_1^2)$ with mean μ_1 and variance σ_1^2 given by $\mu_1 = \sigma_{\rm cn}^2(a_1(K,N) + b_1(K,N)E[x_{\rm TW2}])$ and $\sigma_1^2 = \left(\sigma_{\rm cn}^2 b_1(K,N)\right)^2 V[x_{\rm TW2}]$ respectively. Thus the approximative CDF of λ_1 under $\mathbf{H_0}$ is

$$G_{\rm c}\left(x \,|\sigma_{\rm cn}^2\right) = \Phi\left(\frac{x-\mu_1}{\sigma_1}\right), \tag{17}$$

where $\Phi(\cdot)$ is the CDF of a standard Gaussian random variable. Note that both μ_1 and σ_1^2 are simple functions of the matrix dimensions and noise variance only. Thus the computational complexity is negligible compared with the exact distribution (6). More importantly, by using this closed-form approximation the reliance on numerical calculations from the software package (Dieng, 2006; Perry et al., 2009) is removed.

Under $\mathbf{H_1}$, the covariance matrix $\mathbf{R}$ follows the complex noncentral Wishart distribution. Simple and accurate closed-form approximation for its largest eigenvalue distribution is not available. The first order expansion of λ_1 proposed in (Jin et al., 2008) is unable to capture the detection performance since its accuracy can not be guaranteed except for a threshold around zero. In the following we propose a two-step Gaussian approximation for the λ_1 distribution under $\mathbf{H_1}$. The first step is to establish the relationship between non-central and central Wishart matrices. The results in (Tan & Gupta, 1983) showed that a non-central Wishart matrix $\mathbf{R}$ distributed as $\mathbf{R} \sim \mathcal{W}_K\left(N, \sigma_{\rm cn}^2 \mathbf{I}_K, \mathbf{M}\mathbf{M}^H\right)$, can be well approximated by a correlated central Wishart matrix distributed as $\mathbf{R} \sim \mathcal{W}_K(N, \Sigma_K)$, where the effective correlation matrix Σ_K is given by $\Sigma_K = \sigma_{\rm cn}^2 \mathbf{I}_K + \mathbf{M}\mathbf{M}^H/N$. Since the effective correlation matrix is an identity matrix plus a rank one matrix, the eigenvalues of Σ_K, denoted by ξ_i, can be easily determined as $\xi_1 = \sigma_{\rm cn}^2 + \phi_1/N$, $\xi_2 = \xi_3 \ldots = \xi_K = \sigma_{\rm cn}^2$. The second step is to approximate the largest eigenvalue distribution of a correlated central Wishart matrix by its asymptotic distribution. The results in (Baik & Silverstein, 2005) prove that the largest eigenvalue of a correlated central Wishart matrix converges to a Gaussian distribution $\mathcal{N}(\mu_2, \sigma_2^2)$ with mean $\mu_2 = N\xi_1\left(1 + \frac{K/N}{\xi_1 - 1}\right)$, and variance $\sigma_2^2 = N\xi_1^2\left(1 - \frac{K/N}{(\xi_1-1)^2}\right)$. The convergence occurs when $K \to \infty$, $N \to \infty$, $\frac{K}{N} \to c \in (0,1)$, and in addition ξ_1 must satisfy (Baik & Silverstein, 2005), $\xi_1 > 1 + \sqrt{K/N}$. Thus the approximative CDF of λ_1 under $\mathbf{H_1}$, denoted by $G_{\rm nc}\left(x \,|\sigma_{\rm cn}^2\right)$, is represented as

$$G_{\rm nc}\left(x\,|\sigma_{\rm cn}^2\right) = \Phi\left(\frac{x-\mu_2}{\sigma_2}\right). \tag{18}$$

Based on the approximative distribution of the test statistics under $\mathbf{H_0}$ (17), for a given threshold γ, the false alarm probability can be approximated by

$$P_{\rm fa}(\gamma) \approx 1-\Phi\left(\frac{\gamma-\mu_1}{\sigma_1}\right). \tag{19}$$

Equivalently, for a required false alarm probability the decision threshold can be approximated by $\gamma \approx \mu_1 + \sigma_1\Phi^{-1}(1-P_{\rm fa})$, where $\Phi^{-1}(\cdot)$ is the inverse of a Gaussian CDF. For a given SNR and threshold, the missed detection probability can be obtained by using the approximative distribution under $\mathbf{H_1}$ (18) as

$$P_{\rm m}(\gamma) \approx \Phi\left(\frac{\gamma-\mu_2}{\sigma_2}\right). \tag{20}$$

Similarly for a required missed detection probability and SNR, the approximative decision threshold is $\gamma \approx \mu_2 + \sigma_2\Phi^{-1}(P_{\rm m})$. Finally, by inserting this approximative threshold into (19) we obtain a closed-form approximative receiver operating characteristics for the LE detector

$$P_{\rm m} \approx \Phi\left(\frac{\sigma_1\Phi^{-1}(1-P_{\rm fa})+\mu_1-\mu_2}{\sigma_2}\right). \tag{21}$$

Note that a closed-form ROC expression is not possible by using exact distributions (10) and (11). In the next section we will compare the asymptotic results with the exact distributions in terms of various detection performance metrics.

3.3 A note on computational complexity

The computational complexity discussed here refers to the on-line implementation complexity for the LE detector. If the implementation is based on look-up tables, the computational complexity is negligible when using the approximative distributions. As one can see from (17) and (18), only a 1D table (percentiles of a standard Gaussian CDF) is needed, which is applicable to any K, N and $\sigma_{\rm cn}^2$. It can be seen from (6) and (9) that the look-up table implementation using the exact distributions is more demanding. Under $\mathbf{H_0}$ for each combination of K and N, a 1D table is needed, which is valid for any $\sigma_{\rm cn}^2$. Moreover, under $\mathbf{H_1}$, for each combination of K and N, a 2D table is needed, which is valid for any $\sigma_{\rm cn}^2$. The reason being that the first column of $\widehat{\mathbf{B}}$ is a function of two variables.

Since K and N may be subject to frequent changes in practice, the implementation may rely on realtime computations of the distributions instead of tabulations. In this case the operational complexity when using the exact distribution, which is mainly determined by the number of multiplications, can be shown to be upper bounded by $O(2n^3)$ for both $\mathbf{H_0}$ and $\mathbf{H_1}$ (Borwein & Borwein, 1987), where n is the number of digits needed to represent N, K, $\sigma_{\rm cn}^2$ and the threshold γ. However, the bit-complexity may prevent the use of the exact results. Each multiplication needs to be done with a large n, which is particularly true for $\mathbf{H_1}$. For example when $K=4$, $N=100$ and $\sigma_{\rm cn}^2=1$, by inspecting the distributions (6) and (9), it can be verified that the number of digits n equals 13 and 30 bits for $\mathbf{H_0}$ and $\mathbf{H_1}$, respectively.

4. Detection performance

In this section, several aspects regarding the detection performance are addressed. Firstly we will show the accuracy of the derived asymptotic results in characterizing the detection performance. Then we will compare the performance of the largest eigenvalue based detection to other detection methods, such as the eigenvalue ratio based detection (Penna et al., 2009B; Penna & Garello, 2010) and the energy detection. Finally we will discuss the robustness of the LE detector under noise uncertainty.

4.1 Exact versus asymptotic

The exact characterization versus the asymptotic approximation is basically a trade-off between accuracy and complexity. Here the accuracy means the degree of control in determining the performance metrics, especially the decision threshold. The complexity refers to the computational complexity in calculating various performance metrics from the test statistics distributions. When using the exact distributions (6) and (9), the false alarm probability and the missed detection probability can be determined exactly, thus we have complete control over the decision threshold. However the computational complexity in this case is non-trivial, since the exact distributions (6) and (9) involve matrix determinants with special function as entries. On the other hand, the asymptotic test statistics distributions (17) and (18) provide a trade-off between accuracy and complexity. Since both the approximative false alarm probability (19) and the approximative missed detection probability (20) are Gaussian distributions, the computational complexity in characterizing the detection performance is negligible. In Figure 1, we plot the false alarm probability as a

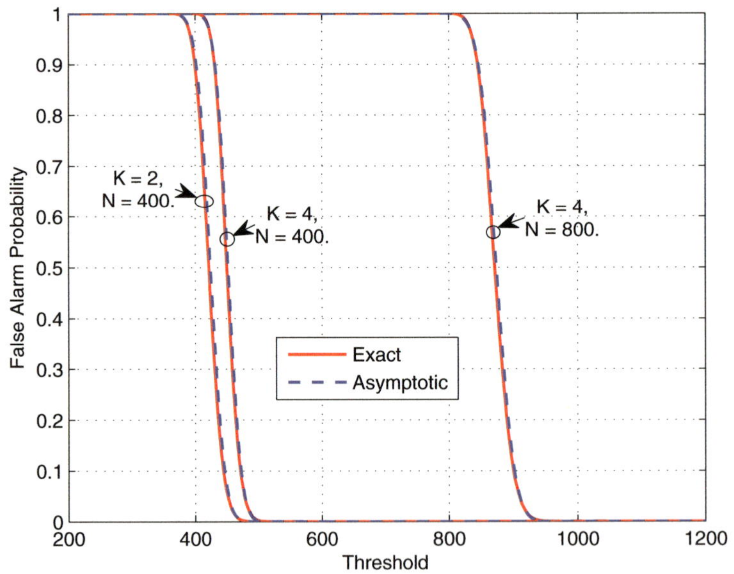

Fig. 1. False alarm probability as a function of threshold: exact v.s asymptotic.

function of the decision threshold for various K and N. The exact curves and the asymptotic curves are obtained from (10) and (19) respectively. We can see from this figure that the asymptotic approximation matches well with the exact characterization in all the parameter settings considered. Note that the considered parameter values (K, N) are realistic in practical spectrum sensing scenarios. The number of samples N can be huge due to the high sample rate. For example, in Digital Television (DTV) signal detection problem studied in (Tawil, 2006), 100 thousand samples corresponds to only 4.65 ms sensing time. The number of receive antennas K can be safely chosen to be less or equal to eight, since nowadays it is possible to have a device with eight antennas. In Figure 2 and Figure 3, we show the missed detection

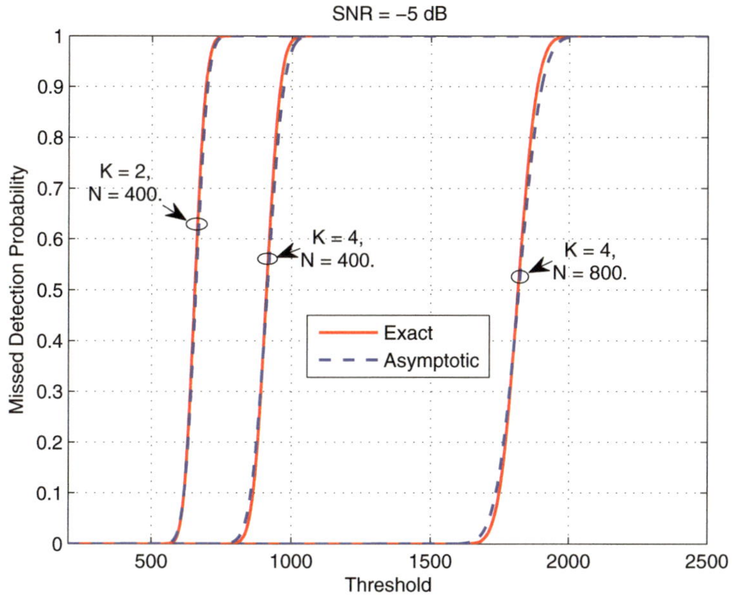

Fig. 2. Missed detection probability as a function of threshold: exact v.s asymptotic.

probability as a function of the decision threshold when SNR equals -5 dB and -10 dB respectively. The (K, N) pairs considered here are the same as in the previous figure. The exact P_{m} plots are obtained by using (11) and the approximative P_{m} plots are obtained from (20). It can be observed from these two figures that the approximation to the missed detection probability (20) is close to the exact result for different (K, N) pairs and SNR values.

4.2 Performance comparison

In general the detection performance is a function of sensing parameters, i.e., the sample size, the number of sensors and the SNR. We can compare the performances by looking at the detection probability for a fixed false alarm probability. Alternatively, the detection performance can be seen from the receiver operating characteristics curve. The ROC curve shows the achieved probability of missed detection as a function of the target false alarm probability. Thus the ROC curve illustrates the overall detection performance for a given detector.

In this subsection, we will compare the LE detector with the classical energy detector (Digham et al., 2003) and the recently proposed eigenvalue ratio (condition number) based detector (Penna et al., 2009A;B; Penna & Garello, 2010) by means of the ROC curves. Specifically, we show how the different sensing parameters (number of sensors, sample size and SNR) affect the detection performance. Here we investigate the case when the noise variance is known

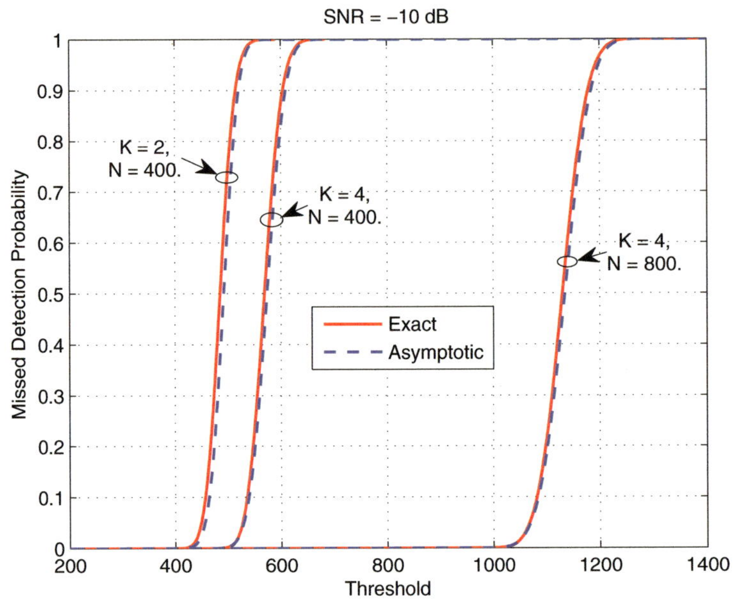

Fig. 3. Missed detection probability as a function of threshold: exact v.s asymptotic.

exactly. The case when there is uncertainty about the noise variance estimates is studied in the next subsection.

The cooperative energy detector will collaborate in the same way as the LE detector in that the decision statistics is a function of the collaboratively formed the received data matrix $\mathbf{Y}$ (1). In the case of the energy detector the test statistics is the norm of received data matrix $\mathbf{Y}$, for example, the Frobenius norm $||\mathbf{Y}||_F^2$. The decision rule is to choose $\mathbf{H_0}$ when $||\mathbf{Y}||_F^2 \leq \gamma$ and choose $\mathbf{H_1}$ when $||\mathbf{Y}||_F^2 > \gamma$. Under $\mathbf{H_0}$, the test statistics $||\mathbf{Y}||_F^2$ follows the central Chi-square distribution with 2KN degrees of freedom (Digham et al., 2003; Proakis, 2001). Under $\mathbf{H_1}$, the test statistics follows the non-central Chi-square distribution with 2KN degree of freedom. In addition to the LE detector, another possible eigenvalue based detector is the eigenvalue ratio based detector. With the test statistics is $T_{\text{ER}} = \lambda_1/\lambda_K$. Asymptotic approximations of T_{ER} distribution under both hypotheses are studied in (Penna et al., 2009B; Penna & Garello, 2010) and the exact distribution of T_{ER} under $\mathbf{H_0}$ is studied in (Penna et al., 2009A). It can be shown that the test statistics of the ER detector does not depend on noise variance asymptotically. Therefore the ER detector is immune to the noise uncertainty problem.

Without loss of generality, we set variance of the complex noise to 1 ($\sigma_{\text{cn}}^2 = 1$). In the following figures, we will compare the detection performance of LE detector with that of the energy detector and the ER detector. In Figure 4, we consider a case where the number of sensors (receive antennas) is 4 , the sample size is 600 per sensor and the SNR is -10 dB. For LE detector, the exact ROC curve is obtained from (10) and (11) while the approximative ROC is drawn by using (21). For the ER detector, the ROC curve is obtained by simulation. From this figure we can see that the LE detector uniformly outperforms both the energy detector and the ER detector since its probability of missed detection is lower for all false alarm probabilities. Moreover we observe that the asymptotic approximative ROC represents the detection performance rather accurately. In Figure 5, we consider a different sensing

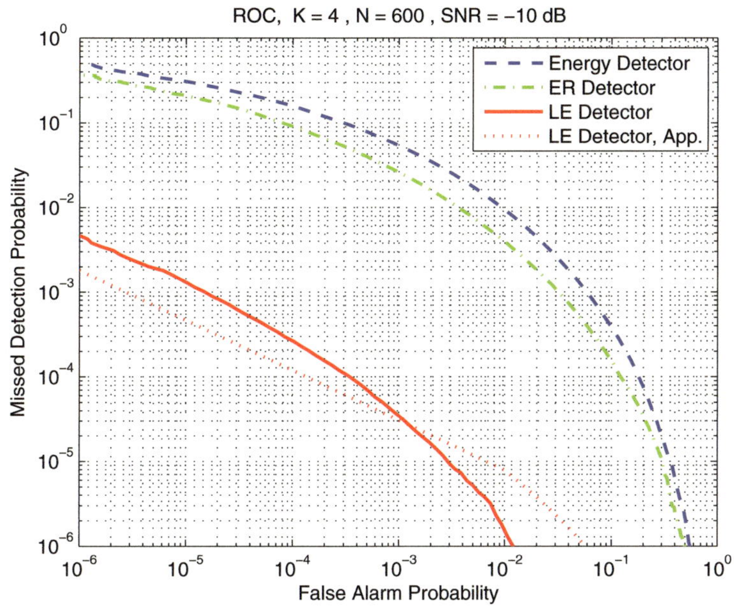

Fig. 4. Receiver operating characteristics: $K = 4$, $N = 600$, SNR $= -10$ dB.

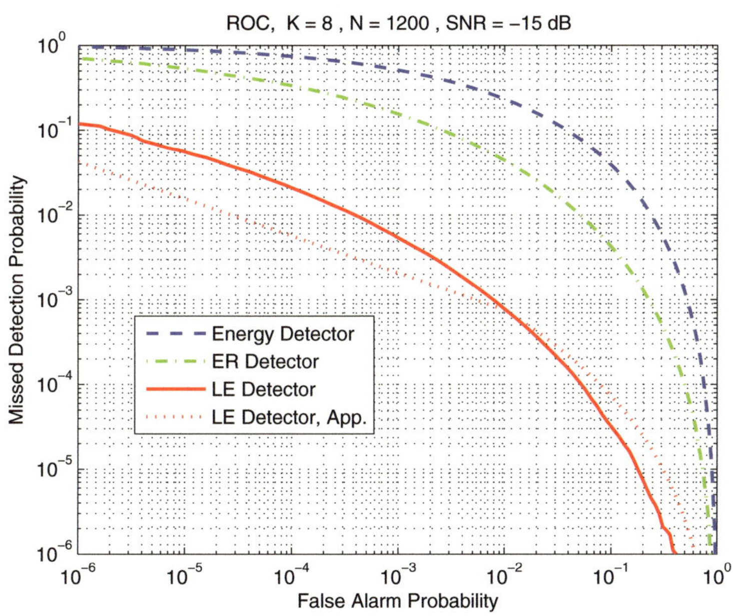

Fig. 5. Receiver operating characteristics: $K = 8$, $N = 1200$, SNR $= -15$ dB.

parameter setting where the number of sensor is 8 with 1200 samples per sensor and SNR -15 dB. In this setting, we again observe that the LE detector performs best among the detectors considered and the loss in characterizing the performance by the approximate ROC is tolerable.

The superior performance of the LE detector over the energy detector can be understood as follows. For cooperative energy detection, the test statistics $||\mathbf{Y}||_F^2$, by definition, equals $||\mathbf{Y}||_F^2 = \text{tr}\{\mathbf{Y}\mathbf{Y}^H\} = \text{tr}\{\mathbf{R}\} = \sum_{i=1}^{K} \lambda_i$, where λ_i is the ith eigenvalue of the received covariance matrix $\mathbf{R}$. Therefore, the test statistics of the energy detector blindly sums up all the K eigenvalues from the covariance matrix $\mathbf{R}$. On the other hand, for the LE detection the test statistics only involves λ_1. In other words, the LE detector will intentionally pick up only the largest eigenvalue as decision statistics. This is an optimal statistical test when there is only one primary user present (matrix $\mathbf{M}$ being rank one) (Roy, 1953). Recall also the implication from Table 1 that blindly adding more eigenvalues as test statistics is unnecessary. When summing all eigenvalues one obtains a more heavy-tailed distribution than the largest eigenvalue distribution. This-heavy tailed distribution will lead to worse detection performance of the energy detector, which is the main motivation behind the LE detector.

4.3 Noise uncertainty analysis

In the analysis done so far we assume the noise variance is known exactly. This is an ideal scenario considering that in any practical system modeling of noise uncertainty is unavoidable. It is especially true for detection problems in CR networks, where robustness to noise uncertainty is a fundamental performance metric (Tandra & Sahai, 2005; 2008). Uncertainty in noise variance may arise due to noise estimation error in the receiver or noise variations during the sensing time or interference caused by other primary users. Note that noise uncertainty analysis may be generalized to incorporate interference uncertainty as well (Zeng et al., 2009).

We consider a situation where there is uncertainty about the noise variance. Let μ be the value in dB of the noise uncertainty. Then the noise power will fall in the interval $\Omega = [\sigma_{\text{cn}}^2/\rho, \rho\sigma_{\text{cn}}^2]$, where $\rho = 10^{\mu/10}$. Naturally, as the uncertainty μ increases the interval that the noise power could fall into will be larger. We would like to see the worst case of performance degradation due to this uncertainty. Thus we need to check all the possible noise power from the interval Ω such that the PDFs under both hypotheses will overlap most. As a result of which, we have the worst case performance for a given uncertainty level μ. Due to the monotonic tails of the largest eigenvalue distributions (6), (9) the noise variance under $\mathbf{H_0}$ is now $\rho\sigma_{\text{cn}}^2$ and the corresponding distribution becomes

$$F_{\text{c}}\left(x \,|\rho\sigma_{\text{cn}}^2\right) = |\det \widehat{\mathbf{A}}|, \tag{22}$$

with $\widehat{\mathbf{A}}_{i,j} = \binom{N-j+i-1}{i-1}\gamma_R(N+i-j, \frac{x}{\rho\sigma_{\text{cn}}^2})$. Similarly, in order to obtain worst case performance, the noise variance under $\mathbf{H_1}$ has to be $\sigma_{\text{cn}}^2/\rho$. The resulting distribution becomes

$$F_{\text{nc}}\left(x \,|\frac{\sigma_{\text{cn}}^2}{\rho}\right) = |\det \widehat{\mathbf{B}}|, \tag{23}$$

with elements in the first column $\widehat{\mathbf{B}}_{i,1} = \frac{\Gamma(K)}{(N-K)\Gamma(N-i)} \int_0^{\rho x/\sigma_{\text{cn}}^2} \beta^{N-i}\, \mathrm{e}^{-\beta}\, {}_0F_1\left(N-K+1; \frac{\rho\phi_1\beta}{\sigma_{\text{cn}}^2}\right) \mathrm{d}\beta$ and in the other $K-1$ columns $\widehat{\mathbf{B}}_{i,j} = \binom{N-i+j-2}{j-1}\gamma_R(N-i+j-1, \frac{\rho x}{\sigma_{\text{cn}}^2})$, $j = 2, \ldots, N$. Notice that in the case of no noise uncertainty ($\rho \to 1$), the distributions (22) and (23) become (6) and (9) respectively. One example to illustrate the effect of noise uncertainty on the LE detector is

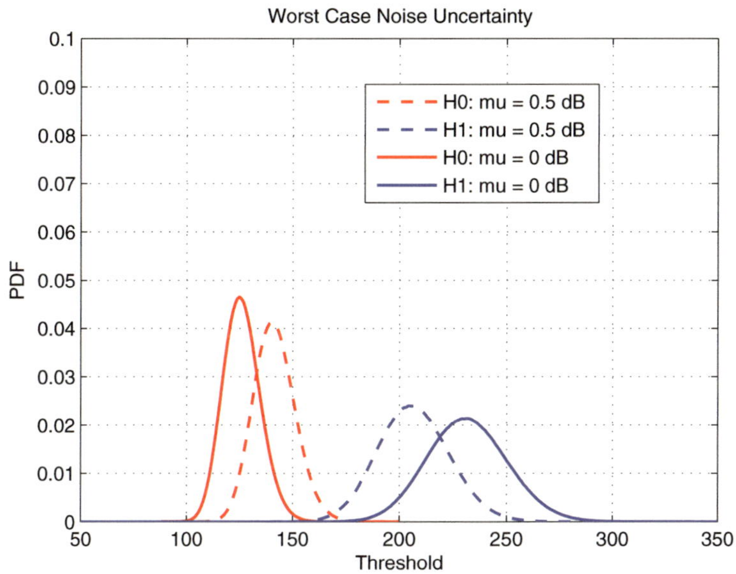

Fig. 6. Impact of worst case noise uncertainty: $K = 4$, $N = 100$, SNR $= -5$ dB, $\mu = 0.5$ dB.

presented in Figure 6, where $K = 4$, $N = 100$, SNR $= -5$ dB. We choose the noise variance $\sigma_{\mathrm{cn}}^2 = 1$ for the case of no noise uncertainty and the uncertainty level $\mu = 0.5$ dB when considering noise uncertainty. Therefore in the worst case scenario the noise variance is 1.122 under $\mathbf{H_0}$ and is 0.891 under $\mathbf{H_1}$. In Figure 6 we plot the PDFs of the test statistics with and without noise uncertainty. We observe that for the case of noise uncertainty, we indeed obtain the worst case of distributions where the curves overlap most. Intuitively, this corresponds to the situation that the two hypotheses are most difficult to distinguish when noise uncertainty exists.

Similarly, for a given uncertainty level μ the worst case approximative test statistics distribution under $\mathbf{H_0}$ is

$$G_{\mathrm{c}}\left(x \mid \rho\sigma_{\mathrm{cn}}^2\right), \tag{24}$$

and under $\mathbf{H_1}$ is

$$G_{\mathrm{nc}}\left(x \mid \frac{\sigma_{\mathrm{cn}}^2}{\rho}\right). \tag{25}$$

In the following figures we will first show the impact of the noise uncertainty on both the false alarm probability and the missed detection probability of the LE detector. In the meanwhile we will illustrate the accuracy of the approximative test statistics distributions under noise uncertainty. Then we will compare the detection performance of the LE detector with that of the energy detector and the ER detector in the case of noise uncertainty. In Figure 7, we show the false alarm probability of the LE detector as a function of the decision threshold for various (K, N) pairs under noise uncertainty. We assume to have 0.4 dB uncertainty in the noise variance. The exact curves and the asymptotic curves are obtained by using (22) and (24) respectively. Comparing this figure with Figure 1 we see that in the case of noise uncertainty the false alarm probability will increase for any given threshold and any (K, N) pairs considered. We can also observe from this figure that the asymptotic approximation

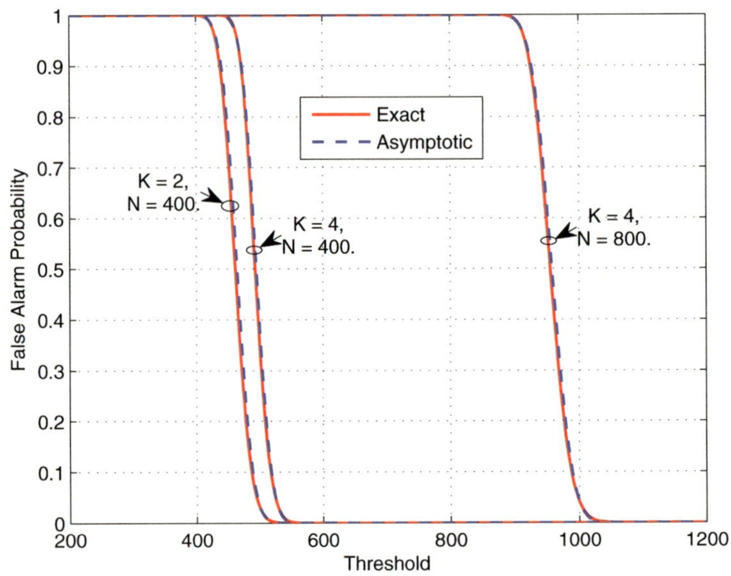

Fig. 7. False alarm probability under noise uncertainty, $\mu = 0.4$ dB: exact v.s asymptotic.

matches well with the exact characterization under noise uncertainty. In Figure 8 and Figure 9, we plot the missed detection probability as a function of the decision threshold when SNR equals -5 dB and -10 dB with 0.4 dB uncertainty in the noise variance. The exact P_{m} plots are obtained by using (23) and the approximative P_{m} plots are obtained

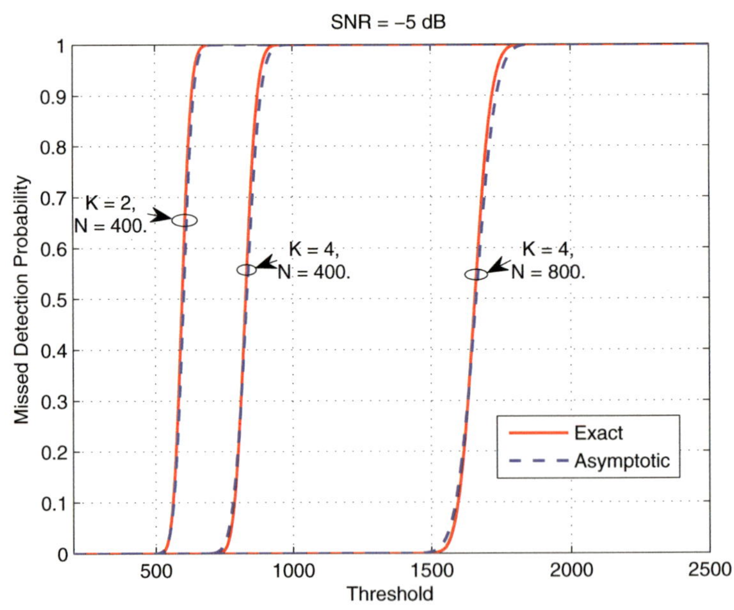

Fig. 8. Missed detection probability under noise uncertainty, $\mu = 0.4$ dB: exact v.s asymptotic.

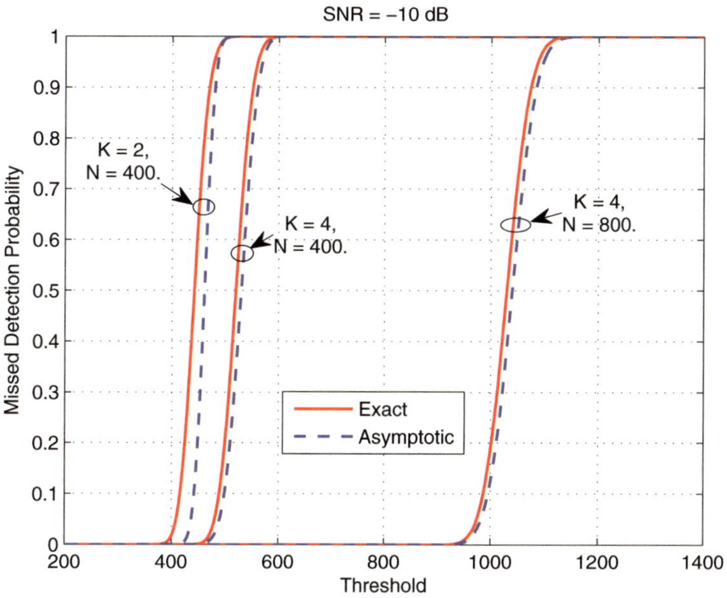

Fig. 9. Missed detection probability under noise uncertainty, $\mu = 0.4$ dB: exact v.s asymptotic.

from (25). By comparing these two figures with Figure 2 and Figure 3 respectively, it can be observed that in the case of noise uncertainty the missed detection probability will also increase for any given threshold and any (K, N) pairs considered. Therefore, we indeed obtain the worst case performance due to noise uncertainty where both the false alarm and missed detection probabilities increase for any given threshold. It can be also observed from these two figures that the asymptotic distribution of the missed detection probability provides a useful approximation for finite size (K, N) pairs. In Figure 10 we compare the impact of noise uncertainty of the LE detector with that of the energy detector and the ER detector by means of the ROC plot. The sensing parameters here are the same as in Figure 4 except that there is now 0.2 dB uncertainty in the noise variance. By comparing this figure with Figure 4, we can see that the ER detector performs better than the LE detector and the energy detector in the case of noise uncertainty. The reason is that the test statistics of the ER detector is not a function of the noise variance, thus its performance will not degrade regardless of the degree of noise uncertainty. On the other hand, the test statistics of both the LE detector and the energy detector depend on the noise variance, thus their detection performances rely on accurate estimation of the noise variance. However, we can observe that the performance degradation is much more severe for the energy detector than that of the LE detector. At $\mu = 0.2$ dB the detection performance of the LE detector and the ER detector are on the same level comparable, but the detection performance of the energy detection becomes too poor to be useful. We also observe that the implementation complexity and accuracy tradeoff reflected by the exact and approximate ROCs is affordable in practice. Finally, in Figure 11 we consider the same sensing parameter setting as in Figure 5 with the exception that we now have 0.2 dB uncertainty in the noise variance. By comparing this figure with Figure 5, we again observe that the impact of noise uncertainty on the ER detector is negligible. However, the energy detector fails in this case with the false alarm probability and the missed detection probability approaching 1. Although the LE detector still works in the case, the performance degradation

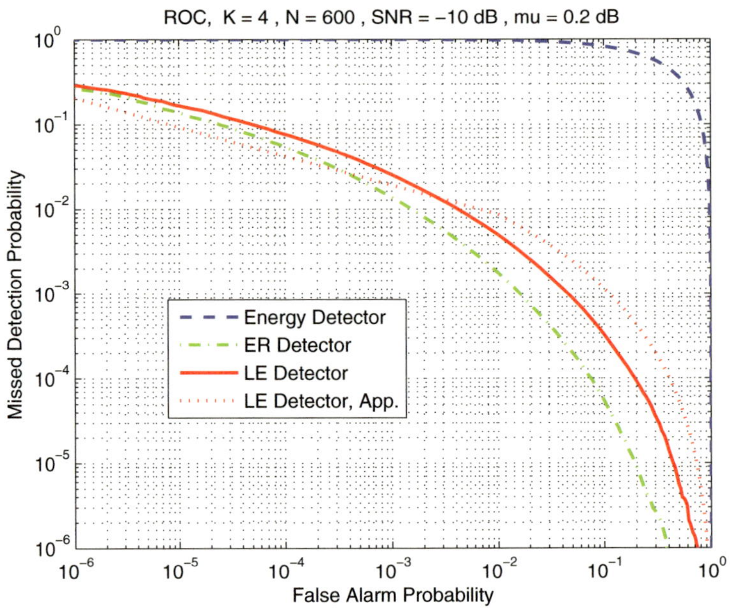

Fig. 10. Receiver operating characteristics under noise uncertainty: $K = 4$, $N = 600$, $\text{SNR} = -10$ dB, $\mu = 0.2$ dB.

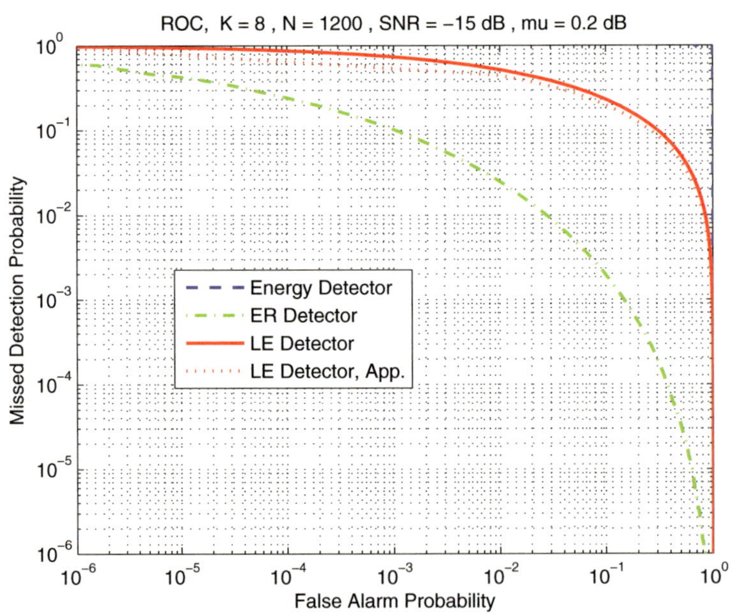

Fig. 11. Receiver operating characteristics under noise uncertainty: $K = 8$, $N = 1200$, $\text{SNR} = -15$ dB, $\mu = 0.2$ dB.

is non-trivial. We can see also that in this case the approximate ROC is able to capture the detection performance almost exactly.

5. Conclusion

In this chapter, we perform both non-asymptotic and asymptotic analysis on the performance of the largest eigenvalue based detection. Analytical formulae have been derived for various performance metrics in realistic spectrum sensing scenarios. It has been shown that the LE detector is more efficient than the energy detector and the ER detector in terms of sample size, number of sensors and SNR requirement. Our analytical framework has also been applied to investigate the detection performance in the presence of noise uncertainty, where we conclude that the superiors performance of the LE detector relies on the accurate estimation on the noise power. From implementation perspective, we studied the computational complexity and accuracy tradeoff which is resolved by the derived tight approximate ROC.

6. References

Anderson, T. W. (1963). Asymptotic theory for principal component analysis. *Ann. Math. Statist.*, 34, 122-148, 1963.

Baik, J. & Silverstein, J. (2005). Eigenvalues of large sample covariance matrices of spiked population models. *Journal of Multivariate Analysis*, 97, pp. 1382-1408, 2006.

Borwein, J. & Borwein, P. (1987). *Pi and the AGM: A Study in Analytic Number Theory and Computational Complexity*, John Wiley, 1987.

Dieng, M. (2006). RMLab, a MATLAB package for computing Tracy-Widom distributions and simulating random matrices. http://math.arizona.edu/ momar/research.htm, 2006.

Digham, F. F., Alouini, M. & Simon, M. K. (2003). On the energy detection of unknown signals over fading channels. *IEEE International Conference on Communications*, May 2003.

Dighe, P. A., Mallik, R. K. & Jamuar, S. R. (2003). Analysis of trasmit-receive diversity in Rayleigh fading. *IEEE Tran. Commun.*, vol. 51, no. 4, pp. 694-703, Apr. 2003.

Gardner, W. A. (1991). Exploitation of spectral redundancy in cyclostationary signals. *IEEE Sig. Proc. Mag.*, vol. 8, pp. 14-36, 1991.

Gupta, A. K. & Nagar, D. K. (2000). *Matrix Variate Distributions*. CRC Press, 2000.

James, A.T. (1964). Distributions of matrix variates and latent roots derived from normal samples. *Ann. Inst. Statist. Math.*, 35, 475-501, 1964.

Jin, S., McKay, M. R., Gao, X. & Collings I. B. (2008). MIMO multichannel beamforming: SER and outage using new eigenvalue distributions of complex noncentral Wishart matrices. *IEEE Tran. Commun.*, vol. 56, no. 3, pp. 424-434, Mar. 2008.

Johansson, K. (2000). Shape fluctuations and random matrices. *Comm. Math. Phys.*, 209:437-476, 2000.

Kang, M. & Alouini, M. S. (2003). Largest eigenvalue of complex wishart matrices and performance analysis of MIMO MRC systems. *IEEE J. Selec. Areas Commun.*, vol. 21, no. 3, pp. 418-426, Apr. 2003.

Karian, Z. A. & Dudewicz, E. J. (2000). *Fitting Statistical Distributions: The Generalized Lambda Distribution and Generalized Bootstrap Methods.* Chapman and Hall/CRC, 2000.

Kay, S. M. (1993). *Fundamentals of Statistical Signal Processing: Estimation Theory.* Englewood Cliffs, NJ: Prentice-Hall, 1993.

Khatri, C. G. (1964). Distribution of the largest or the smallest characteristic root under null hyperthesis concerning complex multivariate normal populations. *Ann. Math. Stat.*, 35, 1807-1810, 1964.

Kritchman, S. & Nadler, B. (2009). Non-parametric detections of the number of signals: hypothesis testing and random matrix theory. *IEEE Trans. Sig. Proc.*, vol. 57, no. 10, pp. 3930-3941, 2009.

Penna, F., Garello, R., Figlioli, D., & Spirito, M. A. (2009). Exact non-asymptotic threshold for eigenvalue-based spectrum sensing. *IEEE International Conference on Cognitive Radio Oriented Wireless Networks and Communications,* Jun. 2009.

Penna, F., Garello, R., & Spirito, M. A. (2009). Cooperative spectrum sensing based on the limiting eigenvalue ratio distribution in Wishart matrices. *IEEE Comm. Letters,* vol. 13, issue 7, pp. 507-509, Jul. 2009.

Penna, F. & Garello, R. (2010). Eigenvalue ratio detection: identifiability and missed-detection probability. *arXiv: 0907.1523.*

Perry, P. O., Johnstone, I. M., Ma, Z. & Shahram, M. (2009). RMTstat: Distributions and Statistics from Random Matrix Theory. 2009, R software package version 0.1.

Proakis, J. G. (2001). *Digital Communications,* Boston, MA: McGraw-Hill, 4th edition, 2001.

Roy, S. N. (1953). On a heuristic method of test construction and its use in multivariate analysis. *Ann. Math. Stat.,* vol. 24, no. 2, pp. 220-238, 1953.

Ruttik, K., Koufos, K. & Jantti, R. (2009). Spectrum sensing with multiple antennas. *IEEE International Conference on Systems, Man, and Cybernetics,* Oct. 2009.

Sahai, A. & Cabric, D. (2005). Spectrum sensing: fundamental limits and practical challenges. *IEEE International Symposium on New Frontiers in Dynamic Spectrum Access Network,* Nov. 2005.

Tan, W. Y. & Gupta, R. P. (1983). On approximating the non-central Wishart distribution with Wishart distribution. *Commun. Stat. Theory Method,* vol. 12, no. 22, pp. 2589-2600, 1983.

Tandra, R. & Sahai, A. (2005). Fundamental limits on detection in low SNR under noise uncertainty. *WirelessCom.,* Jun. 2005.

Tandra, R. & Sahai, A. (2008). SNR walls for signal detection. *IEEE J. Select. Topic in Sig. Proc.,* vol. 2, no. 1, Feb. 2008.

Tawil, V. (2006). 51 captured DTV signal. May 2006, http://grouper.ieee.org/groups/802/22/.

Tracy, C. & Widom, H. (1996). On orthogonal and symplectic matrix ensembles. *Comm. Math. Phys,* vol.177, pp.727-754, 1996.

Wei, L. & Tirkkonen, O. (2009). Cooperative spectrum sensing of OFDM signals using largest eigenvalue distributions. *IEEE International Symposium on Personal, Indoor and Mobile Radio Communications,* Sep. 2009.

Zeng, Y., Koh, C. L. & Liang, Y.-C. (2008). Maximum eigenvalue detection: theory and application. *IEEE International Conference on Communications,* May 2008.

Zeng, Y. & Liang, Y.-C. (2008). Eigenvalue based spectrum sensing algorithms for cognitive radio. *IEEE Tran. Commun.,* vol. 57, no. 6, pp.1784-1793, Jun. 2009.

Zeng, Y., Liang, Y.-C. & Peh, Edward C. Y. (2009). Reliability of spectrum sensing under noise and interference uncertainty. *IEEE International Workshop on the Network of the Future,* 2009.

2

A Practical Demonstration of Spectrum Sensing for WiMAX Based on Cyclostationary Features

Gianmarco Baldini[1], Raimondo Giuliani[1], Diego Capriglione[1] and Kandeepan Sithamparanathan[2]

[1]*Joint Research Centre - European Commission*
[2]*RMIT University*
[1]*Italy*
[2]*Australia*

1. Introduction

Wireless communication systems rely on the use of radio frequency spectrum. The advent of new wireless applications and services and the increasing demand for higher data rates and broadband wireless connectivity have worsen the problem of "'spectrum scarcity"'. It is more and more difficult for spectrum regulators to identify new spectrum bands for new wireless services, because the existing radio frequency spectrum is already allocated to the licensed services. This is also a consequence of the traditional spectrum licensing scheme, where spectrum bands are statically allocated to wireless services in a specific region.

At the same time, recent studies have shown that the spectrum bands are significantly underutilized in time or space. The FCC Spectrum Policy Task Force has reported in (FCC, 2002) vast temporal and geographic variations in the usage of allocated spectrum with utilization ranging from 15% to 85%. Cognitive Radio (CR) technology ((Mitola, 1999), (Haykin, 2007) and (Bhargava, 2007)) offers an alternative to the current system of static spectrum allocation policy by allowing an unlicensed user to share the same radio spectrum resources with the primary user.

To perform the sharing of spectrum resources, CR devices must be able to sense the environment over huge swaths of spectrum to detect spectral holes and to expediently use frequency bands that are not used by primary users, without causing harmful interference to legacy systems. A potential application for CR technology is the "'White Spaces"'concept. The CR nodes opportunistically utilize the spectrum as a secondary user by identifying the 'gaps' in the spectrum known as the 'white space'. The white space arises from partial occupancies of the incumbent users of the spectrum known as the primary users (PU) (e.g. Digital TV broadcasters). The secondary communication by the CR can be performed as long as the white spaces are identified in the spatio-temporal domain (FCC, 2010). The radio spectrum regulatory bodies around the world have also shown great interest in CR technology to improve spectrum utilization (FCC, 2003), (EC, 2007), but the risk of wireless interference to licensed users remains an important concern.

One of the key challenges of CR technology is to reliably detect the presence or absence of primary users at very low signal-to-noise ratio. There are various spectrum sensing techniques available such as the energy detector based sensing, waveform-based sensing, cyclostationarity-based sensing and others (Arslan, 2009). The energy detection method performs the signal measurements and determines the unoccupied spectrum bands by comparing the estimated power to predetermined threshold values. However this method does not perform well under low signal-to-noise ratio conditions.

In this paper we adopt the cyclostationarity-based sensing by considering practical demonstrations and experimentations. Wireless transmissions in general show very strong cyclostationarity features depending on their modulation type, data rate and carrier frequency etc., especially when excess bandwidth is utilized. Therefore the identification of the unique set of features of a particular radio signal for a given wireless access system can be used to detect the system based on its cyclostationarity features. Spectrum sensing based on cyclostationarity performs very well with very low signal-to-noise ratio as described in (Cabric, 2007), (Jondral, 2004).

Spectrum sensing based on cyclostationarity features has received considerable attention from the academic community from the initial papers by Gardner (Gardner, 1991) and (Gardner, 1975), which highlighted that most of the communication signals can be modeled as cyclostationary that exhibits underlying periodicities in their signal structures.

Cyclostationary spectrum sensing has been investigated in (Hosseini, 2010) which addresses the problem that in many applications, for a specific signal, the statistical characteristics are not the same in two adjacent periods, but they change smoothly. So, the periodicity which appears in the aforementioned processes, does not necessarily lead to a pure cyclostationary process, but leads to an almost cyclostationarity which causes limitation on using cyclostationary features. The authors suggests a new estimator for almost cyclostationary signals.

In most cases, spectrum sensing is based on first order cyclostationary analysis but higher orders can be used to improve the detection probability.

Reference (Giannakis, 1994) defined algorithms to detect presence of cycles in the kth-order cyclic cumulants or polyspectra. Implementation aspects and explicit algorithms for $k < 4$ were discussed. Computationally, algorithms for $k < 3$ are more efficient in the time-domain, while algorithms in the frequency domain are simpler to implement for $k \geq 4$.

Spectrum sensing can be implemented on a single CR device or various CR devices, which collaborate to improve the detection probability. (Lunden, 2009) proposes an energy efficient collaborative cyclostationary spectrum sensing approach for cognitive radio systems, which is also applicable for detecting almost cyclostationary signals where the cyclic period may not be an integer number.

The performance of a detector for OFDM signals based on cyclostationary features is described in (Axell, 2011). The detector exploits the inherent correlation of the OFDM signal incurred by the repetition of data in the cyclic prefix, using knowledge of the length of the cyclic prefix and the length of the OFDM symbol. The authors show that the detection performance improves by 5 dB in relevant cases.

A limited number of papers have described the implementation of spectrum sensing based on cyclostationary analysis. In (Baldini, 2009), the authors present experimental results on the cyclostationarity properties of the IEEE 802.11n Wi-Fi transmissions.

In (Sutton, 2008), the authors describe the implementation of a full OFDM-based transceiver using cyclostationary signatures. The system performance was examined using experimental results.

This book chapter provides the following contributions: we perform experimental analysis to study the performance of detecting Worldwide Interoperability for Microwave Access (WiMAX) 802.16e transmissions through its cyclostationarity features as well as energy detection through the computation of the power spectral density (PSD). The experiment is conducted in an anechoic chamber emulating an Arbritary Waveform Generator (AWG) channel for the communications. We describe the implementation of the demonstrator, which uses cyclostationary signatures on a real CR test platform implemented with Software Defined Radio (SDR) technology. This book chapter describes the main constraints and trade-offs, which influenced the design of the demonstrator. Cyclostationary analysis is computationally intensive and the processing resources of the SDR may be limited for the needed signal processing tasks. We present the results for the estimate of the false alarms and missed detection probabilities for different sets of receiver parameters and for different channel conditions.

This book chapter is organized as follows: in section 2 we present the theoretical background for the cyclostationary spectral analysis followed by the description of the spectrum sensing and detection technique in section 3. In section 4 we present the software defined radio platform used to implement the demonstrator. In section 5, we present the experimental analysis.

2. Cyclostationary signal analysis

A random process $x(t)$ can be classified as wide sense cyclostationary if its mean and autocorrelation are periodic in time with some period T_0. Mathematically they are given by,

$$E_x(t) = \mu(t + mT_0) \tag{1}$$

and

$$R_x(t,\tau) = \pi(t + mT_0, \tau) \tag{2}$$

where, t is the time index, τ is the lag associated with the autocorrelation function and m is an integer. The periodic autocorrelation function can be expressed in terms of the Fourier series given by,

$$R_x(t,\tau) = \sum_{\alpha=-\infty}^{\infty} R_x^{\alpha}(\tau)\exp(2\pi j\alpha t) \tag{3}$$

where,

$$R_x^{\alpha}(\tau) = \lim_{T_0\to\infty} \frac{1}{T_0}\int_T x(t - \frac{\tau}{2})x(t + \frac{\tau}{2})\exp(-2\pi j\alpha t)dt \tag{4}$$

The expression in (4) is known as the cycle autocorrelation, and for a cyclostationary process with a period T_0, the function $R_x^{\alpha}(\tau)$ will have component at $\alpha = 1/T_0$. Using the Wiener relationship, the Cyclic Power Spectrum (CPS) or the spectral correlation function can be defined as,

$$S_x^{\alpha}(f) = \lim_{\mathcal{T}\to\infty}\int_{-\mathcal{T}}^{\mathcal{T}} R_x^{\alpha}(\tau)\exp(-j2\pi f\tau)d\tau \tag{5}$$

The CPS in (5) is a function of the frequency f and the cycle frequency α, and any cyclostationarity features can be detected in the cycle frequency domain. An alternative expression for (5), for the ease of computing the CPS, is given by,

$$S_x^{\alpha}(f) = \lim_{T\to\infty} \lim_{T_0\to\infty} \frac{1}{T_0 T} \int_{-T/2}^{T/2} X_{T_0}(t, f + \frac{1}{\alpha})$$
$$\tilde{X}_{T_0}(t, f - \frac{1}{\alpha})dt \quad (6)$$

where, $\tilde{X}_{T_0}(t,u)$ is the complex conjugate of $X_{T_0}(t,u)$, and $X_{T_0}(t,u)$ is given by,

$$X_{T_0}(t,u) = \int_{t-T_0/2}^{t+T_0/2} x(v)\exp(-2j\pi f v)dv \quad (7)$$

Expression in (6) is also known as the time smoothed CPS which theoretically achieves the true CPS for $T >> T_0$. Figure 1 depicts the CPS of a WiMAX signal generated by means of the theoretical presented in this section. In the following section we present the detector based on the CPS considering the cyclostationarity features of the signal.

3. Energy and cyclostationarity feature based detectors

We use the cyclostationarity feature to detect the presence of WiMAX systems in the radio environment. Based on the sensed noisy signal, the binary hypothesis test to perform the decision is given by,

$$H_0^u : r_u(t) = \nu_u(t); H_1^u : r_u(t) = hs(t) + \nu_u(t), \quad (8)$$

where we have H_0^u when signal is not present and H_1^u when signal is present.

$r_u(t)$ is the signal sensed in the uth frequency cluster, $\nu_u(t)$ is the zero mean band limited Gaussian noise at the receiver front end with a noise power of σ_u^2, and $s(t)$ is the WiMAX signal.

The signal to noise ratio (SNR) can be defined as $SNR = P_s^u/\sigma_u^2$ where P_s^u is the received signal power. We consider the channel h to be slowly varying and hence ignore its statistics in our modeling process below, we also assume that $h \approx 1$ in order to make valid comparisons between different experiments and techniques. Since we use the CPS function to detect WiMAX, we can re-write (8) in terms of the CPS as

$$H_0^u : S_r^{\alpha}(f) = S_\nu^{\alpha}(f) H_1^u; S_r^{\alpha}(f) = S_s^{\alpha}(f) + S_\nu^{\alpha}(f), \quad (9)$$

where $S_\nu^{\alpha}(f)$ is the CPS of the AWGN noise ν, and $S_s^{\alpha}(f)$ is the CPS of the WiMAX signal s. In theory, since ν is not a cyclostationary process, the CPS of ν for $\alpha \neq 0$ is zero. Therefore, by using the CPS, one can detect s when it is present. However, for a finite time duration T, or equivalently a finite length of data in the discrete domain with length $N = T/T_s$, where $f_s = 1/T_s$ is the sampling frequency, noise can be present in $S_r^{\alpha}(f)$ for $\alpha \neq 0$. Based on these arguments, we derive the test statistic for the detector as,

$$Z = \sum_{\alpha} \int_{-f_s/2}^{f_s/2} S_r^{\alpha}(f)\tilde{S}_r^{\alpha}(f)df \quad (10)$$

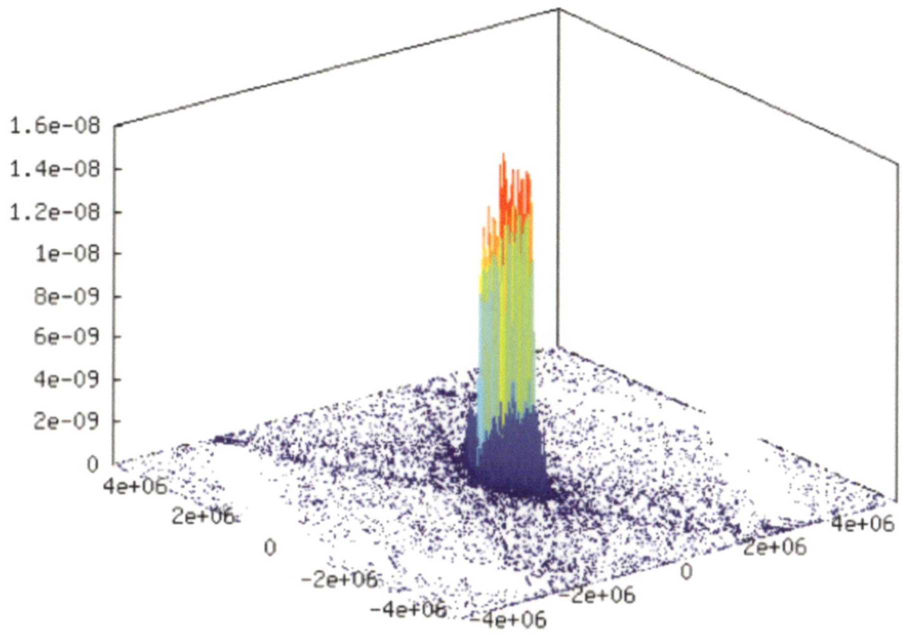

Fig. 1. Cyclostationarity features of a WiMAX signal.

where $\tilde{S}_r^\alpha(f)$ is the conjugate of $S_r^\alpha(f)$. The detector is then given by,

$$H_0^u : Z < \lambda; H_1^u : Z \geq \lambda$$

where λ is the detection threshold. Finding the optimum threshold is the most crucial aspect of the detector and is generally used to target a particular performance criteria for the false alarm probability and the miss detection probability. In general, knowing the noise variance will allow us to have better threshold values and is also feasible in many practical situations. In our work however, we present the receiver operating characteristic curves for possible values of λ in order to study the detection performance under various conditions and also compare the cyclostationary feature based detection with the classical energy based detection technique, which we present subsequently. For the energy detector (Urkowitz, 1967) based spectrum sensing technique the received signal is passed through an energy detector to compute the test statistics Z which is compared with the threshold λ to make a binary decision on the presence of the WiMAX signal. The test statistic Z for the energy detector are mathematically given by

$$Z = \int_0^T r_u^2(t)dt = T_s \sum_{n=1}^{N} |r_u[n]|^2 \tag{11}$$

where, T is the signal observation time window, T_s is the sampling time of the signal, and N is the total number of samples in time T. Then by using equation (11) we perform the detection process using the threshold λ, which is applied to both energy and cyclostationarity feature based detectors.

4. Software defined radio platform

GNU Software Radio (GSR) is an open source project, which provides a real-time digital signal processing software toolkit to develop SDR and CR applications. It is developed for Linux and usable on many other operating systems (OS) on standard PCs (Gnuradio, 2008). While GSR is hardware-independent, it directly supports the so-called Universal Software Radio Peripheral (USRP) front end designed by Ettus et al. A top-down description of the combined GSR and USRP platform is provided in figure 2. The programming environment

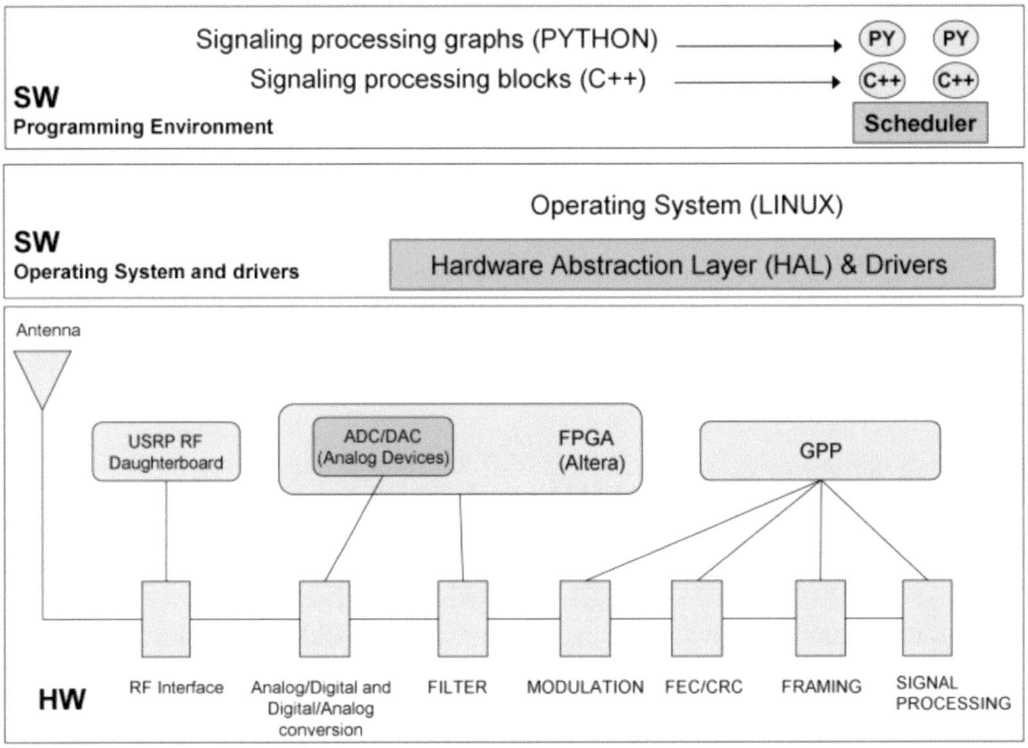

Fig. 2. GNU Radio and USRP architecture

is based on an integrated runtime system composed by a signal-processing graph and signal processing blocks. The signal-processing graph describes the data flow in the SDR platform and is implemented using the object-oriented scripting language Python. Signal processing blocks are functional entities implemented in C++, which operate on streams flowing from a number of input ports to a number of output ports specified per block. Simplified Wrapper and Interface Generator (SWIG) is used to create wrappers for Python around the C++ blocks.

The USRP consists of one main board and up to two receivers (Rx) and two transmitter (Tx) daughterboards. While the main board performs Analog-to-Digital Converter (ADC) and Digital-to-Analog Converter (DAC) conversion, sample rate decimation/interpolation, and interfacing, the daughterboards contain fixed Radio Frequency (RF) front ends including Programmable Gain Amplifiers (PGA) available to adjust the input signal level in order to maximize use of the ADC dynamic range.

This configuration allows an high degree of flexibility because daughter-boards can be connected depending on the type of communications and RF spectrum usage. The ADC/DAC inside the USRP implements sampling and quantization functionality. The analog interface portion contains four ADCs and four DACs. The ADCs operate at 64 million samples per second (Msps) and the DACs operate at 128 Msps. Since the USB bus operates at a maximum rate of 480 million bits per second (Mbps), the Field-Programmable Gate Array (FPGA) must reduce the sample rate in the receive path and it must increase the sample rate in the transmit path to match the sample rates between the high speed data converter and the lower speeds supported by the USB connection. The bottleneck of the system is the USB 2.0 connection to the computer which has, at most, a data rate of 32 MByte/s, thus resulting in a maximum of 8 MS/s of complex signals (16-bit I and 16-bit Q channel).

The ADC/DAC chip is implemented with a AD9862. The AD9862 provides several functions. Each receive section contains four ADCs. Before the ADCs there are PGAs available to adjust the input signal level in order to maximize use of the ADCs dynamic range. The transmit path provides an interpolator and upconverter to match the output sample rate to the DAC sample rate and convert the baseband input to a low IF output. There are PGAs after the DACs. Most of signal processing in the receiver path is performed by the FPGA. The standard FPGA

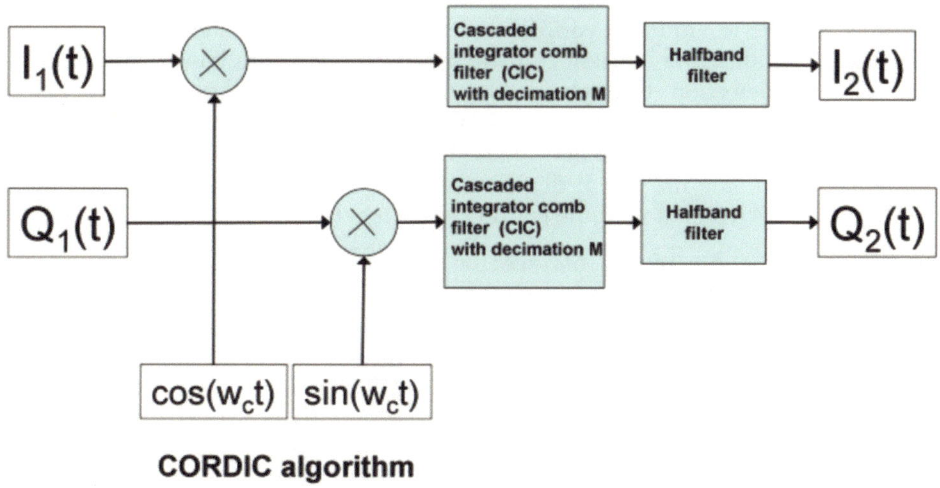

Fig. 3. Block diagram of the digital down-conversion and decimation stage

firmware provides two Digital Down Converters (DDC). The FPGA uses a multiplexer to connect the input streams from each of the ADC's to the inputs of the DDC's. This multiplexer allows the USRP to support both real and complex input signals. The DDC operate as *real* down converters using the data from one ADC fed into the real channel or as *complex* DDCs where the data from one ADC is fed to the real channel and the data from another ADC is fed to the complex channel via the multiplexer.

Figure 3 provides a description of the digital down-conversion and decimation stage to generate the down-converted signal I_2+jQ_2 from the received I_1+jQ_1. W_c represents the frequency for the down-conversion based on the local oscillator, M is the decimation parameter in the Cascaded Integrator Comb (CIC) filter . M is an important parameter for the trade-offs described in the following paragraphs. USRP uses the CORDIC algorithm in the down-conversion (see (Vankka, 2000) for a description of the CORDIC algorithm).

5. Experimentation

5.1 Test setup

This section describes the test-bed setup used for the detection and analysis of the Worldwide Interoperability for Microwave Access (WiMAX) signal. The test-bed setup is described in figure 4. The main components of the test-bed are:

- model RSA3408 spectrum analyzer by Tektronix. It has a frequency range from DC to 8 GHz, 36 MHz bandwidth and a very low background noise floor (-150 dBm/Hz at 2 GHz and a frequency range).
- Rohde & Schwarz SMBV100A signal generator to generate the WiMAX signal based on different standards (e.g. 802.16d and 802.16e). The signal generator has a frequency range from 9 kHz to 6 GHz and very low SSB phase noise (less than -122 dBc at 1 GHz).
- USRP platform already described in section 4.
- variable attenuator with range 0-81 dB to simulate different distances between the WiMAX signal generator and the receiver (i.e. USRP platform).

The signal generator is connected to the receivers (i.e. spectrum analyzer and the USRP platform) through low-loss cables (i.e. 0.327 dB of attenuation at 3.63 GHz). The variable attenuator is needed to simulate various distances between the signal generator and the receivers and to validate the efficiency of the detection algorithm based on cyclostationary features. Signal detection through cyclostationary features is particularly efficient in comparison to signal detection through power sensing when the received signal is at level of the noise floor or below the noise floor. Measurements were repeated with increasing levels of attenuation to validate the signal detection algorithm. Experiments were conducted by generating the WiMAX signal with the features shown in table 1.

Modulation scheme	OFDM
Standard version	802.16 2004-2005
Burst Type	FCH
Modulation	16QAM
Frame	ETSI
Burst length	*3 ms*

Table 1. Features of the WiMAX signal.

5.2 Implementation of the detectors

As described in section 4, the limitation of USB 2.0 to 8 MS/s can increase the spectrum scanning time for a wide frequency band (e.g. 100 MHz). The spectrum scan is implemented by dividing the spectrum in portions of maximum 8 MHz. The algorithm works like a Sliding Spectral Window Technique to cover the entire desired radio frequency spectrum. The algorithm scans the radio frequency spectrum of interest, collecting and processing the data. The same algorithm is used both for *energy detector* based sensing and for *cyclostationary detector* based sensing.

The block diagram for the *energy detector* unit is presented in figure 5. The block diagram for the *cyclostationary detector* is presented in figure 6. Each of the blocks represents a specific signal processing stage programmed in C++. The blocks, which are identified as "Variable",

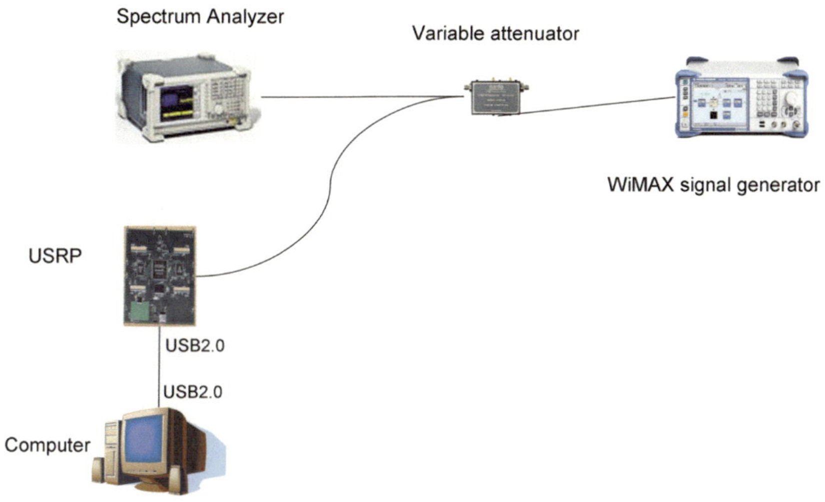

Fig. 4. Test bed configuration

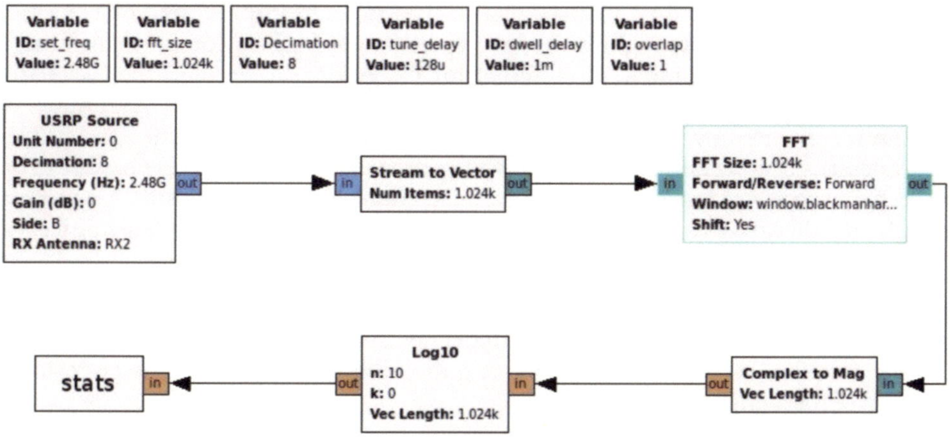

Fig. 5. Block diagrams of the power spectrum algorithm implemented in the *GNU Radio* to calculate the PSD

allow us to dynamically modify the system variables and they will be further discussed in section 5.3.

The first block *USRP source* implements the tuning functionality to tune the receiver to a specific frequency with a bandwidth defined by the decimation value, which is chosen to optimize the processing of the received radio signal. For example, by choosing the maximum decimation value (i.e. 8), we will have a bandwidth of 8 MHz and a sampling frequency of 8

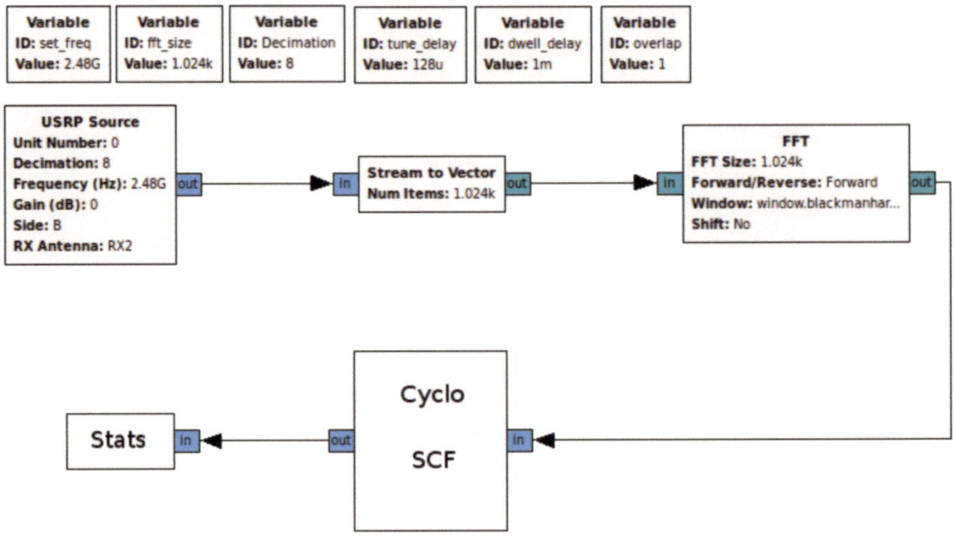

Fig. 6. Block diagrams of the cyclostationary algorithm implemented in the *GNU Radio* to calculate the PSD

MS/s. The received signal is then converted and sampled by the USRP and processed by the *Stream to Vector* block to convert the complex data flow to a complex vector flow. The signal is then multiplied for a window of Blackman-Harris type, to improve the efficiency of the overall algorithm, and then brought in the frequency domain by Fast Fourier Transform (FFT). At this point, the flow diagrams are different for the *energy detector* and the *cyclostationary detector*. In the *energy detector* the *Complex to Mag* block calculates the signal magnitude, which is logarithmic converted by the Log_{10} block and converted to the the Power Spectral Density (PSD). As described in section 3, by choosing an appropriate threshold, it is possible to detect, in real time, the presence/absence of the communication signal. The threshold are computed empirically by calibrating the receiver on continuous wave signals, with defined power levels, transmitted by a signal generator.

In the *cyclostationary detector* flow, the *Cyclo SCF* block calculates the Spectral Correlation Function (SCF) for all of the values of cyclic frequency and the Z on the basis of the theoretical framework described in section 3. By choosing the optimum threshold, the cyclostationary components of the signal can be identified.

Finally, in both flows, the *Stats* block is responsible for the management of the "spectrum scanning" functionality. Specifically, it is responsible for the USRP tuning and the sampling of the received data to be analyzed or discarded.

The real time estimation has a trade-off between the smaller resolution request to estimate the SCF and the USRP's bandwidth. Therefore decimation and length of the FFT are two parameters, which can be used to improve the performance of both detectors. This choice defines the length of processing time: in the case of the energy detector it is proportional to the decimation and to the number of samples on which FFT is calculated; instead for the cyclostationary detector it is proportional to the decimation and to the square number of samples on which FFT is calculated.

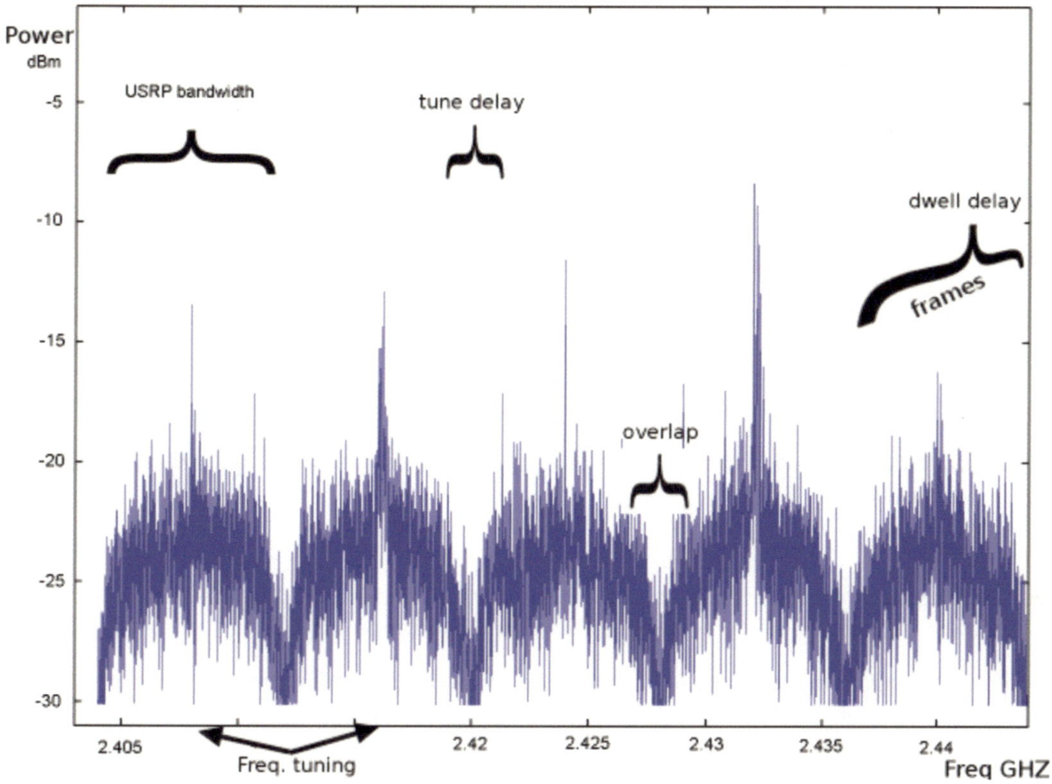

Fig. 7. Spectrum scanning of a band of 40 MHz by using windows of 8 MHz each

5.3 Sensing parameters and related trade-offs

The selection of sensing parameters is an important element to improve the sensing algorithm. The sensing time represents a major trade-off: the increase of the sensing time will reduce the probability of false alarms, but it will also increase the overall the time for scanning the radio frequency spectrum.

To detect the presence of a pulse signal of a short time duration (i.e. burst), we have to scan the spectrum as swiftly as possible. The scanning is carried out by sliding the USRP's Radio Frequency (RF) front-end in shifted frequency given by the relation between the fixed sampling rate of the USRP's ADC/DAC (64 MSa/s) and the decimation choice. Our purpose is to find the best combination between these parameters to improve the probability of detection and to keep the sensing time the shortest possible in order to operate the cyclostationary detection algorithm in real time on the USRP platform. In this book chapter, we will use the experimental results to support the choice of the parameters.

In figure 7, we describe the spectrum scanning concept for a band of 40 MHz and we identify the most important parameters, which contribute to the performance of the spectrum scanning.

The combination of the parameters identified in figure 7 and figures 5 and 6 (i.e. implementation of the detection algorithms) define the overall set of parameters, which drive the performance of the SDR receiver:

1. The ***tuning frequency***: the selection of the center band frequency where the USRP operates;
2. The ***decimation*** (M): define the bandwidth of the RF front-end of the USRP, the sampling frequency on the basis of the following definition:

$$USRP_{rate} = \frac{ADCrate}{M} \tag{12}$$

3. the ***fft size*** where (NFFT) is the number of sampling on which FFT is calculated;
4. the ***tune delay*** (*Td*) is the needed time for the RF front-end to synchronize to a specific frequency. The USRP has a limited throughput so a number of samples must be eliminated on the basis of the following formula:

$$N_{Td} = \frac{Td \cdot USRP_{rate}}{NFFT} \tag{13}$$

5. The ***dwell delay*** (*Dd*) is the time where USRP is synchronized to a specific frequency (i.e. signal observation time).

$$N_{Dd} = \frac{Dd \cdot USRP_{rate}}{NFFT} \tag{14}$$

6. The ***overlap*** is the overlapping of two consecutive spectral windows; it is expressed as a fraction. A value of 100% means that there is no overlapping.

In order to improve the probability of detection and to optimize the sensing time for real-time operation the choices of these parameters are essential.

To serve such purpose we will conduct an analysis of the parameters identified above on the basis of experimental tests. The first consideration is based on the general assumption of the absence of a receiver filter. As described in Section 4, the USRP is not equipped with a receiver filter and additional signal processing techniques are needed for the filtering function. The *overlap* parameter is used to address the absence of the receiver filter in the USRP. The *overlap* works like a band pass filtering centered to the USRP's frequency tuning. The trade-off of the *overlap* parameter is a better filtering and suppression of signal repetitions at the cost of loss of samples, during the acquisition process. For example, if the *overlap* is selected to a value of 0.75, 25% of the data received for every window may be discarded, while the frequency distance between two adjacent windows will be reduced by 25%, reducing the risk of signal's repetitions in the adjacent bands in which the signal is sensed.

The *Td* is the time to manage the incoming data between two adjacent frequency windows. USRP has the limitation that it is not possible to block the flow graph, while it is processing the received data. The processing time and the sampling rate should be synchronized. This is not always possible with complex signals like WiMAX. Before processing the data at the frequency at which the RF front-end is tuned in, the USRP may discard the received data from the previous frequency window still present in the USB pipeline. A reasonable short time to minimize the discarded data is chosen to be equal to *10 ms*, which for *M* equal to 8 and *NFFT* equal to 1024 results in 78 records discarded at each frequency change. On the basis of these considerations, we decided to define the *Td* and the *overlap* constant for all the measurements, and equal respectively to *10 ms* and to *0.75*.

The *Dd* or dwell delay is the signal observation time for each frequency window the RF front-end is tuned in. It is selected based on the hardware elaboration speed. It depends

*NFFT**M*	128	256	512	1024	2048	Units
			3906,25	7812,5	15625	r in Hz
8			39	78	156	S_s in $\sharp$
			64	128	256	T_p in ms
		3906,25	7812,5	15625		r in Hz
16		39	78	156		S_s in $\sharp$
		64	128	256		T_p in ms
	3906,25	7812,5	15625			r in Hz
32	39	78	156			S_s in $\sharp$
	64	128	256			T_p in ms

Table 2. Frequency resolution (r), number of records discarded (S_s) and elaboration time (T_p) needed for a single window reference to M and $NFFT$.

upon the choice of the value of decimation: for M equal to 8, the time to collect 1024 samples is equal to 128 ms.

As illustrated in table 2, the decimation and number of samples on which FFT is calculated determines the value of:

- frequency resolution: r in Hz.
- number of records discarded for each frequency change: S_s in $\sharp$.
- minimum hardware time needed to process data from a single window: T_p in ms.

In order to examine the performance of the receiver implementation outlined in the previous section, we executed measurement campaigns with the WiMAX parameters defined in table 1.

Figure 8 shows a spectrum scanning from 2.4 GHz to 2.5 GHz using the Energy Detector for 35 seconds of sensing time. The frequency peak at 2.405 GHz represents a WLAN signal, which was present in the measurement environment but it not part of the experiment. The WLAN signal was filtered out in the signal analysis phase. The WiMAX signal generated with 0 dBm power appears at 2.484 GHz. Figure 9 describes the SCF estimates for the same experiment. The SCF is presented at the central frequency of 2.484 GHz for a bandwidth of 4 MHz and using a 256-bin Fourier transform. From the figure we clearly identify all the WiMAX signal sub-carriers at the cycle frequencies $\alpha = 4MHz$.

We have generated WiMAX signal with an increasing number of channels (from 1 to 8). Examples of the resulting SCFs are shown in figure 10 (two channels) and figure 11 (eight channels).

5.3.1 Detection performance in relation to FFT-bin size and decimation

A number of tests were executed with changing values of the USRP's decimation, bandwidth and FFT-bin size, to investigate the performance of the detector. From section 3, we can obtain the equations for the percentages of false alarm and detections, which are given by:

$$\begin{aligned} P_{FA} &= P(Z > \lambda | H0) \\ P_D &= P(Z > \lambda | H1) \end{aligned} \tag{15}$$

By using the AWGN source generator, we set the threshold to reach the percentage of false alarm, P_{FA}, of 10^{-3}. The WiMAX signal is transmitted at a range of power levels, centered at 2.484 GHz. In the test, we used a signal observation time equal to *1,28 ms*. For every

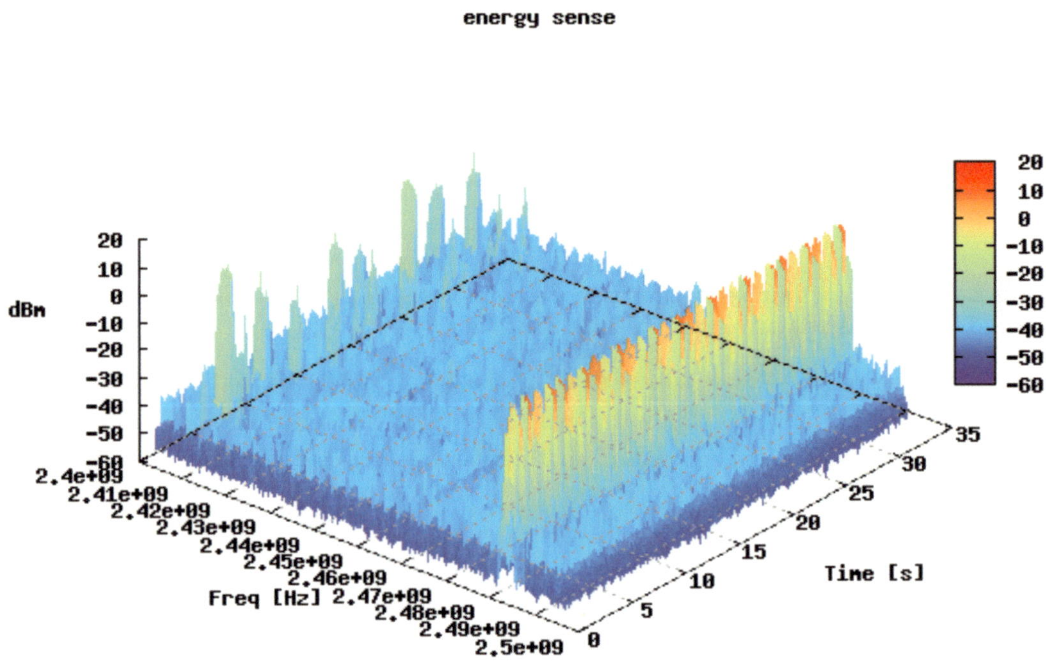

Fig. 8. PSD for *35 s* of spectrum sensing from 2,4 GHz to 2,5 GHz; WiMAX signal

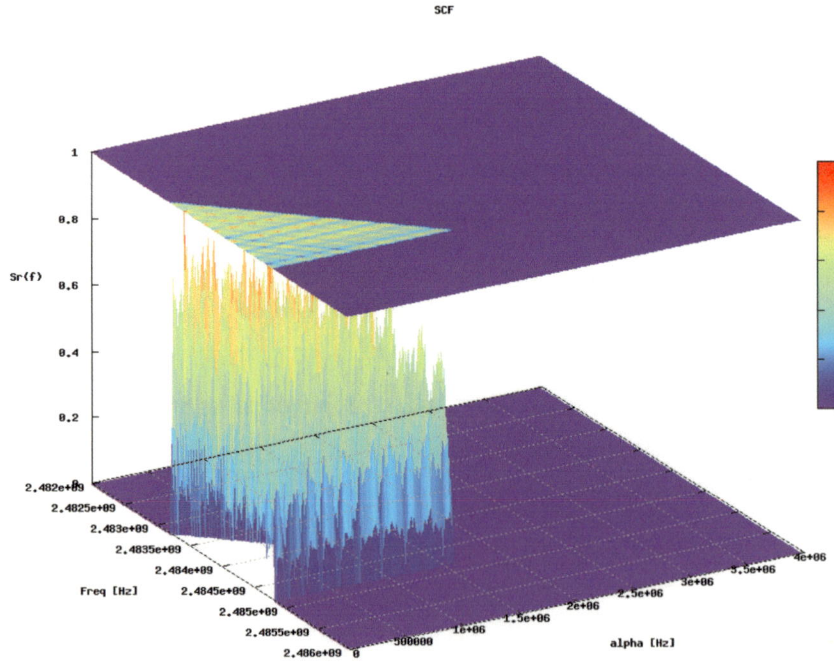

Fig. 9. SCF at the frequency of 2,484 GHz; WiMAX signal, which uses all the available channels

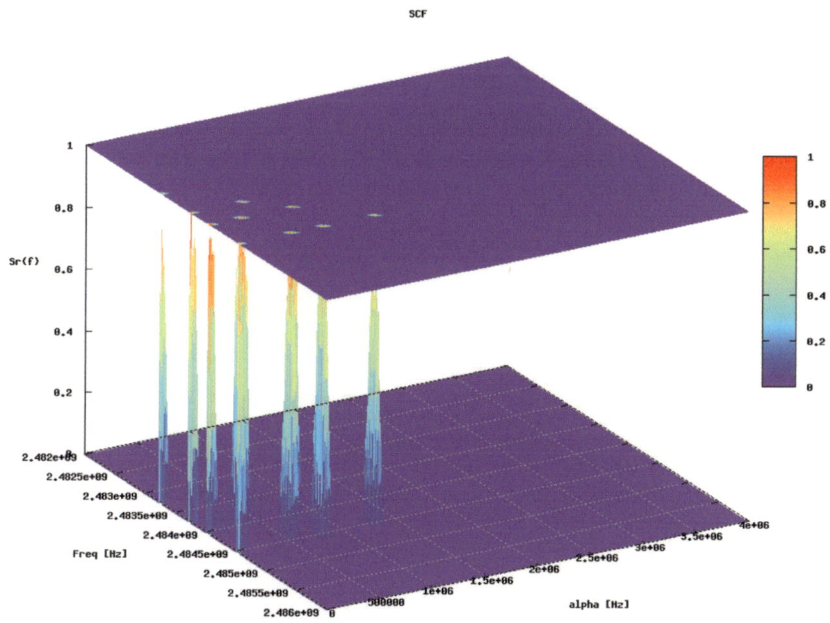

Fig. 10. SCF at the frequency of 2,484 GHz; WiMAX signal, which uses only two channels.

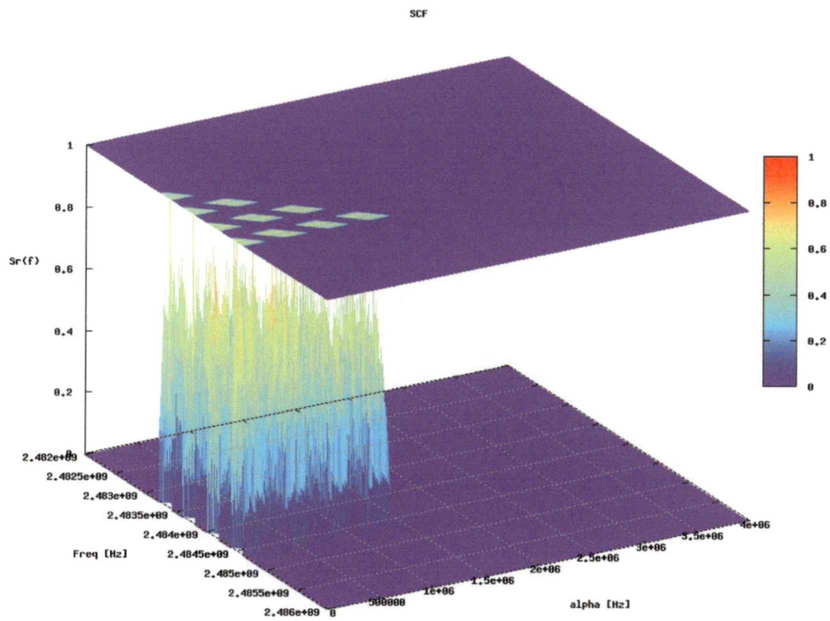

Fig. 11. SCF at the frequency of 2,484 GHz; WiMAX signal, which uses only eight channels.

combination of $NFFT$ and M, the percentage of detection is estimated. The results are illustrated in figures 12 and 13 respectively for *energy detector* and for the *cyclostationary detector*. For both detection techniques the performances improved using the smallest USRP's bandwidth. This is caused by the inability of the processor to manage the incoming data from the USRP. This effect is greater in the *cyclostationary detector* than the *energy detector* due to the higher complexity of the cyclostationary detection algorithm. The best detection performance is for values of decimation set to 32 and FFT-bin size to 128.

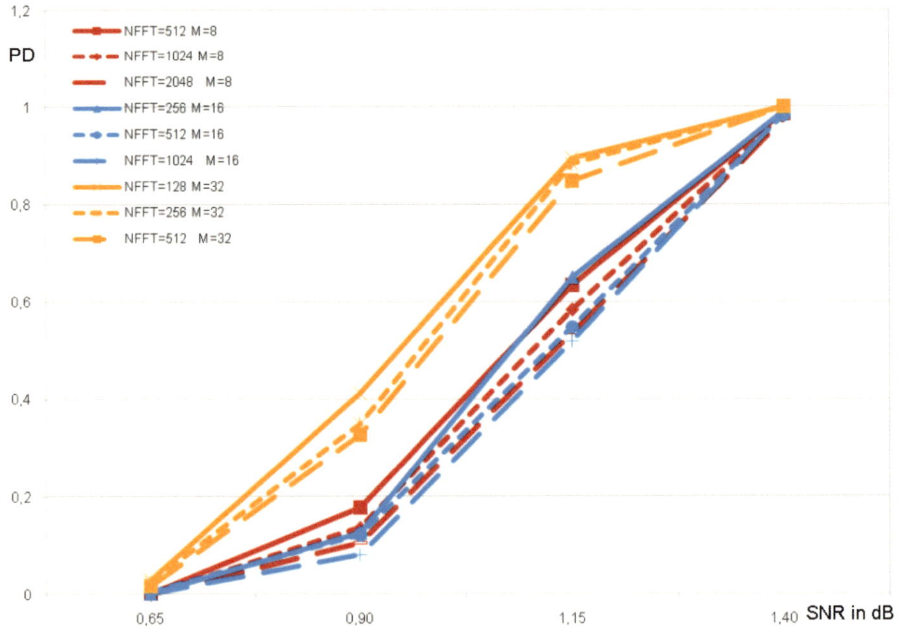

Fig. 12. PD for the power detection technique in relation to Signal/Noise ratio for combinations of NFFT and M.

5.3.2 Detection performance in relation to observation time

Ten thousands tests were executed to examine the performance for both detectors with increasing signal observation time *Td*. Naturally, this value must be chosen in relation to the requested speed of the spectrum scanning and the type of signal to be detected (e.g. WiMAX). Using a pulse signal type the observation time have to be lower than the duration of the single burst. Using an AWGN source generator we set the threshold to reach the percentage of false alarm, P_{FA}, of 10^{-3}. WiMAX signal is transmitted at a range of power levels, centered at 2.484 GHz. Figure 14 describes the percentage of detection for received SNR. Performance is seen to improve considerably with increased observation time from 0.064 ms to 0.36 ms for both detection techniques.

Table 3 displays the processing times required to scan 100 MHz of spectrum for *energy detector* (PSD) and *cyclostationary detector* (SCF). The energy detector algorithm spends *0.36 ms* for every combination of *NFFT* and *M*. Real time elaboration is possible only if the time spaces between the signal bursts are shorter than *3,6 ms*. As expected the *cyclostationary detector* has longer processing times and for some combinations of *NFFT* and *M*, real time elaboration is not possible.

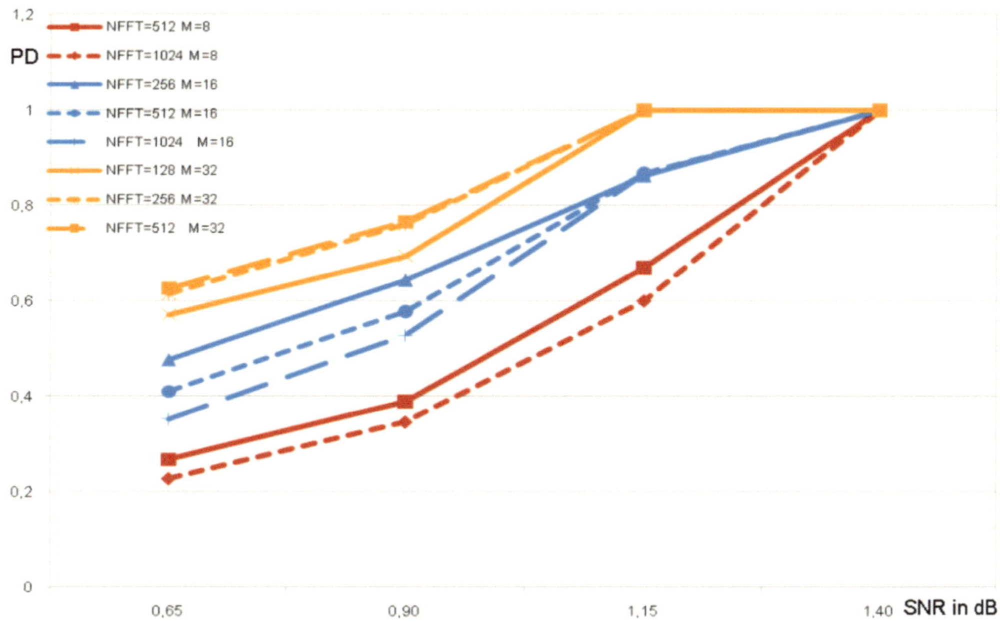

Fig. 13. PD for the cyclostationarity detection technique in relation to Signal/Noise ratio for combinations of NFFT and M.

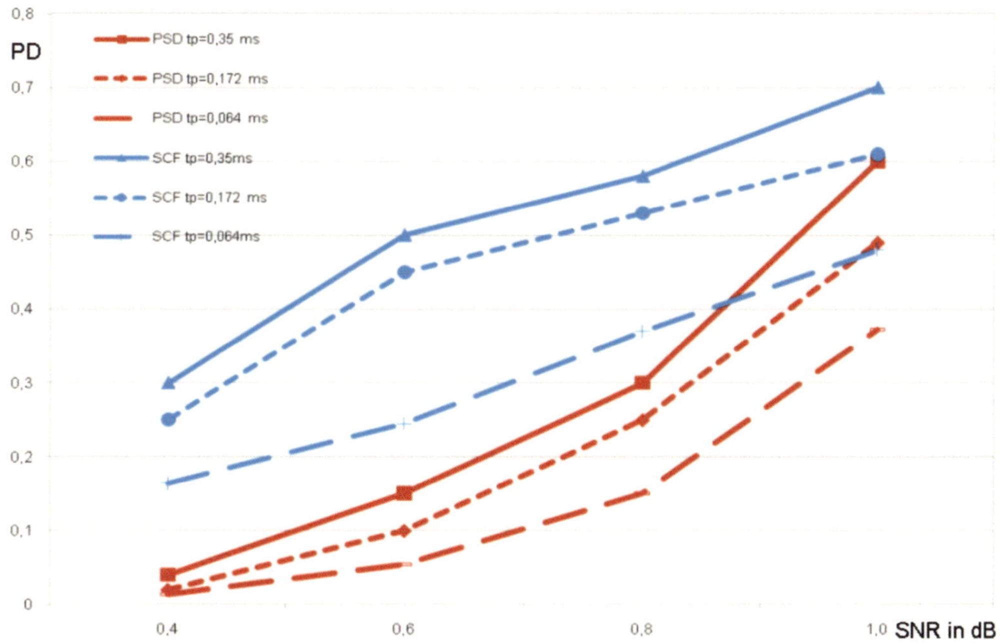

Fig. 14. Time Performance, percentage of detection PD for both detection techniques increasing observation time.

NFFT\M	128	256	512	1024	2048	Detector
8			0.35	0.36	0.39	$T_p(PSD)$
			3.12	11.69	44.32	$T_p(SCF)$
16		0.37	0.36	0.38		$T_p(PSD)$
		1.5	6.08	22.68		$T_p(SCF)$
32	0.36	0.37	0.38			$T_p(PSD)$
	0.84	3.13	11.04			$T_p(SCF)$

Table 3. Time Performance, processing times (T_p) required to scan 100 MHz of spectrum.

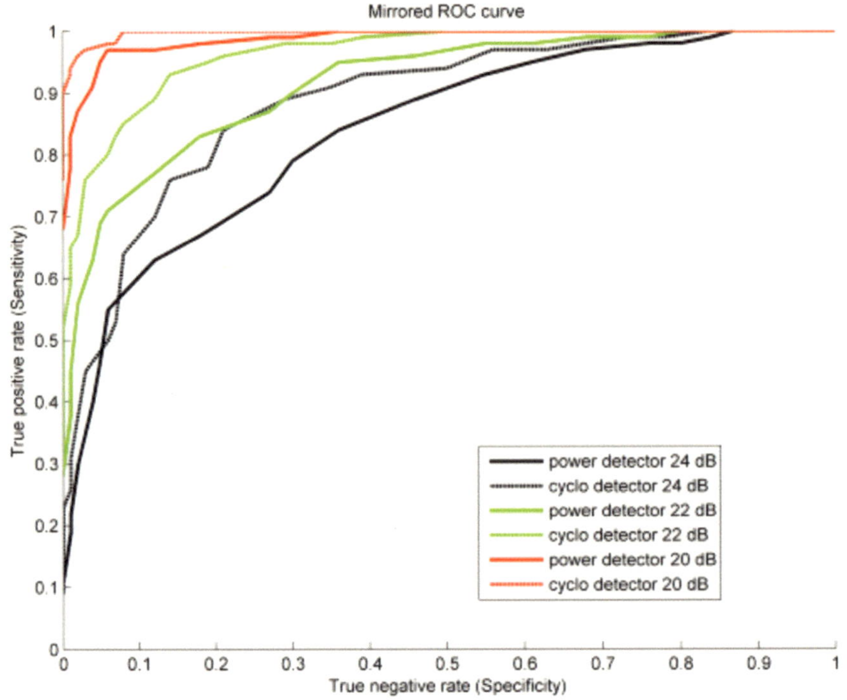

Fig. 15. Receiver operation characteristic (ROC) performance for various SNR levels, energy based detector against cyclostationary based detector.

5.3.3 Receiver operating characteristic (ROC)

Figure 15 describes the receiver operating characteristic (ROC) curves for ten thousand measurements performed for both detectors for various SNR levels. The signal is received using USRP, which was configured to process a spectrum band of 2 MHz bandwidth centered to 2.484 GHz. The experimental results show a significant increase in gain in the curves with increased SNR. The figure 15 also shows that the cyclostationary based detector perform better than the energy based detector in discriminating against the additive sensing noise. This confirms the robustness of the cyclostationary detection algorithm with low SNR in comparison to the energy detector. The trade-off is that cyclostationary detection is computationally more complex and requires significantly longer signal observation time to get a precise estimate of SCF to perform the detection.

6. Conclusion and future developments

This book chapter presents an experimental evaluation of the cyclostationary and energy based spectrum sensing techniques for cognitive radios, implemented on a functional software defined radio platform. The sensing techniques were tested against a WiMAX signal generated in an anechoic chamber emulating an AWGN communications channel. The parameters that affect the real-time detection performance of the cognitive radios were evaluated. Moreover, we presented the trade-offs among the main parameters (e.g. FFT size, decimation) in relation to different values of SNR and observation time. Our results show that the cyclostationarity based technique performs well in comparison to the energy based detector at the expense of additional computational complexity. In particular, the cyclostationary based technique requires more time samples to generate the cyclic spectral density with the needed level of accuracy.

Our future developments will include the design and implementation of an adaptive sensing algorithm, which can tune automatically to the described parameters as well as selecting the best detection techniques for any wireless communication systems (i.e. not only WiMAX) in any radio environment. The adaptive scanning approach will thus require more sophisticated and efficient cyclostationary detection methods. The authors will investigate and apply other computationally efficient algorithms such as the Strip Spectral Correlation Algorithm (SSCA) described in (Roberts, 1991).

7. References

FCC, Spectrum Policy Task Force Report, ET Docket No. 02-155, Nov 02, 2002.

FCC White Spaces Access Rules: "Unlicensed Operation in the TV Broadcast Bands", Final Rules (http://edocket.access.gpo.gov/2009/pdf/E9-3279.pdf). Last Accessed 10 September 2010.

J. Mitola and G. Maguire Jr., Cognitive Radio: Making Software Radios More Personal, *IEEE Personal Communications*, vol. 6, no. 4, pp. 13 8, Aug. 1999.

S. Haykin, Cognitive Radio: Brain-Empowered Wireless Communications, *IEEE Journal on Select Areas in Communications*, vol. 23, no. 2, pp. 201 - 220, Feb. 2005.

Ed Arslan, Cognitive Radio, Software Defined Radio, and Adaptive Wireless Systems, Springer, Netherlands, 2007.

V. Bhargava, Cognitive Wireless Communication Networks, Springer, New York, 2007.

Federal Communications Commission, Facilitating Opportunities for Flexible, Efficient, and Reliable Spectrum Use Employing Cognitive Radio Technologies, NPRM and Order, ET Docket no. 03-322, Dec. 2003.

The Commission of the European Communities, Commission Decision 2007/131/EC on allowing the use of the radio spectrum for equipment using ultra-wideband technology in a harmonised manner in the Community, Official Journal of the European Union, Feb. 21, 2007.

W.A. Gardner, Exploitation of spectral redundancy in cyclostationary signals, *IEEE Signal Processing Magazine*, Vol 8,April 1991,pp 14-36.

W. A. Gardner and L. E. Franks, ŞCharacterization of cyclostationary random processes,Ť *IEEE Trans. Inform. Theory*, vol. 21, pp. 414, 1975.

R.S. Roberts, W.A. Brown, H.H. Loomis, Computationally efficient algorithms for cyclic spectral analysis, *IEEE Signal Processing Magazine*, April 1991, Vol 8, Iss 2, pp 38-49.

S. Kandeepan; G. Baldini; R. Piesiewicz; , Experimentally detecting IEEE 802.11n Wi-Fi based on cyclostationarity features for ultra-wide band cognitive radios, *Proceedings of the 2009 IEEE 20th International Symposium on Personal, Indoor and Mobile Radio Communications*, vol., no., pp.2315-2319, 13-16 Sept. 2009.

A. Tkachenko, A.D. Cabric, R.W Brodersen, Cyclostationary Feature Detector Experiments Using Reconfigurable BEE2, *Proceedings of the 2007 IEEE International Symposium on New Frontiers in Dynamic Spectrum Access Networks (DySPAN)*, 17-20 April 2007, pp 216-219, Dublin.

Kyouwoong Kim Akbar, I.A. Bae, K.K. Jung-sun Urn Spooner, C.M. Reed, J.H., Cyclostationary Approaches to Signal Detection and Classification in Cognitive Radio, *Proceedings of the 2007 IEEE International Symposium on New Frontiers in Dynamic Spectrum Access Networks (DySPAN)*, 17-20 April 2007 pp 212-215, Dublin.

S. Enserink, D. Cochran, "A cyclostationary feature detector", *Proceedings of the Twenty-Eighth Asilomar Conference on Signals, Systems and Computers*, 31 Oct-2 Nov 1994 Vol 2, pp806-810, Pacific Grove, CA.

M. Oner; F. Jondral, "Cyclostationarity based air interface recognition for software radio systems", *Proceedings of the IEEE Radio and Wireless Conference*, 19-22 Sep. 2004 pp 263-266.

Eric Blossom, Exploring GNU Radio. http://www.gnu.org/software/gnuradio/doc/exploring-gnuradio.html. Last Accessed 16 September 2008.

H. Urkowitz, Energy Based Detection of Unknown Deterministic Signals, *IEEE Proceedings*, Vol.55, No.4, page:523-531, April 1967.

T. Yucek; H. Arslan; "A survey of spectrum sensing algorithms for cognitive radio applications," *IEEE Communications Surveys & Tutorials*, vol.11, no.1, pp.116-130, First Quarter 2009

J. Vankka; M. Kosunen; I. Sanchis; K.A.I. Halonen; "A multicarrier QAM modulator," *Circuits and Systems II: IEEE Transactions on Analog and Digital Signal Processing*, vol.47, no.1, pp.1-10, Jan 2000

A. V. Dandawatk and G. B. Giannakis, "Statistical Tests for Presence of Cyclostationarity", *IEEE Transactions on signal processing*, vol. 42, no. 9, September 1994.

Hosseini, S.M.A.T.; Amindavar, H.; Ritcey, J.A.; , "A new cyclostationary spectrum sensing approach in cognitive radio," *Proceedings of the 2010 IEEE Eleventh International Workshop on Signal Processing Advances in Wireless Communications (SPAWC)*, vol., no., pp 1-4, 20-23 June 2010.

Lunden, J.; Koivunen, V.; Huttunen, A.; Poor, H.V.; , "Collaborative Cyclostationary Spectrum Sensing for Cognitive Radio Systems," *IEEE Transactions on Signal Processing*, vol.57, no.11, pp 4182-4195, Nov. 2009.

Sutton, P.D.; Nolan, K.E.; Doyle, L.E.; , "Cyclostationary Signatures in Practical Cognitive Radio Applications," *IEEE Journal on Selected Areas in Communications*,vol.26, no.1, pp 13-24, Jan. 2008.

Axell, E.; Larsson, E.G.; , "Optimal and Sub-Optimal Spectrum Sensing of OFDM Signals in Known and Unknown Noise Variance," *IEEE Journal on Selected Areas in Communications*, vol.29, no.2, pp.290-304, February 2011

3

Modulation Classification in Cognitive Radio

Adalbery R. Castro, Lilian C. Freitas, Claudomir C. Cardoso,
João C. W. A. Costa and Aldebaro B. R. Klautau
*Signal Processing Laboratory (LaPS) and Applied Electromagnetism Laboratory (LEA) –
Federal University of Pará (UFPA), Belém – PA
Brazil*

1. Introduction

The automatic modulation classification (AMC) problem aims at identifying the modulation scheme of a given communication system with a high probability of success and in a short period of time. AMC has been used for decades in military applications in which friendly signals should be securely transmitted and received, whereas hostile signals must be located, identified and jammed (Gardner, 1988). More recently, the interest for AMC has been renewed by the research on cognitive radios (Haring et al., 2010; Wang & Wang, 2010; Xu et al., 2011), where AMC plays an important role in spectrum sensing (Haykin et al., 2009).

The AMC approaches are typically organized in *likelihood* and *feature-based* methods (Dobre et al., 2007). Alternatively, the AMC methods are distinguished in this chapter by the corresponding learning algorithm: *generative* (also called *informative*) or *discriminative* (Rubinstein & Hastie, 1997). One advantage of this nomenclature is to benefit from the insights accumulated in the machine learning community with respect to the generative and discriminative approaches (Long & Servedio, 2006).

Generative algorithms perform the classification based on probabilistic models that are typically constructed by estimating probability distributions for each class separately. Examples are Naïve Bayes (Tan et al., 2006), hidden Markov models obtained with maximum likelihood estimation (Bremaud, 2010) and methods that uses likelihood ratio tests such as the average likelihood ratio test (ALRT) (Su et al., 2008), generalized likelihood ratio test (GLRT) (Xu et al., 2011) and the hybrid likelihood ratio test (HLRT) (Polydoros, 2000).

Discriminative algorithms are used to learn classifiers that focuses in class boundaries, not on modeling distributions. Examples include support vector machines (SVM) (Cortes & Vapnik, 1995) and neural networks (Krose & van der Smagt, 1996).

There are research of AMC (Dobre et al., 2007; Xu et al., 2011) and spectrum sensing (Haykin et al., 2009; Yucek & Arslan, 2009). Instead of providing an overview of alternatives, this chapter compares two AMC methods: one that uses the well-established features based on cyclostationary analysis (Gardner & Spooner, 1992; Haykin et al., 2009) and the recently proposed *CSS* (concatenated sorted symbols) front end (Muller et al., 2011). The idea is to contrast the characteristics of these methods and emphasize practical aspects. In the experiments, SVM is adopted as the learning algorithm given its performance in many classification tasks. Moreover, the implementation on a field-programmable gate array

(FPGA) of an AMC system using CSS and SVM is discussed and suggested the hardware requirements of an AMC module.

2. The modulation classification problem

An AMC system consists of a *front end* and a *back end* or *classifier*. The *front end* converts the received signal $r(t)$ to a vector $\mathbf{x}[k], k = 1, \ldots, N$ composed of N elements. Having $\mathbf{x}[k]$ as input, the *classifier* decides the class $y \in \{1, \ldots, C\}$ among C pre-determined modulation schemes. The process is depicted in the diagram below:

$$r(t) \text{ (signal)} \rightarrow \boxed{\text{front end}} \rightarrow \mathbf{x}[k] \text{ (parameters)} \rightarrow \boxed{\text{classifier}} \rightarrow y \text{ (class)}$$

There are several options to implement the front end (Mishali & Eldar, 2011), but in order to be concrete, the following alternative will be assumed:

$$r(t) \rightarrow \boxed{\text{FILT/DOWN}} \rightarrow c(t) \rightarrow \boxed{\text{A/D}} \rightarrow c[n] \rightarrow \boxed{\text{DSP}} \rightarrow s[n] \rightarrow \boxed{\text{parameter extraction}} \rightarrow \mathbf{x}[k]$$

where FILT/DOWN denotes operations such as filtering, down-conversion and signal conditioning, $c(t)$ is the input to the analog-to-digital (A/D) converter, $s[n]$ is obtained from $c[n]$ via digital signal processing (DSP) and the last block converts the time-domain signal $s[n]$ into the parameters of interest.

A classifier in a likelihood-based AMC (Su et al., 2008) assumes that $s[n]$ is a sequence of N received *symbols* (Proakis, 2001) and $\mathbf{x}[k] = s[n]$. Hence, these classifiers are considered here to be a special case of feature-based classifiers. The adoption of symbols as features is restricted to the case where the modulation is digital and linear, such as QAM, PSK, etc. (Proakis, 2001). Hence, the likelihood-based AMC can be implemented as

$$r(t) \rightarrow \boxed{\text{FILT/DOWN}} \rightarrow c(t) \rightarrow \boxed{\text{A/D}} \rightarrow c[n] \rightarrow \boxed{\text{DSP}} \rightarrow \mathbf{x}[k],$$

where $c[n]$ is the *complex envelope* (Proakis, 2001). In this case, the DSP block samples the complex envelope according to the signaling rate. This kind of model is widely adopted in the AMC literature (Mak et al., 2007).

In practice, several impairments must be mitigated to have $\mathbf{x}[k]$ as a reasonable approximation of the transmitted symbols, such as lack of synchronism, carrier frequency offset and channel noise. The front end that perfectly recovers the transmitted symbols in a digital modulation is called here *canonical*. Several works assume such front end and then contaminate $\mathbf{x}[k]$ by additive white Gaussian noise (AWGN) and other impairments. Alternatively, a front end based on cyclostationarity can extract features other than the symbols. This chapter focuses on the two distinct front ends for AMC and one classification technique, which are discussed in the sequel.

2.1 Front end: CSS

The CSS front end has been recently proposed by Muller et al. (2011) and uses the symbols of the constellations (Proakis, 2001) as input parameters for the classifier. The CSS front end takes the magnitude and the phase of the received symbols, normalizes them and sorts them separately. The two ordered vectors (magnitude and phase) are concatenated, generating a new vector with length $D = 2N$, which should reflect an individual *signature* of the corresponding constellation. Recall that N is the number of symbols.

For example, Fig. 1(a) and Fig. 1(b) represent the constellations of noise-free (ideal) 16QAM and 8PSK modulations, respectively. An example of two possible vectors of parameters

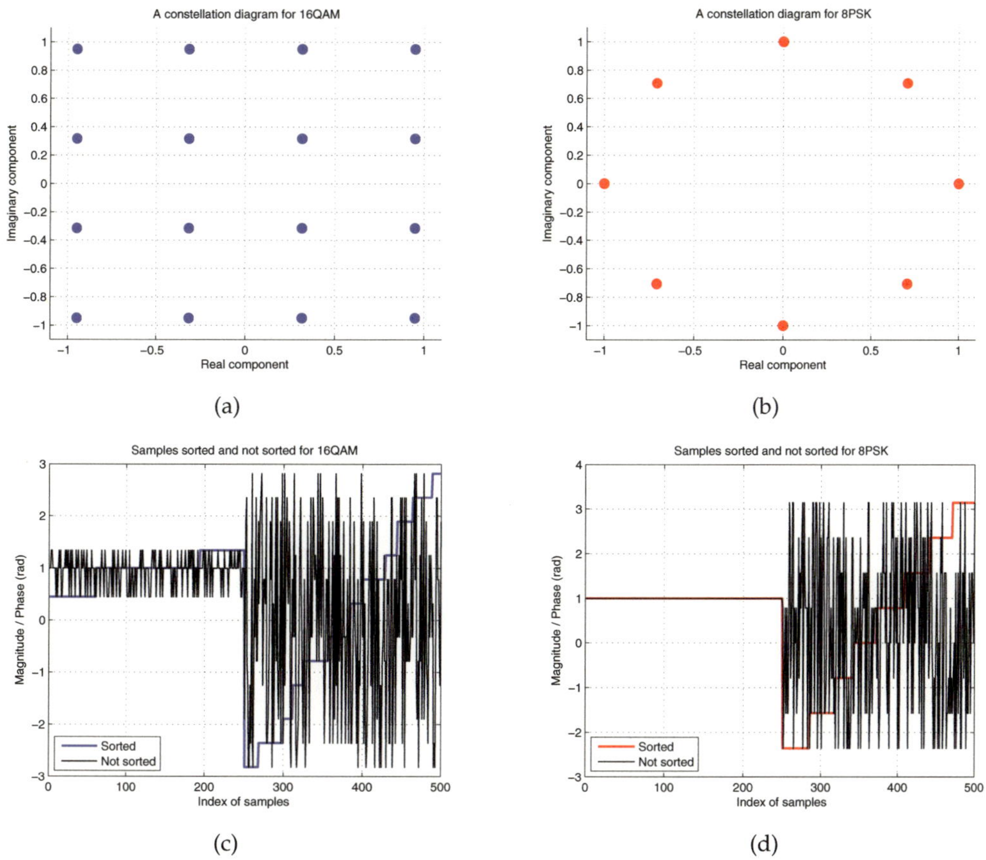

Fig. 1. Examples of vector of parameters with $D = 2N = 500$ for a 16QAM and 8PSK modulation, without noise. (a) A constellation diagram for 16QAM. (b) A constellation diagram for 8PSK. (c) Samples sorted and not sorted for 16QAM. (d) Samples sorted and not sorted for 8PSK.

representing a 16QAM and 8PSK modulations is also illustrated. There is no noise and both curves of received symbols, before and after ordering, are shown in Fig. 1(c) and Fig. 1(d). Each vector is composed of $D = 2N = 500$ features (corresponding to 250 magnitudes and 250 phases). It is observed that ordering creates a pattern that can be used to identify the type of modulation used in the generation of the received signal. For example, one can note from Fig. 1(d) that all first $N = 250$ normalized and ordered symbols corresponding to the magnitude of the 8PSK modulation are equal to one. For improved clarity, Fig. 2 summarizes the information in Fig. 1 by showing the comparison between the signatures of the 16QAM and 8PSK modulations provided by the CSS front end. For obtaining Fig. 3 and Fig. 4, white

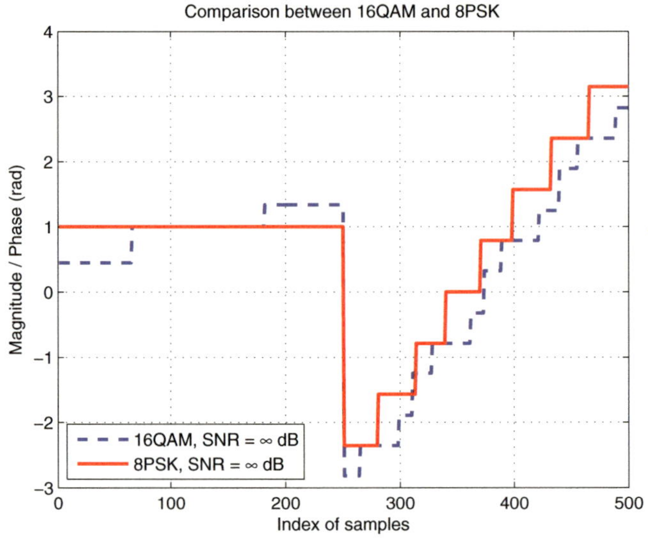

Fig. 2. Examples of vector of parameters with $D = 2N = 500$ for comparison between 16QAM and 8PSK under ideal conditions.

Gaussian noise was added to achieve signals with SNR = 15 dB. It can be observed that the signatures are modified with respect to the ideal case and their differences are less visible. However, in Muller et al. (2011) it was shown that even with noise, it is possible to distinguish these two modulations when the SNR is large enough by using, for example, an SVM classifier.

The next subsection presents an alternative front end, which will be used for performance comparisons.

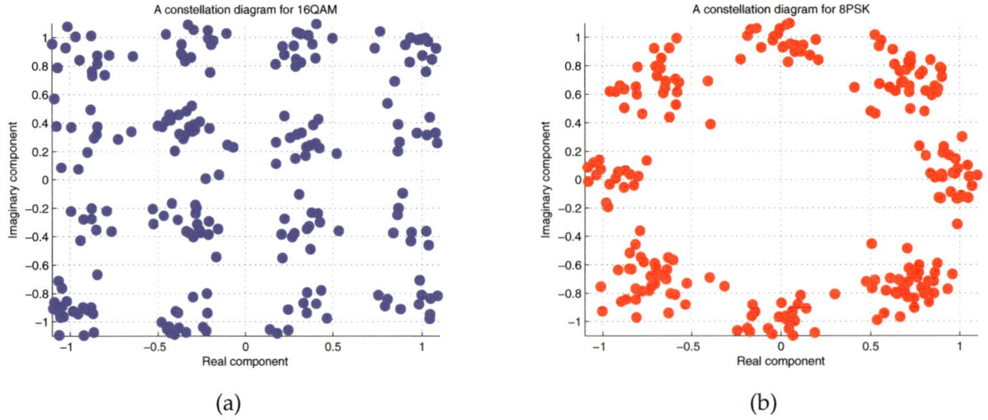

Fig. 3. 16QAM and 8PSK modulation, with SNR = 15 dB. (a) A constellation diagram for 16QAM. (b) A constellation diagram for 8PSK.

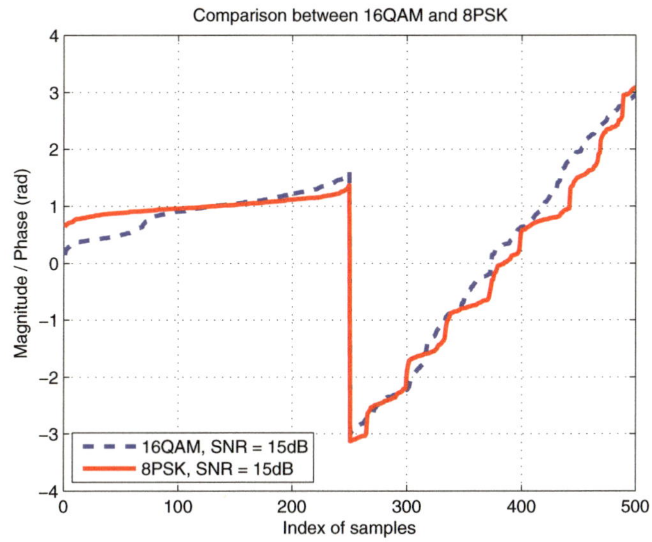

Fig. 4. Examples of vector of parameters with $D = 2N = 500$ for comparison between 16QAM and 8PSK. SNR = 15 dB.

2.2 Front end: Cyclostationarity

Cyclostationary analysis has been increasingly considered for use in a large range of applications, including signal detection, classification, synchronization and equalization. In Gardner & Spooner (1992), a large number of advantages of cyclostationary analysis are identified when compared to radiometric approaches (which are based in the measured energy of the received signal). Among its advantages are the reduced sensibility to noise and interfering signals, and also its ability to extract signal parameters such as the carrier frequency and the symbol rate.

In spite of its robustness, it is well-known that the computational cost of a cyclostationary analysis is high. According to the work of Lin & He (2008), a discrete-time signal $x[n]$ is defined to be cyclostationary if its autocorrelation function is invariant in relation to time shifts to all integer m multiple of T_o, that is

$$R_x(n + mT_0, 1) = R_x(1). \tag{1}$$

Two mathematical functions are used to characterize cyclostationary signals. The first is the cyclic autocorrelation function (CAF) (Castro, 2011), defined by

$$R_x^{\alpha}(1) = \lim_{N \to \infty} \frac{1}{2N+1} \sum_{n=-N}^{N} \{x[n+l]e^{-j2\pi\alpha(n+l)}\}\{x[n]e^{-j2\pi\alpha n}\}^* \tag{2}$$

where R_x^{α} denotes the CAF for a discrete-time signal $x(n)$ and α is referred to as the cyclic frequency. When the signal $x(n)$ is cyclostationary, its CAF is nonzero at some time delay 1 and cyclic frequency $\alpha \neq 0$ (Wang et al., 2010).

The second function, the spectral correlation density (SCD), or cyclic spectral density, is calculated from the Fourier transform of the CAF:

$$S_x^\alpha(k) = \sum_{l=-\infty}^{\infty} R_x^\alpha(l) e^{-j2\pi kl}, \tag{3}$$

where k is the frequency. When $\alpha = 0$, the CAF and SCD represent the autocorrelation and power spectral density functions, respectively.

2.2.1 Estimation of the SCD

When the AMC is based on cyclostationary features, estimating the SCD is fundamental to extract features that should distinguish modulated signals. Two algorithms to estimate the SCD were proposed in Schnur (2009): FFT Accumulation Method (FAM) and Strip Spectral Correlation Algorithm (SSCA). The FAM algorithm is considered to be more computationally efficient than the SSCA and is adopted in the sequel.

Assuming discrete-time processing, the FAM algorithm calculates

$$S_x^\alpha(k) \approx S_x^\alpha(n,k) = \frac{1}{N}\sum_{n=0}^{N-1}\left[\frac{1}{N'} X_{N'}(n, k+\frac{\alpha}{2}) X_{N'}^*(n, k-\frac{\alpha}{2})\right] \tag{4}$$

where N is the number of time samples within the range of observation of the signal, k and α are the frequency and cyclic frequency, respectively. The parcels $X_{N'}(n, k \pm \frac{\alpha}{2})$ represent the complex envelope of the spectral component of $x[n$ and can be computed in the following way:

$$X_{N'}(n,k) = \sum_{r=-N'/2}^{N'/2} a[r]x[n-r]e^{-j2\pi k(n-r)T_s}, \tag{5}$$

where $a[r]$ is the data taper window (for instance Hamming window) and T_s is sampling period. In this method, the complex envelopes are estimated by means of sliding N' FFT points, followed by a downshift in frequency to baseband. For improved clarity, Fig. 5 represents the diagram implementation of this method. Fig. 5 pictorially represents the steps for estimating the SCD, which are:

- The input sample sequence $x[n]$ of length N is divided into P blocks, where each block containing N' samples, and L is the overlap factor;
- A Hamming window is applied across each block;
- The FFT of each block is computed;
- The complex envelopes $X_{N'}(n,k)$ are downshift in frequency to baseband;
- The SCD function is estimated by multiplying $X_{N'}(n,k)$ by its complex conjugate;
- The smoothing operation of the product sequences is executed by means of P-points FFT.

The value of L is configured to be equal to $N'/4$, because this value is a good trade off between computational efficiency and minimizing cycle leakage and cycle aliasing. The value of N' is

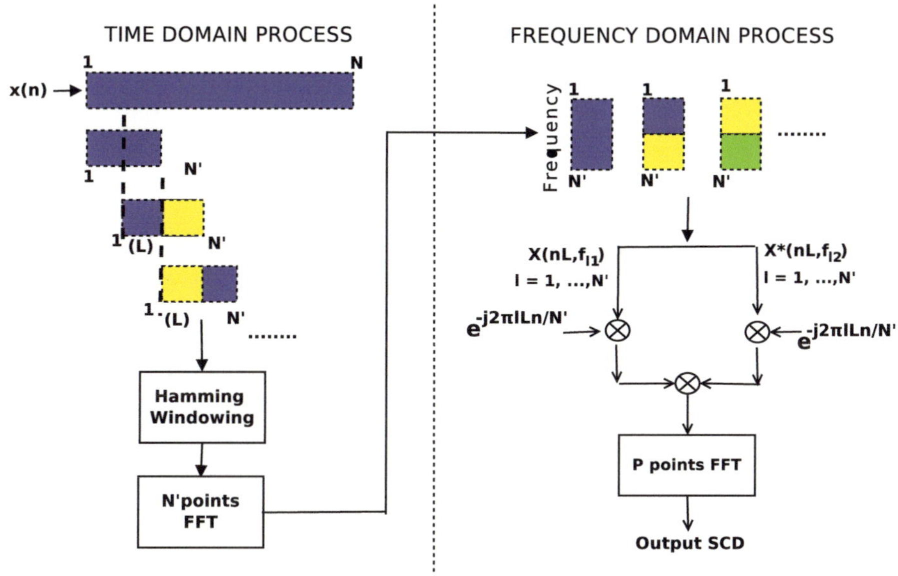

Fig. 5. Implementation of the method FAM.

computed according to the desired resolution Δk, and is defined by:

$$N' = \frac{f_s}{\Delta k}. \tag{6}$$

The value of P is determined according to the desired cyclic frequency resolution $\Delta\alpha$, and is given by:

$$P = \frac{f_s}{L\Delta\alpha}. \tag{7}$$

To perform AMC by means of cyclostationarity features it is typical to normalize the SCD. This normalization can be obtained by:

$$C_x^{\alpha}(k) = \frac{S_x^{\alpha}(k)}{[S_x^0(k+\alpha/2)S_x^0(k-\alpha/2)]^{1/2}}, \tag{8}$$

where $C_x^{\alpha}(k)$ is the spectral autocoherence function.

Fig. 6 shows the estimation of the SCD for BPSK and QPSK modulations respectively. This example adopted a sampling frequency $f_s = 8192$ Hz, carrier frequency $K = 2048$ Hz, cyclic frequency resolution $\Delta\alpha = 20$ Hz and frequency resolution $\Delta k = 80$ Hz. It can be noticed that the SCD is three-dimensional. The features that will distinguish each modulation, that is, the cyclic domain profile (CDP), is obtained from the SCD. The CDP $I(\alpha)$ uses only the peak values in the SCD and is obtained by

$$I(\alpha) = \max_k |C_x^{\alpha}(k)|. \tag{9}$$

The CDP for the BPSK and QPSK modulations of Fig. 6 are shown in Fig. 7.

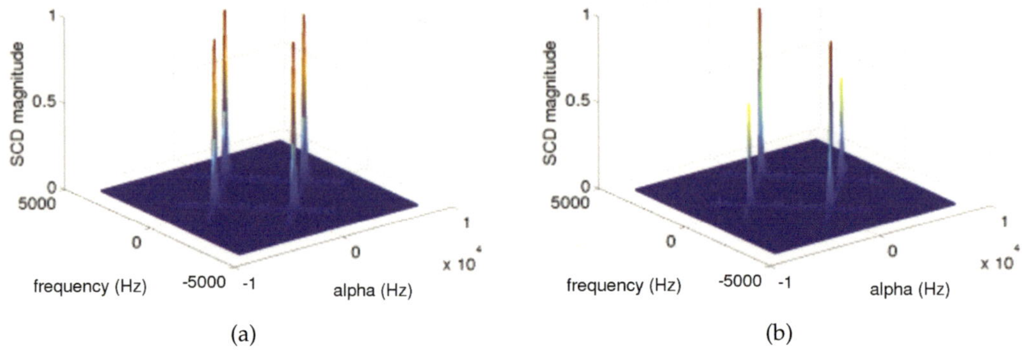

Fig. 6. Spetral Cyclic Density. (a) BPSK. (b) QPSK

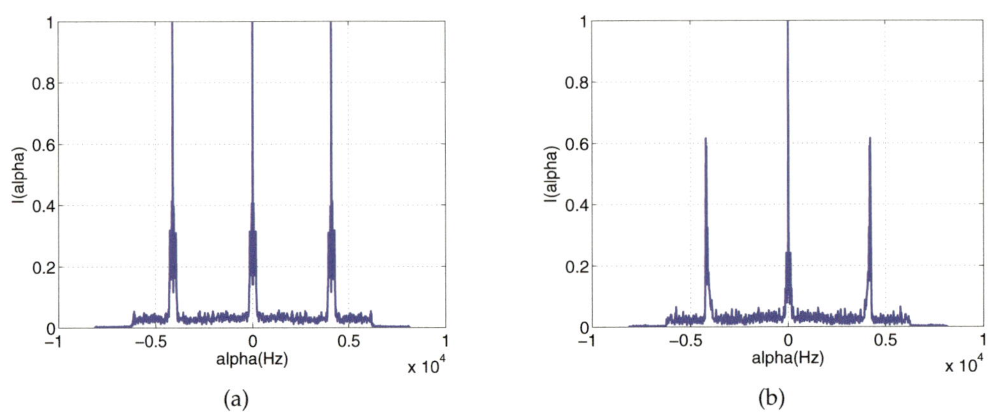

Fig. 7. Cyclic Perfil. (a) BPSK. (b) QPSK

2.3 Back end: SVM classifier

Support vector machine (SVM) is a class of learning algorithms based on the statistical learning theory, which implements the principle of the structural risk minimization (Vapnik, 1998). A basic idea of SVM is to map the input space into a feature space. This mapping can be done linearly or not, according to the kernel function used for the mapping. In the literature, various possibilities for SVM kernels are presented in applications involving pattern recognition such as: linear kernel, polynomial kernel, gaussian kernel and radial basis network (Burges, 1998).

In the feature space, an SVM builds a *maximum margin* hyperplane **w** to separate classes while minimizing the classification error. The hyperplane can be written as a combination of few points (training examples) in the feature space, called the *support vectors* of the optimal hyperplane.

Maximum margin is defined as the shortest distance that separates the training examples of different classes in relation to the hyperplane, as seen in Fig. 8. The distance of any point $\mathbf{x}_i$ to the hyperplane is given by Equation 10, where $||\mathbf{w}||$ is the norm of the vector.

$$d = \frac{\langle \mathbf{w},\mathbf{x}\rangle + \mathbf{b}}{||\mathbf{w}||} \tag{10}$$

An SVM is a binary classifier given by

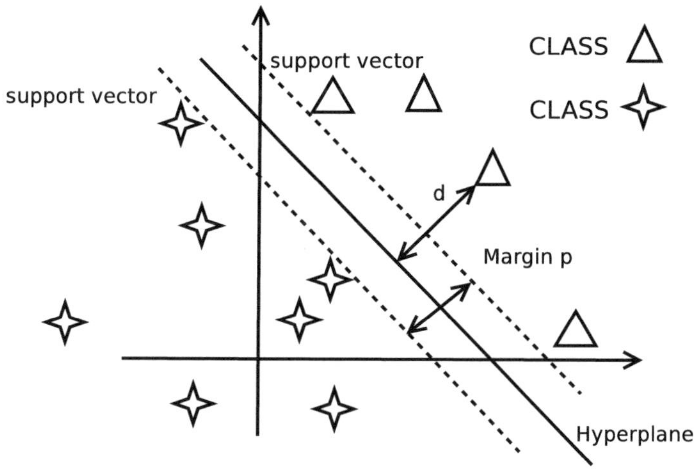

Fig. 8. Examples of the margin and support vectors.

$$f(\mathbf{x}) = \sum_{m=1}^{M} \gamma_m \mathcal{K}(\mathbf{x}, \mathbf{x}_m) + c, \tag{11}$$

where $\mathcal{K}(\mathbf{x}, \mathbf{x}_m)$ is the kernel function between the test vector $\mathbf{x}$ and the m-th training example $\mathbf{x}_m$, with $c, \gamma_m \in \Re$. The effectively used examples have $\gamma_m \neq 0$ and are called *support vectors*. The number $V \leq M$ of support vectors can be large and impact the computational cost. An SVM with a linear kernel $\mathcal{K}(\mathbf{x}, \mathbf{x}_m) = \langle \mathbf{x}, \mathbf{x}_m \rangle$ given by the inner product between $\mathbf{x}$ and $\mathbf{x}_m$, can be converted to a perceptron $f(\mathbf{x}) = \langle \mathbf{a}, \mathbf{x} \rangle + c$, where $\mathbf{a} = \sum_{m=1}^{M} \gamma_m \mathbf{x}_m$ is pre-computed.

Therefore, linear SVMs were adopted in this chapter due to their lower computational cost when compared to non-linear SVMs with kernels such as the Gaussian (Cristianini & Shawe-Taylor, 2000). To combine the binary SVMs $f_b(\mathbf{x}), b = 1, \ldots, B$, to obtain $F(\mathbf{x})$ this work adopted the *all-pairs* error-correcting output code (ECOC) matrix with Hamming decoding (Allwein et al., 2000), where the winner class is the one with the majority of "votes". Note that an alternative to all-pairs, which uses $B = 0.5C(C-1)$ SVMs, is the one-vs-all ECOC that uses $B = C$ SVMs (Klautau et al., 2003).

3. Results

3.1 Simulation results

Simulations were performed to compare the performance of the CSS and cyclostationarity front ends to classify the modulations BPSK, 4-PAM, 16-QAM and 8-PSK in channels with additive white Gaussian noise (AWGN). For these simulations, linear SVM classifiers were trained using several integer SNR values in the range [-5, 15] dB. The training and test sets used in all simulations were made disjoint and each had 500 examples. All the constellations were normalized to have unitary energy. For the CSS front end, $N = 250$ symbols were used per training / test instance. For the cyclostationarity front end, all signals were generated with sampling frequency $f_s = 8192$ Hz, carrier frequency $K = 2048$ Hz, cyclic frequency resolution $\Delta\alpha = 20$ Hz and frequency resolution $\Delta k = 80$ Hz.

The results are shown in Fig. 9. Note that the approach based on the CSS front end considers that the received signal was properly demodulated, whereas the cyclostationarity front end does not require demodulation and, consequently, demand less knowledge about the input signal. In fact, the cyclostationary analysis itself can be used to estimate important parameters for the demodulation task, such as the carrier frequency and symbol rate. Because of these aspects, a complete AMC may use more than one complementary front ends. It is observed

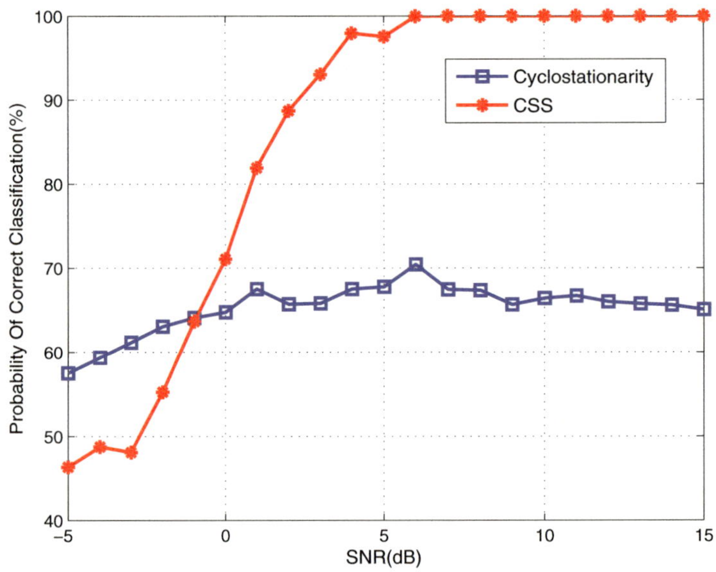

Fig. 9. Probability of correct classification of the cyclostationarity and CSS front ends.

from Fig. 9 that for the adopted AMC problem, the CSS outperformed the cyclostationarity front end. This is due to the difficulty of correctly separating cyclostationarity pair of modulations [BPSK, 4PAM] and [16QAM, 8PSK]. For both pairs, the cyclical characteristics are similar, as shown by the confusion matrix in Table 1 and discussed in A. Fehske (2005). For SNR = -1 dB, the performance of both techniques is similar. The confusion matrix shown in Table 2 and Table 3 illustrate that both front ends do not distinguish the same pairs of modulations. For SNR values larger than -1 dB, the CSS presents better results than the cyclostationarity. Although the cyclostationarity front end has not presented good results for the set of adopted modulations (BPSK, 4PAM, 8PSK and 16QAM), it is capable of producing good results for others (A. Fehske, 2005; da Silva et al., 2007; Haykin et al., 2009). Also, the CSS is restricted to linear digital modulations while the cyclostationarity is not. In spite of

classified as ->	BPSK	4PAM	16QAM	8PSK
BPSK	291	209	0	0
4PAM	157	344	0	0
16QAM	0	1	331	168
8PSK	0	0	164	336

Table 1. Confusion matrix for cyclostationarity. SNR = 15 dB

classified as ->	BPSK	4PAM	16QAM	8PSK
BPSK	391	109	0	0
4PAM	219	279	2	0
16QAM	6	1	275	218
8PSK	5	0	144	351

Table 2. Confusion matrix for cyclostationarity. SNR = -1 dB

classified as ->	BPSK	4PAM	16QAM	8PSK
BPSK	436	64	0	0
4PAM	204	295	0	1
16QAM	0	0	316	184
8PSK	0	0	274	226

Table 3. Confusion matrix for CSS. SNR = -1 dB

these two aspects, the presented results confirm that CSS is a competitive technique for AMC. Thus, the CSS front end was implemented in a FPGA for investigating its real-time processing capabilities, as described in the sequel.

3.2 Implementation results

Most of the AMC research is based on computer simulations and does not target the hardware implementation. In addition, there are few commercial devices aimed at detecting empty channels or classify modulations. But the available equipment is proprietary, directed generally to the military. Thus, there is great interest in academia and also in the industry to implement and test algorithms for AMC and spectrum sensing (Mishali & Eldar, 2011) to be easily embedded in devices such as FPGA or DSP (digital signal processor).

The next subsection describes the implementation of a CSS-SVM classifier for AMC starting by the classifier.

3.2.1 Architecture of the programmable SVM classifier

For the implementation of an SVM classifier in FPGA using the all-pairs ECOC, an architecture was designed in which the test instance (input parameters $\mathbf{x}[k]$) is continuously classified and the coefficients of the SVMs can be changed on-the-fly. This is a programmable architecture proposed for multiclass classification, using binary classifiers. The training of the classifiers is performed offline.

Equation 12 represents the function for the decision problem between two classes used by the binary SVM. Where **w** and **b** are the coefficients of the classifier and **x** is the vector test. The sign of **f(x)** indicates the result of the classifier, with **f(x)** $= 0$ the decision threshold between the two classes for which the classifier was trained.

$$\mathbf{f(x)} = \sum_{i=1}^{n} \mathbf{w}_i \mathbf{x}_i + \mathbf{b}. \tag{12}$$

To combine binary SVMs it was used matrix all-pairs ECOC with decoding Hamming, where the winner class was chosen by having most of the "votes". In other words, each one of binary SVM is trained to distinguish a pair of classes and, in the test phase, the chosen class was the one that had the largest number of binary SVMs, indicating the winner (Muller et al., 2011).

Fig. 10 shows the process of programming the FPGA with the SVM coefficients and reading the input test data. The structure uses a demultiplex followed by groups of shift registers. The value of the input **i** is stored in a shift register, and several registers are disposed in sequential arrangements with the information moved by the circuit until all the registers are updated. In Fig. 10, n is the number of features of the test data. Considering y classes, recall that the number B of binary classifiers for the all-pairs ECOC is $B = (y(y-1))/2$. The registers **w** and **b** store the coefficients of the classifiers and **x** stores the data for testing. After updating the coefficients, the test values are stored and classification can be started. If necessary to

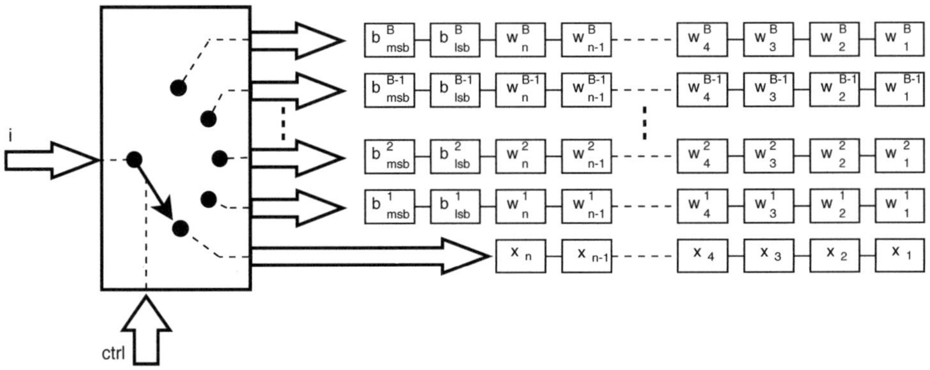

Fig. 10. Diagram representing the on-the-fly programming of SVM coefficients and input data. The row at the bottom corresponds to the input data and the other rows to SVM coefficients. The ctrl input controls the demultiplexer.

update the SVM coefficients, the flow of test data will be suspended for the time necessary. In these cases, the input **i** will be directed to the corresponding SVM register set and not to the registers that store the test data.

The proposed SVM architecture uses four steps, which are depicted in Fig. 11. The implementation uses a state machine and can be understood as:

- **The first step** makes the multiplication of elements of the vectors **w** and **x** shown in Equation 12 and stores the results in other sets with the same number of registers. However, to be able to store the multiplication results without roundoff errors, the registers that receive the results have twice the number of bits of the registers that store the vectors **w** and **x**. For each SVM classifier, a new set of registers is necessary to receive the results of multiplications.
- **The second step** performs, for each classifier, the sum of all values from the first step, added to the coefficient **b** (the "y-intercept" of a linear SVM).

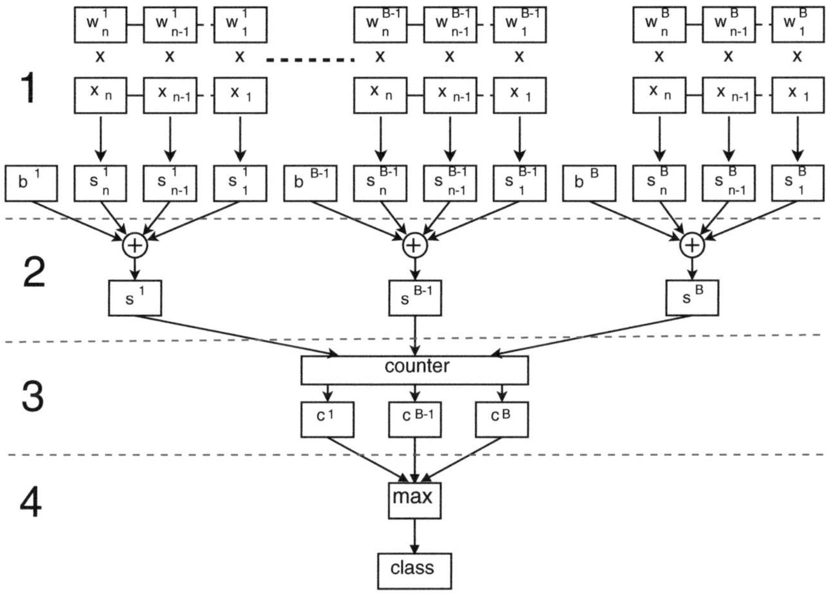

Fig. 11. Diagram that performs the classification.

- **The third step** checks the result of each classifier. As binary classifiers, the result of a classifier can be considered as a vote given to the one of the classes for which it was trained. At the end of the third step, we have totaled the number of votes for each of the classes.
- **The fourth step** verifies the class that received the largest number of votes counted in the third step. The result of this step is the output of the classification.

It is important to note that this architecture is tailored to implementation in a FPGA, where it is possible to describe hardware that performs various operations on each clock cycle (clock). In this case, each of the steps to perform the calculation of the SVM should occur in one clock period. In the first step, for example, in a single clock cycle, all the multiplications needed to compute the inner product between the test vector and the coefficients of binary classifiers are performed. The disadvantage of this architecture is the large amount of FPGA resources used, which make this approach feasible only when the number of coefficients is small. This is because the architecture, as described, uses only registers and logic elements to execute several multiplications in one clock period. A better use of the FPGA resources is obtained by using its RAM memory, as will be described together with the implementation of the CSS.

3.2.2 Architecture of the CSS-SVM modulation classifier

The front end CSS proposed by Muller et al. (2011) uses the symbols represented by magnitude and phase. For a set of received symbols, an ordered vector with the magnitude is concatenated with a vector with ordered phases, generating a third vector x with twice the number of samples.

For implementing the CSS front end in a FPGA, the VHDL language (*IEEE Standard VHDL Language Reference Manual*, 2009) was adopted within the Altera Quartus II development platform (Altera, 2011). In order to explain the proposed architecture, first a high-level description will be provided, which treats the CSS-SVM as a processing block. Emphasis is placed on its inputs (symbols for being classified and SVM coefficients) and output (the classification result). Fig. 12 depicts the implemented block. The symbols can be entered in the same sequence they are received. The implementation assumes that demodulation has been performed and the CSS implementation corresponds to ordering the magnitudes and phases of the symbols. This ordering and the SVM-based classification are performed by the cited block. The signals in Fig. 12 have the following role:

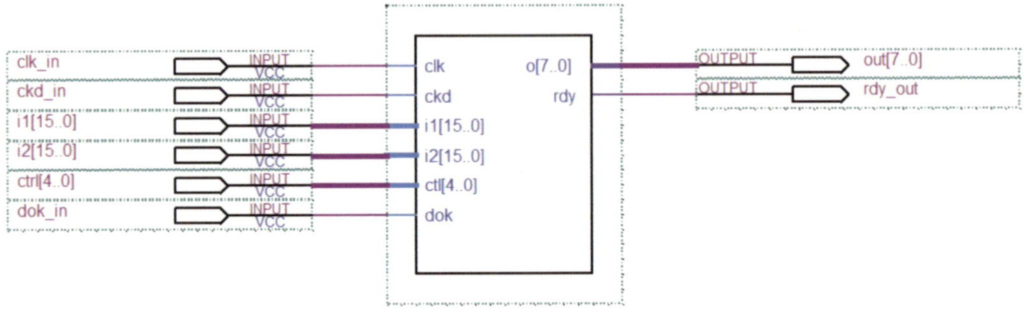

Fig. 12. Representation of the implemented CSS-SVM block with its inputs and outputs signals.

- **clk** represents the clock and is the base time system.
- **ckd** indicates the arrival of a new sample for classification or a new coefficient for the classifiers.
- **i1** and **i2** are the inputs that can be used both for input symbols that will be classified and for the SVM coefficients. The symbols or coefficients values should reach the block with a rate defined by the signal **ckd**. In this implementation, the rising clock edge is used to update the signals **i1** and **i2**. For these signals it was used the two's complement notation, with 16 bits for magnitude and 16 bits for phase for each of symbols that will be classified.
- **dok** indicates that a whole set of N test symbols has been informed and a new classification can begin.
- **Ctrl** is a control signal used with the inputs **i1** and **i2** that can be used to provide the symbols or the SVM coefficients. Obviously, all the SVM coefficients should be informed before symbols are received. When zero is the value given in **Ctrl**, **i1** and **i2** must contain, respectively, the magnitude and phase of a new symbol. For other values, it is understood that the value given in **i1** is a coefficient. In this case, the value of **Ctrl** tells to which classifier the coefficient corresponds and **i2** the position of the coefficient.

As explained, mapping all the processing into registers, logic elements and multipliers consume too many resources of the FPGA, because all values of samples and coefficients need to be available for accomplishing many multiplication operations in a single clock cycle. Another way to implement the CSS-SVM classifier is to direct all coefficients and symbols to be stored in RAM, which are available in FPGAs such as Altera's Cyclone II.

In this improved architecture, the SVM coefficients are separated into different blocks in the FPGA memory. The disadvantage is that only one position of each memory block can be accessed in a clock cycle and, consequently, a greater number of clock cycles is required for performing a classification. In the architecture described in Section 3.2.1, the rate of symbols input could be the same as the clock, but in the improved architecture, the clock rate must be higher than the rate of input symbols. For $N = 250$, the clock rate should be approximately 130 times the symbol rate and for clk = 50 MHz (1/clk = 20 ns) took 627.62 us to perform the classification of a set of symbols.

Fig. 13 represents the different memory blocks used in the classifier. The upper lines (**w**) represent the memory blocks for the SVM coefficients and they have $D + 2$ positions due to the fact that storing the coefficient **b** consumes twice the number of bits used by the other coefficients. The explanation is that, during the simulations, the coefficients **b** were an order of magnitude larger than the other coefficients. Therefore, they are represented using two memory positions. The lower (smaller) blocks represent the memory space to store the

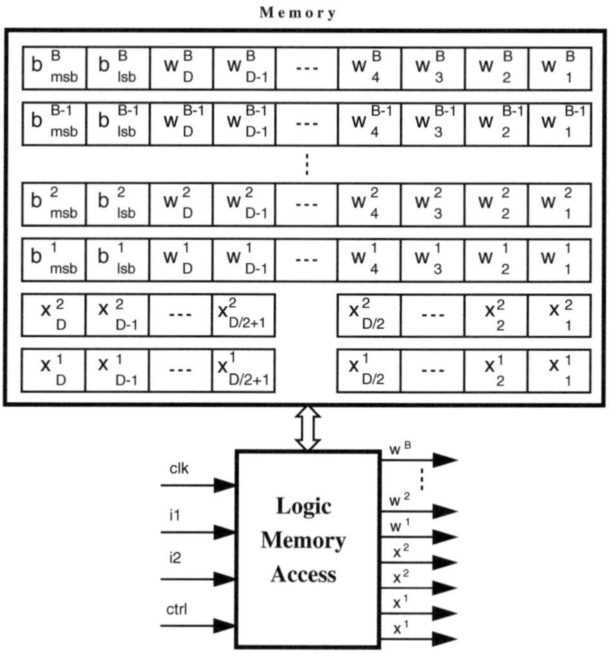

Fig. 13. Representation of the memory blocks.

symbols to be classified and that will compose the vector (**x**). As the FPGA design allows accessing memory blocks one element at a time, the set of test symbols are stored in two separate memory blocks: one for the magnitude and another block for the phase.

For supporting a steady stream of symbols, there are two sets with two memory blocks each. The classification process works on a set of previously received symbols while new symbols are received and stored.

The signal **dok** indicates that a new classification should start. A sorting algorithm is used to sort the values of magnitude and phase before being classified through the SVM classifier. During this process, new symbols received are directed to another memory block.

After sorting, the multiplication of vectors (inner product between) **w** and **x** is performed according to Equation 12, for each classifier. This stage requires D clock cycles because each cycle is used to perform the multiplication of a value of vector **x** with a coefficient of each classifier, which means that each cycle makes B multiplications. There are B registers with the SVM output values **f(x)**, one for each SVM classifier. The next step is to count the votes of the classifiers. Finally, the result of the CSS-SVM classifier is the modulation that received the majority of votes.

Table 4 shows the used resources of an Altera Cyclone II FPGA model EP2C20F484C7 to program a CSS-SVM classifier, assuming $N = 250$ symbols for classification.

Resources	QTY	%
Estimated total logic elements	1,824	10%
Total combinational functions	1,586	8%
Dedicated logic registers	704	4%
Total memory bits (RAM)	64,000	27%
Embedded multiplier 9-bit elements	24	46%

Table 4. Used resources of an Altera Cyclone II FPGA model EP2C20F484C7 to implement the CSS-SVM classifier.

It is observed that the proposed architecture is relatively efficient with respect to the use of FPGA resources. It uses memory and multipliers available in the FPGA, which releases registers and logic elements to other algorithms such as demodulation and functions other than AMC. If this is not the case, the first architecture can be adopted.

4. Conclusions

This chapter discussed the AMC task in cognitive radio. Two front end techniques were contrasted: CSS and cyclostationarity. The adoption of SVMs as the base classifiers for AMC systems is showed and experimental results were presented.

The results shown in this chapter were obtained for classification of the BPSK, 4PAM, 8PSK and 16QAM modulations. The CSS front end combined with the linear kernel SVM classifier achieved good results for the set of modulations adopted, and proved to be feasible for implementation in a FPGA, processed in real time. In addition, a synthesizable CSS-SVM architecture was proposed, which presented satisfactory results for AMC with a relatively efficient use of the available FPGA resources.

5. References

A. Fehske, J. Gaeddert, J. R. (2005). A new approach to signal classification using spectral correlation and neural networks, *DySPAN* pp. 144–150.

Allwein, E., Schapire, R. & Singer, Y. (2000). Reducing multiclass to binary: A unifying approach for margin classifiers, *Journal of Machine Learning Research* pp. 113–141.

Altera (2011). Altera corporation. `http://www.altera.com/`.

Bremaud, P. (2010). *Markov Chains: Gibbs Fields, Monte Carlo Simulation, and Queues*, Springer.

Burges, C. J. (1998). A tutorial on support vector machines for pattern recognition, *Data Mining and Knowledge Discovery* 2(2): 1–43.

Castro, M. E. (2011). *Cyclostationary detection for ofdm in cognitive radio systems*, Master's thesis, Faculty of The Graduate College at the University of Nebraska.

Cortes, C. & Vapnik, V. (1995). Support-vector networks, *Machine Learning* 20(3): 273–297. URL: *citeseer.nj.nec.com/cortes95supportvector.html*

Cristianini, N. & Shawe-Taylor, J. (2000). *An introduction to support vector machines and other kernel-based learning methods*, Cambridge Univ. Press.

da Silva, C. R. C., Choi, B. & Kim, K. (2007). Distributed spectrum sensing for cognitive radio systems, *Information Theory and Applications Workshop*, pp. 120–123.

Dobre, O. A., Abdi, A., Bar-Ness, Y. & Su, W. (2007). Survey of automatic modulation classification techniques: Classical approaches and new trends, *IET Commun.* pp. 137–156.

Gardner, W. (1988). Signal interception: a unifying theoretical framework for feature detection, *IEEE Transactions on Communications* 36, Issue 8: 897 – 906.

Gardner, W. A. & Spooner, C. M. (1992). Signal interception: Performance advantages of cyclic-feature detectors, *IEEE Transactions on Communications* 40: 149–159.

Haring, L., Chen, Y. & Czylwik, A. (2010). Automatic modulation classification methods for wireless OFDM systems in TDD mode, *IEEE Transactions on Communications* 58: 2480 – 2485.

Haykin, S., Thomson, D. & Reed, J. (2009). Spectrum sensing for cognitive radio, *Proceedings of the IEEE* 97: 849–877.

IEEE Standard VHDL Language Reference Manual (2009). *IEEE Std 1076-2008 (Revision of IEEE Std 1076-2002)* pp. c1 –626.

Klautau, A., Jevtić, N. & Orlitsky, A. (2003). On nearest-neighbor ECOC with application to all-pairs multiclass SVM, *J. Machine Learning Research* 4: 1–15.

Krose, B. & van der Smagt, P. (1996). *An introduction to Neural Networks*. URL: *"citeseer.nj.nec.com/article/schapire97using.html"*

Lin, Y. & He, C. (2008). Subsection-average cyclostationary feature detection in cognitive radio, *Neural Networks and Signal Processing, 2008 International Conference on*, pp. 604 –608.

Long, P. M. & Servedio, R. A. (2006). Discriminative learning can succeed where generative learning fails, *19th Annu. Conf. Learning Theory*, pp. 319–334.

Mak, P.-I., U, S.-P. & Martins, R. P. (2007). Transceiver architecture selection: Review, state-of-the-art survey and case study, *IEEE Circuits and Systems Magazine*, pp. 6–25.

Mishali, M. & Eldar, Y. C. (2011). Wideband spectrum sensing at sub-nyquist rates, *IEEE Signal Processing Magazine* 28: 102–135.

Muller, F. F., Cardoso-Jr., C. & Klautau, A. (2011). A front end for discriminative learning in automatic modulation classification, *IEEE Communications Letters (in Press)* .

Polydoros, P. P. A. A. A. (2000). Likelihood ratio tests for modulation classification, *21st Century Military Communications Conference Proceedings MILCOM 2000*, Vol. 2, pp. 670 – 674.

Proakis, J. G. (2001). *Digital Communications*, 4th edn, McGraw-Hill.

Rubinstein, Y. & Hastie, T. (1997). Discriminative vs informative learning, *Proc. 3rd Int. Conf. Knowledge Discovery Data Mining*, pp. 49–53.

Schnur, S. R. (2009). *Identification and Classification of OFDM Based Signals Using Preamble Correlation and Cyclostationary Feature Extraction*, PhD thesis, Naval Postgraduate School.

Su, W., Xu, J. L. & Zhou, M. (2008). Real-time modulation classification based on maximum likelihood, *IEEE Commun. Lett.* 12: 801–803.

Tan, P.-N., Steinbach, M. & Kumar, V. (2006). *Introduction to Data Mining*, Addison-Wesley, chapter Classification: Basic Concepts, Decision Trees, and Model Evaluation.

Vapnik, V. (1998). *Statistical learning theory*, John Wiley and Sons.

Wang, F. & Wang, X. (2010). Fast and robust modulation classification via kolmogorov-smirnov test, *IEEE Trans. Commun.* 58(8): 2324–2332.

Wang, R., Hou, C. & Chen, D. (2010). Blind separation of instantaneous linear mixtures of cyclostationary signals, *Image Analysis and Signal Processing (IASP), 2010 International Conference on*, pp. 492 –495.

Xu, J. L., Su, W. & Zhou, M. (2011). Likelihood-ratio approaches to automatic modulation classification, *IEEE Transactions on Systems, Man, and Cybernetics, Part C: Applications and Reviews* 41: 455–469.

Yucek, T. & Arslan, H. (2009). A survey of spectrum sensing algorithms for cognitive radio applications, *Communications Surveys Tutorials, IEEE* 11(1): 116 –130.

4

Link Quality Prediction in Mobile Ad-Hoc Networks

Gregor Gaertner and Eamnn O'Nuallain
School of Computer Science and Statistics,
Trinity College Dublin,
Ireland

1. Introduction

Cognitive radio is the radio revolution of our time. With it comes the promise of huge swathes of unused bandwidth which will provide the fuel for sophisticated wireless applications and fast and reliable wireless internet use. It is as yet unclear precisely what form cognitive networks will take. They may be infrastructure-based, ad-hoc or some combination of the two resulting from a trade-off between technological ideals and economic imperatives. With regard to ad-hoc networks it is clear that they will, at the very least, form an important part of cognitive radio networks for without them co-operative sensing would be considerably less effective. Cooperative sensing is an important tool in guaranteeing non-interference. Indeed in areas or at times where there is no radio infrastructure ad-hoc networks are essential. Communication between ad-hoc nodes is a vexed problem. The propagation environment may be harsh and/or changeable. Furthermore it is very difficult to predict accurately what the signal strength at an intended receiver is. Of course propagation models abound but none can accurately predict signal strength without recourse to topological data and even then these are slow to compute. This is why researchers have focused on path-loss models or statistical methods with which to make their predictions. The following is a literature review of the most prominent methods which have been published over the last fifteen years which will convey for the reader the coalface of research in this area.

Wireless mobile ad hoc networks (MANETs) (Schiller 2000, pp. 275–286) are self-organizing communication networks without any infrastructure. Peer nodes work collaboratively to transport packets through the network in a store and forward fashion since the limited transmission radius of nodes necessitates multi-hop communication. MANETs are appropriate in scenarios in which an infrastructure is either not feasible due to economic constraints or not available due to physical constraints such as natural disasters or battlefield deployments. Applications range from the communication-enabled soldier, disaster recovery and Voice over IP to mobile gaming.

The mobility of nodes and radio propagation effects cause frequent changes in the topology of a MANET; link failures and link recoveries, which are infrequent events in wired networks, occur frequently in MANETs. Consequently, reliable high-bandwidth communication is a challenge that can only partially be addressed by existing methods for wired networks. One successful strategy to alleviate the impact of topology changes is to predict them such that corrective actions can be taken before the change occurs (e.g. Goff et al. 2001). In the literature the effectiveness of such proactive operations has been demonstrated in the area of multicast communications and routing, in which packet latency is significantly reduced by discovering new routes before links fail. Similar performance enhancements are envisioned for group communications with partition prediction. Link quality prediction (LQP) is the foundation for proactive operations and is therefore a key technology for the efficient operation of MANETs.

2. Link quality prediction algorithms

LQP algorithms can be classified using a variety of criteria. One criterion is the type of input that the algorithm uses to make predictions. The inputs can be signal-power measurements, location measurements, or other measurements such as the ratio of transmitted to received packets within some time interval. In fact, the type of input is the most significant criterion by which to classify LQP algorithms since it typically represents the focus of the algorithms' creators. Predictions based on signal-power measurements focus on the radio propagation model while predictions based on location measurements concentrate on the mobility model. Another criterion by which to classify existing algorithms is the output of the LQP. If the output is subset of a finite set of states, we classify the prediction as being deterministic. Most deterministic prediction algorithms surveyed have only two possible output states; packets are predicted to be either lost or received for some time into the future. If the output of an LQP algorithm is a subset of an infinite set of states, we classify this prediction as stochastic since the output is in that case usually a probability. For example, an algorithm that estimates the probability of successfully receiving future transmitted packets is stochastic. Another possible classification criterion is the application of LQP. Successful applications are found in the area of routing, group communication, multicast communication and clustering. A final possible classification criterion is the radio propagation model on which a LQP algorithm is based. Since link quality is primarily dependent on the radio propagation model, even algorithms that focus on location measurements and mobility models are explicitly or implicitly derived with respect to a specific radio propagation model. In the literature we commonly found the use of the simple Radial Propagation Model, the Free Space Propagation Model (Rappaport 2002, pp. 107–109) and the Two-Ray Ground Propagation Model (Rappaport 2002, pp. 120–125). In the simple Radial Propagation Model a node 'i' can communicate with a node 'j' if the distance between 'i' and 'j' is less than or equal to a threshold; otherwise 'i'does not receive packets that are sent by 'j' and vice versa.

A graphical overview of the taxonomy of the surveyed literature based on these criteria is presented in Figure 1. Note that Roman et al.'s (2001) and Killijian et al.'s (2001) works are of a conceptual nature only such that we could only assess them with respect to their applications. Apart from these two studies all other works have been classified according to their input parameters, output parameters, radio propagation model and application area. Punnoose et al.'s (1999) study included two different radio propagation models as discussed later in this chapter. We found that LQP algorithms used various input parameters with signal power and location measurements being the most popular. Older LQP algorithms

Fig. 1. Taxonomy of LQP algorithms by different criteria. The surveyed algorithms use a variety of input parameters. Older LQP algorithms tend to provide deterministic predictions while newer algorithms focus on stochastic predictions. Routing has been by far the most important application area for LQP. The underlying radio propagation model is unspecified for a considerable number of algorithms.

tend to provide deterministic predictions while newer algorithms favour stochastic predictions. The most important application area for LQP in the literature was found in routing. For a considerable number of LQP algorithms the radio propagation model, on which these algorithms were based and evaluated, was unspecified.

All LQP algorithms are based on radio propagation models that characterize the propagation effect of path loss but do not account simultaneously for large-scale and wideband small-scale fading effects. However, the latter effects affect the radio propagation significantly (see, for example, Neskovic et al. 2000) and path-loss-only models do not consequently represent the real-world behavior well. To avoid the detailed repetition of this crucial critique, we refer to LQP algorithms based on such radio propagation models as LQP algorithms being based on 'path-loss-only' radio propagation models.

In this chapter, we refer to 'indirect' and 'direct' evaluations of LQP algorithms. We refer to an indirect evaluation if the accuracy of a LQP algorithm is evaluated by demonstrating that some metric, which refers to a higher-layer protocol, has changed due to the use of LQP. A higher-layer protocol might be a routing protocol or a multicast protocol and typical metrics are packet latency, packet delivery ratio and data throughput. In contrast, an evaluation is direct if the accuracy of a LQP algorithm is assessed by comparing its prediction of some quantity, such as the probability of receiving future transmitted packets, directly with future measurements of this quantity.

3. Notation and nomenclature

In order to unify the various notations used to describe LQP algorithms, we introduce the following notation inspired by Harvey (1993, p. 33). Let t denote the variable that represents time and T represent the point in time at which a prediction is made. Then τ is the prediction horizon that represents the amount of time into the future for which this prediction is made. We call the interval $[\mathrm{T},\mathrm{T}+\tau]$ the 'prediction interval' and the interval $[\mathrm{t}_c,\mathrm{T}]$ the 'history interval' where $\mathrm{t}_c \leq \mathrm{T}$ is some point in time. The data collected within the history interval is used at time T to predict a future value of a quantity at $\mathrm{T}+\tau$. The probability of a transmitted packet being successfully received is symbolized by the packet reception probability (PRP) Φ with range $\Phi \in [0,1]$. We denote a predictor for a quantity by a tilde over its symbol. A deterministic prediction of the PRP at time T for $\mathrm{T}+\tau$ is then defined as:

$$\tilde{\Phi}_{\mathrm{T}+\tau|\mathrm{T}} = \begin{cases} 0 & \text{- if the packet is predicted at time T not to be received at time T+}\tau \\ 1 & \text{- if the packet is predicted at time T to be received at time T+}\tau \end{cases} \tag{1}$$

A stochastic PRP prediction is defined as the predicted probability that a packet will successfully be received at time $\mathrm{T}+\tau$:

$$\Phi_{\mathrm{T}+\tau|\mathrm{T}} \rightarrow \{0, \ldots ,1\} \in \mathbb{R} \tag{2}$$

It is intended that, based on the result of the prediction that a decision be made, whether or not to send data. If the prediction warrants it then data would be sent immediately following time T. Given that in MANETS the distances are relatively short it is envisaged that the transmit time is negligible by comparison with the time the link is predicted to be stable. It is also important to discuss the interpretation of the time variables in the context of continuous real-world processes and the sampling by the computer. The stochastic processes of propagation effects investigated in this chapter are continuous. However, computers measure quantities derived from these processes in a discrete fashion. Assume, for example, a continuous stochastic real-world process X that is sampled with a sampling interval of T. Hence, the sample x_t corresponds to the realization of this process at time t. Assume that this sample has been taken at $t = 10:45:00$ hours and that the sampling interval T is 0.5 seconds. If we now refer to a sample that was taken at x_{t+8} this sample was not taken at $10:45:08$ hours but at $10:45:04$ hours since x_{t+8} corresponds in the discrete system to $t + 8T$, which is $10:45:00$ plus 8 times 0.5 seconds. In general, the discrete samples x_{t+c} correspond to samples of the continuous system at the actual time $t + c\tau$ while in the sampled computer representation t and c are some indices with the meaning explained above.

There are several other quantities that may be used or predicted by LQP algorithms. The location of a mobile node is represented by 'o'. Knowledge of successive locations allows for the derivation of the velocity. The magnitude of the velocity vector (i.e. the speed) is denoted by ν and the direction by v'. The signal power of a received packet is denoted by R in Watts and by $R(\mathrm{dBm})$ when measured in dBm; the signal envelope is symbolized by r.

4. LQP algorithms based solely on signal-power measurements

In this section we present LQP algorithms that rely solely on signal-power measurements to predict link quality.

4.1 Dube, Rais, Wang & Tripathi (1997)

LQP in MANETs has been introduced in the context of routing by Dube et al. (1997). Their on-demand routing protocol uses two conditions for route selection that are based on signal-power measurements. These signal-power measurements are assumed to represent future link quality albeit that the prediction horizon is not specified. The first condition requires that signal-power measurements for each link on a route have higher exponential average power than a threshold. The second condition demands that the exponential average signal power for each link has been above its threshold for a specified period of time. The first condition is termed 'signal stability', the second one 'location stability'. Applying both conditions leads to the selection of routes that have longer remaining lifetimes and therefore require less route maintenance. This was confirmed by simulations in which this routing protocol is compared to a simple, imaginary routing protocol that chooses the shortest route without considering signal-power measurements.

The LQP algorithm used to select routes with higher remaining lifetimes is given by the following equation:

$$\tilde{R}_{T+1|T} = c_1\tilde{R}_T + (1 - c_1)R_T \quad (3)$$

where c_1 is an experimentally determined constant, $\tilde{R}_T$ is the previously predicted signal power for some link, and R_T is the last signal power value measured. If $\tilde{R}_{T+1|T}$ is higher than an experimentally determined threshold, the link between the transmitter and receiver of a packet is classified as being 'strong', otherwise as being 'weak'. Since this prediction has only two possible outcomes we classify it as being deterministic. Routes with only strong links are preferred over routes that contain weak links. The expectation is that a route consisting of only strong links has a higher remaining lifetime than a route that also contains weak links. Older routes of strong links are preferred to younger routes.

A few observations are noteworthy. Equation (3) is proposed without derivation or justification. The constant c_1 as well as all thresholds are chosen by experience. The performance evaluation is based on simulations with a radio propagation model that considers only the distance between nodes but is not further specified. Hence, we cannot assess the applicability of this LQP algorithm to real-world environments.

4.2 Paul, Bandyopadhyay, Mukherjee & Saha (1999)

Paul et al. (1999) propose an affinity metric that measures the strength of connection between two nodes. Based on this affinity metric a routing protocol for low volume data transmission and another for high volume data transmission are presented. The affinity metric is a value proportional to the predicted remaining lifetime of a link and is intended to identify routes with higher lifetime than without using such a predictive measure.

Let $\Delta_{Avg}\left(R_T(dBm),[T - c_1, T]\right)$ be the average rate of change of the signal-power evaluated at time T . This average is obtained using the samples $R_T(dBm)$ over a preceding interval of c_1 seconds . The 'affinity' quantity $g(t)$ is now defined as:

$$g(t) = \begin{cases} \text{high} & \text{- if } \Delta_{Avg}\left(R_T(dBm),[T - c_1, T]\right) \geq 0 \\ \dfrac{c_2 - R_T(dBm)}{\Delta_{Avg}\left(R_T(dBm),[T - c_1, T]\right)} & \text{- if } \Delta_{Avg}\left(R_T(dBm),[T - c_1, T]\right) < 0 \end{cases} \quad (4)$$

where c_2 is the signal power at the maximum transmission range and $R_T(dBm)$ is the last signal-power sample. If the value of $\Delta_{Avg}\left(R_T(dBm),[T - c_1, T]\right)$ is positive, the affinity between two nodes is increasing, given the authors' assumptions, since the nodes are moving closer to each other. In this case the affinity is classified simply as being 'high' since no link failure is predicted for the future. Otherwise, the affinity is calculated periodically using the above formula. A deterministic link failure prediction can now be given by the following equation:

$$\tilde{\Phi}_{T+\tau|T} = \begin{cases} 0 & \text{if } c_3 g(T) \leq T + \tau \\ 1 & \text{otherwise} \end{cases} \quad (5)$$

where c_3 is an experimentally determined constant that represents the proportionality factor between the affinity metric and the remaining link lifetime.

Paul et al. (1999) evaluated the accuracy of their proposed LQP algorithm indirectly by showing that the number of route breaks can be significantly reduced in a variety of different scenarios when the affinity metric is used as a predictive measure to select routes with expected high remaining lifetimes.

The authors did not motivate or derive their LQP algorithm. Therefore, we do not know what radio propagation model was used as the basis for its derivation and hence, we cannot judge its applicability to real-world environments. Nevertheless, the presented LQP algorithm is interesting since it could provide accurate predictions at least for path-loss-only radio propagation models such as the Two-Ray Ground Propagation Model and the Free Space Propagation Model. These path-loss models are linear if the path loss is expressed in decibels versus the logarithmic distance. Consequently, the linear equations (4) and (5) are appropriate for these models, albeit with a certain prediction error if the signal power of nodes with constant speed vectors is measured in linear time intervals.

4.3 Agarwal, Ahuja, Singh & Shorey (2000)

Agarwal et al. (2000) built their Route-Lifetime Assessment Based Routing (RABR) protocol on Paul et al.'s (1999) LQP algorithm discussed in the previous section. Since their work utilizes the affinity metric without changes, we omit a discussion of it here. RABR introduces three new features compared to previous routing protocols. First, route selection for Transmission Control Protocol (TCP) traffic is extended to consider the throughput function of TCP as well as the affinity metric. Since the throughput of TCP traffic can be approximated by a function of the number of hops (Gerla et al. 1999), route selection is based on the product of the weakest link's affinity and the bandwidth of that route. Therefore, a route that has a lower predicted remaining lifetime but higher bandwidth can be more suited to a transmission than the route that has the longest predicted remaining lifetime but with much lower bandwidth. Secondly, the affinity metric is employed to increase the effectiveness of route cache maintenance. Route caching reduces the latency of route discovery in on-demand routing protocols since an intermediate rather than the destination node can answer the route discovery request from some source node, if a suitable route exists in its cache. However, stale routes in the cache can increase latency considerably. Hence, a timeout that equals the expected lifetime of the route is assigned to each route cache entry such that no stale entries exist if the LQP is accurate. Thirdly, the authors introduce a power saving scheme that reduces the transmission power using the affinity metric. However, only between 3% to 5% of power was saved in their simulations.

The effectiveness of Paul et al.'s (1999) LQP ALGORITHM was demonstrated in simulations with NS-2. For a wide range of scenarios the performance of the RABR protocol was compared to other common routing protocols (Agarwal et al. 2000). For medium to large average node speeds the RABR protocol was shown to outperform its competitors in terms of TCP throughput and ratio of delivered to sent packets. However, the suggested LQP algorithm and its evaluation are based on a path-loss-only radio propagation model.

4.4 Goff, Abu-Ghazaleh, Phatak & Kahvecioglu (2001)

Goff et al. (2001) were the first to introduce the concept of proactive routing for on-demand routing protocols. On-demand routing protocols exhibit low traffic overhead since they only update active routes that have been used in the recent past. However, updates are only triggered after a link failure is detected in which case multiple retransmissions have been unsuccessful. Therefore, time passes while a new route is being discovered which results in high mean and variance of packet latency. Proactive routing, however, tries to minimize this latency by initiating route discoveries early enough to have discovered alternative routes by the time links fail. The efficiency of proactive route discovery relies on accurate LQP. If a link is falsely predicted to fail early, unnecessary route discoveries are initiated and a route with lower quality than currently used might be chosen. However, if a link failure is predicted too late or not predicted at all, the route discovery does not finish before the link fails; hence, packet latency is increased.

The authors rely on the concept of preemption for enabling proactive route discovery. A communication task (e.g. route discovery) is initiated preemptively such that it can be completed even if links between two nodes fail when these two nodes move apart at maximum speed. The preemptive region starts at the maximum distance at which two nodes can be separated such that if they move apart at maximum speed, the communication task can still be completed. The preemptive region ends at the maximum transmission radius. The preemptive threshold is the signal power that corresponds to the beginning of the preemptive region and this threshold is intended to determine whether a node has entered the preemptive region with respect to another node.

The proposed LQP ALGORITHM differs from previous ones in that it is more a protocol than a mathematical expression. The idea behind this approach is that a number c_1 of 'ping-pong' rounds are exchanged between two nodes if a packet is received whose signal power is below the preemptive threshold c_2. If the signal power of c_3 or more of these c_2 packets are below the threshold or these packets are lost, the link is predicted to fail. The 'ping-pong' rounds are intended to reduce falsely predicted link failures that are caused by temporary fades due to small-scale fading. The preemptive threshold is calculated such that if two nodes move away from each other at maximum speed, the link failure is predicted early enough to allow a route discovery to complete before the link fails. Therefore, the threshold c_2 corresponds to the signal power at the beginning of the preemptive region that represents the difference between the maximum transmission distance and the product of the maximum relative speed between two connected nodes and the prediction horizon. For their evaluation Goff et al. (2001) set the number of 'ping-pong' rounds c_1 and the threshold c_3 to three. The LQP algorithm was derived considering the Two-Ray Ground Propagation Model and small-scale fading.

Significant improvements due to proactive instead of on-demand routing were shown for the Dynamic Source Routing (DSR) (Johnson & Maltz 1996) and the Ad hoc On-demand Distance Vector (AODV) (Perkins et al. 2001) protocol in an augmented version of the NS-2 simulator. The simulator was augmented with an error model that assigns either a good or bad state to a link. In order to approximate small-scale fading, the signal power was decreased by division by a uniformly distributed random number between 2 and 100 when a bad state was assigned to a link. The mean length of staying in a state was set to 20 000 packets in the good state and to 2 packets in the bad state. It was shown for both routing protocols that the number of broken routes and the packet latency was significantly reduced for constant-bit-rate traffic due to proactive route discovery. A similar reduction in packet latency was also demonstrated for TCP traffic.

This deterministic LQP algorithm relies critically on the assumption of bidirectional links since the 'ping-pong' mechanism to mitigate small-scale fading does not work otherwise. Furthermore, this algorithm only allows for LQP for one future point in time and not for a set of points (interval). For example, the algorithm can be used to predict link failure for 2 seconds ahead but not 0.5, 1.0, 1.5 and 2 seconds ahead simultaneously. This property is based on the protocol's nature. Since the 'ping-pong' mechanism is triggered by the preemptive threshold that is a function of the prediction horizon, it cannot be trigged for a different threshold simultaneously. In terms of evaluation, the accuracy of the proposed approximation model for small-scale fading is questionable. Although small-scale fading affects the signal power of all transmitted packets, only deep fades, which occur very rarely with a probability of 1 in 10 000 packets in their model, are simulated. Goff et al. (2001) do not explain how this probability is obtained. Furthermore, the important propagation effect of large-scale fading has not been considered. For these reasons, the applicability of the proposed LQP algorithm to real-world environments is unclear.

4.5 Qin & Kunz (2002) and (2003)

Qin & Kunz (2002) add proactive route discovery to the DSR protocol similarly to Goff et al.'s (2001) work. However, Qin & Kunz (2002) use a different LQP ALGORITHM and they focus on reducing the number of dropped packets by means of LQP rather than improving the number of broken routes and the packet latency. Packets can be dropped during route discovery if no alternative route has yet been found. This loss of packets causes significant degradation in throughput as shown, for example, for TCP (Holland & Vaidya 2002). However, if alternative routes are discovered proactively before a link, and therefore a route, fails, no packets will be lost.

Qin & Kunz (2002) derive their LQP ALGORITHM trigonometrically using the same principle as Lee et al. (1999) and Jiang et al. (2005) whose works are discussed later in this chapter. However, the latter two approaches use location measurements to predict the time of link failure, while this algorithm uses only signal-power measurements. These signal-power measurements are used to estimate the distance between two nodes by assuming that only the path loss as given by the Two-Ray Ground Propagation Model determines the signal power. The LQP ALGORITHM is derived from the Law of Cosines (Kay 2001, pp. 27–31) under the assumption that nodes maintain their speed and direction during the history and prediction intervals. Therefore, the time of link failure $\tilde{\Phi}_{t>T|T}$ is predicted at time T by:

$$\tilde{\Phi}_{t>T|T} = \frac{\sqrt{g_2^2 - 4g_1g_3} - g_2}{2g_1} \tag{6}$$

where

$$\begin{aligned}
g_1 &= t_{T-1}g_4\sqrt{R_{T-1}c_1} \\
g_2 &= \sqrt{c_1}\left(\left(\sqrt{R_{T-2}} - \sqrt{R_{T-1}}\right) - t_{T-1}^2 g_4\sqrt{R_{T-1}}\right) \\
g_3 &= t_{T-1}\sqrt{R_{T-1}c_1} - t_{T-1}\sqrt{R_{T-2}R_{T-1}} \\
g_4 &= \frac{t_{T-1}\sqrt{R_{T-2}R_{T-1}} + T\sqrt{R_{T-1}R_T} - T\sqrt{R_{T-2}R_T} - t_{T-1}\sqrt{R_{T-1}R_T}}{\sqrt{R_{T-1}R_T}\left(t_{T-1}T^2 - Tt_{T-1}^2\right)}
\end{aligned} \tag{7-10}$$

and c_1 symbolizes the signal-power threshold below which packets are lost, R_t represents the signal power measured at time t, and t_{T-x} denotes the times at which samples are taken. Equation (6) is only applied if three consecutive packets have decreasing signal power. Otherwise the nodes are assumed to be moving closer and the link is predicted not to fail. The deterministic PRP predictor is therefore defined as:

$$\tilde{\Phi}_{T+\tau|T} = \begin{cases} 0 & \text{if } T+\tau \geq \tilde{\Phi}_{t>T|T} \\ 1 & \text{otherwise} \end{cases} \tag{11}$$

Qin & Kunz (2002) also discuss how their LQP ALGORITHM can be made resilient against power fluctuations. They propose to preprocess the sampled signal power by linear regression and showed, for one example situation, that the accuracy of their link failure prediction increased compared to not using any preprocessing. This result stems from simulations in which the signal power varies by random noise of between 5% and 10% from its mean. However, there are a number of problems with this approach. First, Qin & Kunz (2002) do not define the cause of the power variations. Do these variations refer to measurement noise or radio propagation effects? Secondly, the distribution and correlation structure of the noise process is not given. However, linear regression is only an unbiased estimator if the noise process is normally distributed and uncorrelated (Hines et al. 2002, pp. 409–414). These assumptions are invalid, for example, when the noise refers to small-scale and large-scale fading. Thirdly, the authors do not describe why the power varies only between the small values of 5% and 10% from its mean.

Qin & Kunz's (2002) work is one of the few studies that directly evaluated the accuracy of the proposed LQP ALGORITHM. In a simulation with the NS-2 simulator it was shown that more than 90% of lost packets were successfully predicted for different mobility patterns. It was also demonstrated that the number of dropped packets in DSR due to unavailable routes was reduced by between 20% and 44.95% while the overhead of control messages increased by up to 33.5%. This increase in control messages, however, did not cause network congestion or other significant, negative side effects. However, these simulations, as well as the LQP ALGORITHM itself, are based on a 'path-loss-only' radio propagation model and even the previously discussed simple power variations were omitted.

5. LQP algorithms based on signal-power and other measurements

In this section we describe algorithms that use signal power as well as other types of measurements to predict link quality. These other measurements can be, for example, the average packet loss ratio or the hop count.

5.1 Wang, Martonosi & Peh (2006)

Wang et al. (2006) suggest the use of supervised learning to predict the quality of links in static wireless sensor networks. The term 'prediction' refers in this context more to estimation or classification of current link quality than to the forecast of link quality for a specific time into the future. Nevertheless, since Wang et al. (2006) assume similar to De Couto et al. (2003) and Dowling et al. (2005) justifiably that the current link quality represents the future link quality to some degree, we include this work in our survey. The proposed link quality estimator replaces the original link quality metric ETX in the MintRoute (Woo et al. 2003) routing protocol and its impact on different higher-layer metrics (e.g. packet delivery ratio and packet latency) in the original and augmented MintRoute protocol is assessed. The suggested link quality estimator intends to be more accurate in scenarios with high network traffic. In these situations snooping-based methods such as ETX can underestimate link quality due to high packet loss of probe packets that is caused by the hidden and exposed terminal problems as well as the unfairness of the 802.11 medium access control (MAC) protocol (Tobagi & Kleinrock 1975, Xu & Saadawi 2001).

Supervised learning establishes the mapping between an input and an output measure in an offline training phase. The input measure consists of a feature vector that contains features that can be measured by the system under consideration. These features are selected by humans at design time and they should have a meaningful relation to the desired output measure. If the output measure has a finite set of outcomes, the problem is called classification. In the training phase, the learning algorithm tries to minimize the classification error between the input measure and the given output measure. The output measure in Wang et al.'s (2006) work is the link state, which is assigned to be one of the three states 'bad', 'medium' and 'good'. Different metrics are suggested as the feature vector:

- The signal power of received packets.
- The usage of send and forward buffers to detect network congestion.
- The hop count of the sensor node from the base station.
- The average packet loss ratio, which is determined by a snooping-based method such as ETX, for the link in both directions.

Interestingly, the classification accuracy using all features was 80.8% and still around 70% for using the signal power or the average packet-loss ratio alone. Therefore, the classification was only moderately improved by using additional features.

Packet delivery ratio, packet latency and a fairness index were the metrics chosen to compare the performance of the original MintRoute protocol with its augmented version. The fairness index is a summary statistic that captures the packet delivery ratio for all nodes of a network. It has a value between 0 and 1 and is 1 when all nodes have the same packet delivery ratio. The evaluation was conducted in a test bed of 30 wireless sensors nodes that were located across multiple offices in a building. The wireless nodes communicated over 802.15.4 (ZigBee) and the packet size for all experiments was set to 29 bytes. All nodes were static and the link quality in the network varied due to changes in the environment, radio propagation effects and different network loads. The new LQP ALGORITHM led to significantly higher packet delivery ratios, a comparable average packet latency and improved fairness of the MintRoute protocol.

The proposed supervised learning approach was shown to be effective for wireless networks with static nodes only. However, it is not clear if this method is also applicable to MANETs with node mobility. Moreover, the suggested LQP ALGORITHM is suitable for selecting routes in routing protocols leading to the reported performance improvements but is not able to predict link quality for a specifiable prediction horizon. Furthermore, all results were obtained for a small packet size. Since in many application scenarios longer and varying packet sizes can be expected, it is uncertain whether this learning approach will extend to these situations since the packet loss depends on the packet length, as it was, for example, observed in the work of De Couto et al. (2003).

6. LQP algorithms based solely on location measurements

In this section we present LQP algorithms that rely solely on location measurements for their predictions. Location measurements are usually obtained by GPS in outdoor environments. However, GPS has a user-equivalent range error of 19.2 m (Grewal et al. 2001, pp. 128–130) and is therefore fairly inaccurate. The user-equivalent range error summarizes the overall error experienced by a user from the various error sources of the GPS (e.g. transmission delays caused by the atmosphere, inaccuracy of satellite clocks, etc.). Furthermore, GPS requires a line-of-sight (LOS) connection to at least three satellites that are often not in view in dense urban environments.

6.1 Lee, Su & Gerla (1999)

Lee et al. (1999) presented a deterministic PRP predictor to enhance the performance of the On-Demand Multicast Routing Protocol (ODMRP) from Gerla et al. (1998). While the original ODMRP protocol uses a fixed refresh interval for periodic flooding to update each nodes' knowledge about the network topology, in the augmented ODMRP protocol the refresh interval is set to some time before the link with the shortest predicted remaining lifetime is assumed to fail. This anticipation of topology changes reduces latency by updating the network topology before it actually changes. LQP is also employed to select routes with longer remaining lifetimes such that route maintenance is reduced.

Lee et al.'s (1999) predictor is based on location measurements, the resulting speed vector and the Free Space Propagation Model. Although the authors do not give any mathematical derivation of their predictor, it is obvious from the expression that it is derived by trigonometry. If two nodes keep the same speed and direction within the prediction interval, the time until the link fails corresponds to the time at which the nodes are predicted to be further apart than the maximum transmission radius. Formally, let the coordinates of the mobile node at time T be (x_i, y_i). Let v_i be its speed and v_i' its direction. Then the link failure time $\tilde{\Phi}_{t>T|T}$ for the link between two nodes i and j is predicted by:

$$\tilde{\Phi}_{t>T|T} = \frac{-(g_1 g_2 + g_3 g_4) + \sqrt{c_1^2\left(g_1^2 + g_3^2\right) - \left(g_1 g_4 - g_2 g_3\right)^2}}{g_1^2 + g_3^2} \tag{12}$$

Where:

$$\begin{aligned} g_1 &= v_i \cos(v_i') - v_j \cos(v_j') \\ g_2 &= x_i - x_j \\ g_3 &= v_i \sin(v_i') - v_j \sin(v_j') \\ g_4 &= y_i - y_j \end{aligned} \tag{13-16}$$

and c_1 is the maximum transmission radius. If $v_i = v_j$ and $v_i' = v_j'$ then the link failure time is set to infinity without applying (12) since the nodes are maintaining a constant distance. The deterministic PRP predictor is now defined by:

$$\tilde{\Phi}_{T+\tau|T} = \begin{cases} 0 & \text{if } T+\tau \geq \tilde{\Phi}_{t>T|T} \\ 1 & \text{otherwise} \end{cases} \tag{17}$$

The accuracy of the proposed LQP ALGORITHM was shown indirectly in the GloMoSim simulator (Bajaj et al. 1999) by comparing different aspects of the enhanced multicast protocol with the original one. It was shown that the packet delivery ratio of the enhanced ODMRP protocol was considerably higher than the original protocol, especially for high node speeds. Moreover, the packet latency was reduced significantly by anticipation of topology changes. However, the suggested LQP ALGORITHM and its evaluation are based on a 'path-loss-only' radio propagation model.

6.2 McDonald & Znati (1999b) and (1999a)

McDonald & Znati (1999b) suggest a stochastic PRP predictor for a cluster-based routing framework (McDonald & Znati 1999a). The idea behind this routing approach is to divide the mobile network into clusters that guarantee with certain probability for a certain length of time unbroken links for nodes within this cluster. The probability and length of time are specifiable. Table-driven routing is employed within a cluster; on-demand routing is used for inter-cluster routing. Table-driven routing protocols periodically update the routing information throughout the network for all possible routes. Therefore, for highly dynamic topologies the overhead is high while latency is still low since routes have been precomputed. On-demand routing protocols in contrast only maintain previously used routes. Hence, the traffic overhead is low but latency is high if a route between source and destination has to be discovered. Mixing these different routing schemes aims to balance the trade-off between table-driven and on-demand routing based on clustering nodes. The efficiency of this clustering relies on low effort for cluster maintenance that is achieved by LQP.

McDonald & Znati (1999b) derive their LQP ALGORITHM from their proposed mobility model and the simple Radial Propagation Model. Their mobility model assumes that each node's movements consist of a sequence of random length intervals called mobility epochs during which it moves in a constant direction at constant speed. The parameters of this mobility model are termed the 'mobility profile'. Formally, for a node i the epoch length is assumed to be an exponential Independent and Identically Distributed (IID) random variable with mean $1/c_1$. The speed v_i is taken to be an IID random variable with mean c_2 and variance c_3. The direction of movement is also a uniformly distributed IID random variable having a value of between 0 and 360 degrees. This mobility model is similar to the Random Walk Mobility Model (Polya 1921) and McDonald & Znati (1999b) do not explain how their model relates to the commonly used Random Waypoint Mobility Model (Johnson & Maltz 1996) or any other mobility model. The predicted PRP for a link (i,j) between two nodes is given by:

$$\begin{aligned} \tilde{\Phi}_{T+\tau|T} &\approx \frac{1}{2}\left(1 - I_0\left(\frac{-2c_4^2}{g(T+\tau)}\right)\exp\left(\frac{-2c_4^2}{g(T+\tau)}\right)\right) \\ g(t) &= 2t\left(\frac{c_{3,i}^2 + c_{2,i}^2}{c_{1,i}} + \frac{c_{3,j}^2 + c_{2,j}^2}{c_{1,j}}\right) \end{aligned} \tag{18-19}$$

where I_0 is the modified Bessel function of the first kind and zero order, and c_4 is the maximum transmission radius. As evident from (19) the PRP is predicted based on the mobility profile (c_1, c_2, c_3). However, McDonald & Znati (1999a) assume these values to be known by each node. Clearly, this assumption is valid for simulations only. Nevertheless, real-world use can be enabled when mobility profiles are estimated by each node using location measurements. We therefore classify this algorithm as being based on location measurements.

McDonald & Znati (1999b) evaluated the accuracy of their LQP ALGORITHM in a custom-built, discrete event simulator. On average, the LQP was accurate for longer prediction horizons but substantially underestimated link quality for shorter horizons as also observed by Jiang et al. (2005). The accuracy of the suggested LQP ALGORITHM was also assessed indirectly in McDonald & Znati (1999a) where it was shown that a proposed clustering algorithm, which uses (18) and (19), led to clusters that were stable and adaptive to node mobility. Again, these evaluation results and the design of the authors' LQP ALGORITHM stem from a 'path-loss-only' radio propagation model.

6.3 Punnoose, Nikitin, Broch & Stancil (1999)

Punnoose et al. (1999) propose a signal-power predictor for MANETs. Their approach uses location measurements from which the signal power is predicted by using a site-specific three-dimensional ray tracing propagation model with terrain map. The following components are suggested for their model:

- Non-location-based prediction. In the absence of location information this component predicts signal power by extrapolation of past signal-power samples.
- Mobility model. Probable future node locations are predicted using past location and speed measurements.
- Terrain map. The terrain map is used to provide the radio propagation model with necessary data about the environment and is updated when the environment changes.
- Radio propagation and communication model. This model predicts link quality by considering the predicted node locations, the terrain map and feedback about the accuracy of past predictions.
- Decision maker. The decision maker interacts with the network protocol to determine which links should be tracked and to provide LQP for the network.

While the above components have been suggested as a general architecture, these were only partly implemented and evaluated in the work of Punnoose et al. (1999). The non-location-based prediction was, for example, neither explained nor implemented. A terrain map was used for one evaluation scenario; however, it was not employed for a second scenario. The update of the terrain map was omitted. The mobility model for prediction was implemented and past information was used to increase prediction accuracy for cyclic mobility patterns. Nevertheless, the authors give no details about their algorithm for mobility prediction such that we cannot assess this component. They used two different radio propagation models in their work. In one scenario, a three-dimensional "N+2 ray+diffraction" propagation model (Punnoose et al. 1999) was employed that takes into account a direct ray, a ground reflected ray and rays reflected off N objects as well as diffraction from these objects. This propagation model relies on terrain maps that contain the shapes of buildings with reflection and diffraction coefficients of their materials. In order to reduce the computational cost of this ray tracing model, the search for all multipath sources was restricted. Specific details about this proposed enhancement such as the resulting computational effort and the reduction in accuracy were not given by Punnoose et al. (1999). In a second evaluation scenario the Two-Ray Ground Propagation Model with empirically determined, site-specific constants was employed for prediction. No explanation was given as to why two different radio propagation models have been chosen for the two scenarios. Furthermore, the suggested use of past locations was omitted from the prediction model implemented. The decision maker tracked the link quality for all nodes and interacted with the DSR protocol such that routes with higher predicted link quality were preferred.

The LQP accuracy was evaluated within an experimental outdoor test bed. In the first scenario, five mobile nodes in vehicles followed a looped course. Each mobile node was equipped with a differential GPS receiver. For a link between two nodes the predicted and measured signal power was displayed together with the packet loss as three separate graphs in Punnoose et al.'s (1999) paper. Unfortunately, these graphs use a large time scale that precludes a direct evaluation of the prediction accuracy in the order of seconds. Furthermore, the prediction horizon was not given such that the span of the prediction into the future is unknown. In the second scenario, one mobile node and two stationary nodes were used to show that the augmented DSR protocol chose routes with higher link quality due to the proposed LQP ALGORITHM. In summary, the results presented were not sufficiently detailed to allow us a full assessment.

6.4 Jiang, He & Rao (2005)

Jiang et al. (2005) present a stochastic LQP ALGORITHM that has some similarity to McDonald & Znati's (1999b) approach. Both are based on the prediction of node mobility for exponential mobility models such as the Random Walk Mobility Model. Furthermore, both assume a simple Radial Propagation Model such that two nodes are connected if they are within a specified maximum transmission radius. Both approaches predict link quality to assess how reliable the connection between nodes is. While McDonald & Znati (1999b) use this information for clustering, Jiang et al. (2005) employ it for selection of routes in the context of on-demand routing. One major difference between both LQP approaches is that McDonald & Znati's (1999b) LQP ALGORITHM predicts at time T the PRP for $\mathrm{T}+\tau$, while Jiang et al.'s (2005) approach predicts the probability that all packets, which will be sent between T and $\mathrm{T}+\tau$, are received. However, this latter definition of link failure, which requires for an available link that every sent packet will be received, is restrictive. A link might experience brief periods of unavailability due to temporary packet loss, which is caused by small-scale fading, but is available for all times outside this brief period. Another considerable difference between both approaches is that for a link McDonald & Znati's (1999b) LQP ALGORITHM predicts the PRP probability for a specifiable time $\mathrm{T}+\tau$ while Jiang et al.'s (2005) LQP ALGORITHM predicts first the link failure time and subsequently estimates the probability that the link will really last until the predicted link failure time.

More specifically, Jiang et al.'s (2005) LQP ALGORITHM works in the following two stages. First, a measurement-based approach predicts the time of link failure. The underlying assumption is that nodes will keep their past speed and direction within the prediction interval. Secondly, based on this link failure time the probability that the link will be available through the whole interval is calculated. For this prediction the situations are considered in which the speed and direction of the nodes do not change and in which one node or both nodes change their direction and speed.

The time until link failure $\tilde{\Phi}_{t>T|T}$ for two nodes is derived similarly to Lee et al. (1999) and Qin & Kunz (2002) by employing trigonometry. Therefore, $\tilde{\Phi}_{t>T|T}$ is predicted at time T by:

$$\tilde{\Phi}_{t>T|T} = \frac{\sqrt{g_2^2 + 4c_1^2 - 4g_1g_3} - g_1}{2g_3} - T \tag{20}$$

where

$$\begin{aligned} g_1 &= \frac{d_{T-1}^2 T - d_T^2 t_{T-1} - d_{T-2}^2 (T - t_{T-1})}{t_{T-1} T (t_{T-1} - T)} \\ g_2 &= \frac{d_{T-1}^2 T - d_T^2 t_{T-1} - d_{T-2}^2 (T^2 - t_{T-1}^2)}{t_{T-1} T (t_{T-1} - T)} \\ g_3 &= d_{T-2}^2 \end{aligned} \tag{21-23}$$

and c_1 denotes the maximum transmission range, d_t the distance between two nodes measured at time t. t_{T-x} symbolizes the times at which samples are taken.

Jiang et al. (2005) stated that the distance could either be determined by location measurements or be approximated by signal-power measurements. However, since Jiang et al. (2005) used location measurements for their evaluation and they did not elaborate on how signal-power measurements can be employed, we classify this predictor as being based on location measurements. Note that Qin & Kunz (2002) proposed a very similar predictor that corresponds to Jiang et al.'s (2005) predictor with the distance being estimated by signal-power measurements using the Two-Ray Ground Propagation Model.

Let $g(\tilde{\Phi}_{t>T|T})$ denote the predicted probability at time T that all future packets will be received successfully until the link is predicted to fail at $\tilde{\Phi}_{t>T|T}$. Considering that the nodes might change their speed and direction, this probability is now predicted by:

$$g(\tilde{\Phi}_{t>T|T}) \approx \frac{1 - e^{-2c_2\tilde{\Phi}_{t>T|T}}}{2c_2\tilde{\Phi}_{t>T|T} + c_3} + \frac{2c_2\tilde{\Phi}_{t>T|T} e^{-2c_2\tilde{\Phi}_{t>T|T}}}{2} \tag{24}$$

where $1/c_2$ is the mean epoch length of the mobility model and c_3 is a constant that is updated at runtime. The mean epoch length is the amount of time in which, on average, the nodes do not change speed or direction and its value is dependent on the mobility model. In order to reduce the computational cost the predictor (24) is an approximation of the exact solution to the problem as defined by Jiang et al. (2005). More specifically, the predictor only considers changes of the nodes' speed and direction that lead to nodes moving further away but does not consider changes that lead to higher link quality as in the case of nodes moving closer. Consequently, the constant c_3 was introduced to correct for the difference between the approximate and exact solutions by comparing the predicted with the actual probability of link failure at run-time of the algorithm. Since the determination of c_3 is complex, we refer the reader to Jiang et al.'s (2005) paper for further details.

The LQP ALGORITHM was evaluated in the OPNET (MIL 3 Inc. 2006) simulator. It was shown that (24) is, on average, a good predictor for remaining lifetime of links for exponential mobility models such as the Random Walk Mobility Model. However, the accuracy was significantly reduced for non-exponential mobility models such as the Random Waypoint Mobility Model. Furthermore, we noticed that the scale of the presented graphs emphasized long-term prediction horizons such that we cannot comment on the prediction accuracy for short-term horizons of the order of a few seconds. Jiang et al. (2005) also showed that packet latency, packet loss ratio and the ratio of received to delivered packets were all significantly improved for the DSR protocol when routes with longer remaining lifetimes were selected based on their LQP ALGORITHM. However, this improvement could only be observed for exponential mobility models and the improvement was only minor for the non-exponential Random Waypoint Mobility Model. All results and the proposed LQP ALGORITHM were derived by a 'path-loss-only' radio propagation model.

7. LQP algorithms based on other measurements

In this section we introduce LQP algorithms that use measurements other than signal power and location for their predictions. Possible measurement metrics are the link age or metrics that calculate ratios based on the number of transmitted and received packets in the recent past.

7.1 Gerharz, Waal, Frank & Martini (2002)

Gerharz et al. (2002) present two metrics that are intended to select the link with the highest remaining lifetime from a set of links. Both metrics are based on measurements of link age and its distribution. The first metric recommends choosing the link with the highest predicted average remaining lifetime. The second metric advocates choosing the link with the highest time at which this link is predicted to fail with 25% probability. Both metrics led in simulations to link choices that result in links with higher remaining lifetimes than choosing links randomly. For most scenarios examined these metrics also led to links with higher remaining lifetimes than choosing the oldest link.

The proposed metrics for LQP are not intended to accurately predict the remaining lifetime of links; rather they help to select links that are likely to have the longest remaining life. Gerharz et al. (2002) derived their metrics empirically from the link age data that was obtained by conducting a number of simulations for different mobility models. All simulations, including those for evaluation, were done with the Free Space Propagation Model and a maximum transmission radius of 50 m. Therefore, the applicability of the proposed metrics is bound to this specific radio propagation environment and it is unclear if the metrics extend to more realistic radio propagation models or even the Free Space

Propagation Model with a different maximum transmission radius. Furthermore, it was suggested that 30 minutes of link age data are required to use the proposed metrics. It was not explained how LQP can be provided within this initialization interval, however. Moreover, Gerharz et al. (2002) assumed that the mobility patterns (e.g. the parameters of the mobility model) are static with respect to time. Due to all these severe shortcomings it is unclear whether Gerharz et al.'s (2002) method can successfully be applied in more realistic simulations or real-world deployments.

7.2 De Couto, Aguayo, Bicket & Morris (2003)

De Couto et al. (2003) introduce ETX in order to find higher-throughput routes in multi-hop wireless networks than those given by the traditional minimum hop-count metric, which is the most common metric used in ad hoc routing protocols. The Expected Transmission Count (ETX) metric is the expected number of transmissions, including possible retransmissions used in successfully making a unicast. It is assumed that the expected number of future transmissions corresponds to the estimated number of past transmissions. This estimation is realized by counting the number of transmitted and received packets with a subsequent calculation of their ratio. Metrics such as ETX, which use such counting mechanisms, are classified as being snooping-based.

The ETX metric was motivated by observations of link behavior in a wireless test bed of 29 nodes that were statically located on different floors in an office building. It was observed that the minimum hop-count metric often chooses routes with lower throughput than the optimal route. Furthermore, a high variance in packet reception ratios was perceived and around 20% of all links were asymmetric such that forward and reverse packet reception ratios differed by at least 25%. However, it should be noticed that a high degree of asymmetry was observed in an experiment in which all nodes had broadcast packets for 5 seconds consecutively. Therefore, the packet reception ratios of all possible node pairs were determined over a large time span (the worst case is a difference of 140 seconds), such that the degree of asymmetry could be overestimated due to a change of propagation characteristics in the environment during that time span. These observations led to the following definition of the ETX metric g_e for a link:

$$g_e = \frac{1}{g_f g_r} \tag{25}$$

where g_f and g_r are the measured packet reception ratios for the forward and reverse links respectively. The ETX metric for a route is simply the sum of the ETX metrics of the links that constitute the route. The proposed metric has several desirable properties:

- ETX is based on packet reception ratios that are strongly correlated to the throughput.
- ETX considers the asymmetry of links by incorporating the packet loss in both directions.
- ETX penalizes routes with higher hop counts that have naturally a lower throughput.

The snooping interval was set to 10 seconds for all trials. However, an explanation of this value is missing.

De Couto et al. (2003) confirmed the suitability of the ETX metric for routing in their wireless test bed of static nodes. It was demonstrated that the ETX metric leads to route selections with significantly higher throughput than the minimum hop-count metric in the DSDV protocol (Perkins & Bhagwat 1994). Furthermore, it was observed that the throughput of the DSR protocol was only slightly improved by employing the ETX metric since DSR already includes link-layer transmission feedback and is therefore able to avoid links with high asymmetry. Moreover, the packet reception ratio was shown to depend on packet length. Since ETX determines this ratio via probe packets that are significantly longer than acknowledgements, it underestimates the reception probability for these acknowledgements, which leads to a biased estimation of the ETX metric as observed by the authors.

The ETX metric was evaluated in a static ad hoc network. Therefore, it is unclear how the above results translate to scenarios with node mobility, especially since large-scale and small-scale fading are spatial phenomena. Furthermore, the estimation of the ETX metric is representative for one packet length only since an influence of the packet length on the packet reception ratio was shown. Moreover, it is unclear how the ETX metric's performance, which was evaluated indoors, relates to the performance in an outdoor MANET. These LQP ALGORITHMs and the ETX metric are complementary concepts in which ETX is a higher-level metric.

7.3 Dowling, Curran, Cunningham & Cahill (2005)

Dowling et al. (2005) present a framework for collaborative reinforcement learning to solve optimization problems in dynamic, distributed networks. As an example application for this framework, a routing protocol termed 'SAMPLE', which exploits online learning to minimize the cost of routing in MANETs, is proposed and evaluated. The learning goals are to maximize the overall network throughput, to minimize the ratio of delivered to undelivered packets and to minimize the number of transmissions for every packet sent.

Dowling et al. (2005) advocate a stochastic state transition model for links to model the cost of routes. This model predicts the probability of a successful unicast and is therefore a snooping-based method considering unicast packets as is the ETX metric of De Couto et al. (2003). The number of required transmissions for a successful unicast is given by:

$$E\left[\frac{g_s}{g_a}\right] = \frac{g_s + c_1 c_2 \left(g_r + g_b + g_p\right)}{g_a + c_2 \left(g_r + g_b + g_p\right)} \tag{26}$$

where g_s is the number of successful unicast transmissions, g_a is the number of attempted unicast transmissions, g_r is the number of received unicast transmissions, g_b is the number of received broadcast transmissions, g_p is the number of promiscuously received (overheard) unicast transmissions, and c_1 and c_2 are static constants. c_1 represents the belief in the successful transmission of a packet considering only received packets. c_2 controls the weight between received and transmitted packets. Equation (2.26) is based on snooping transmitted and received packets within a fixed time interval that was set by Dowling et al. (2005) to 10 seconds. c_1 and c_2 were set to 0.5 and 0.2 respectively.

The performance of the SAMPLE routing protocol was evaluated against AODV and DSR in the NS-2 simulator. Two scenarios were chosen with the first one representing the design assumptions of AODV and DSR in that a true ad hoc network without any fixed infrastructure was available. In the second scenario, these design assumptions were violated by including a subset of static nodes that provided a backbone. In the first scenario the packet delivery ratio of all routing protocols was high. However, when the NS-2 simulator was augmented such that links experienced random packet loss (without further details given about its statistical description), the SAMPLE protocol showed much better adaptivity. For packet loss of up to 20%, SAMPLE showed packet delivery ratios above 85% while AODV and DSR exhibited only 60% and 10% respectively. In the second scenario it was shown that SAMPLE outperformed its competitors in terms of throughput and packet delivery ratio since it better utilized the backbone. While the accuracy of the proposed LQP ALGORITHM was confirmed indirectly by these simulations, it is unclear how the obtained results relate to real-world radio links since the quality of links is not arbitrarily random but a complicated stochastic process that is governed by diverse propagation effects depending on the environment. Furthermore, the constants in (26) and the observation length were set to values without any further explanation.

8. Related literature

In this section we discuss literature that does not suggest new LQP algorithms for MANETs but is related to LQP and its applications.

8.1 Roman, Huang & Hazemi (2001)

Roman et al. (2001) use the concept of announced disconnections to create a group communication system for MANETs that provides consistency. Consistency is defined such that all messages are sent and delivered in the same group view by all members and no message is lost under the assumption that no node fails and communication failures are announced in advance. These properties simplify the development of mobile applications greatly since even frequent changes in the network topology are masked from the application layer by the principle of transparency. The announcement of disconnection relies on LQP that is provided by Roman et al. (2001) by means of the principle of safe distance. This safe distance is defined as the maximum distance at which one can guarantee that any communication task between two nodes can still be completed before the link between them fails, even if the nodes move away from each other at maximum speed. This concept corresponds to the preemptive region from Goff et al. (2001) with the difference that Goff et al.'s (2001) communication task is the discovery of routes, while Roman et al.'s (2001) task is the delivery of messages to achieve consistency. The proposed group communication system consists of a group discovery and a reconfiguration protocol. The former uses location information to determine if new groups or group members are in the vicinity such that either groups or members join. The latter protocol merges and splits groups, handles joins and leaves of group members and ensures consistency of message delivery. Reconfigurations rely on the principle that all members of a group are within a safe distance such that consistency can be provided due to announcement of disconnections. If one or more group members move outside the safe distance with respect to other members, the group is split into multiple groups.

The work of Roman et al. (2001) focuses on the algorithms in their group communication system and does not provide any detailed LQP ALGORITHM. All the algorithms presented are only briefly outlined and no evidence of an implemented system is provided. Therefore, this work is of conceptual nature only. While we believe that LQP algorithms can be of great benefit for efficient group communication in MANETs, the practical feasibility of the proposed approach is questionable. Roman et al. (2001) assume perfect LQP as a basis for providing consistency. In this way, Fischer et al.'s (1985) impossibility results for achieving consensus in an asynchronous distributed system, which experiences arbitrary link failures, is circumvented and the complicated task of achieving consensus in MANETs with restricted link failure patterns is greatly simplified. However, the assumption of perfect LQP is unrealistic and thus should not be the basis for providing consistency, since this property must be provided in all circumstances, even when the LQP is incorrect.

8.2 Killijian, Cunningham, Meier, Mazare & Cahill (2001)

Killijian et al. (2001) suggest a model for a location-aware group communication system. The key idea is that groups are defined not only by interest but also by proximity. A proximity group therefore consists of members that are within a defined space and that are interested in this group. The space is specified either absolutely or relatively to mobile nodes. For example, an absolute proximity might be defined for a traffic light that is to inform cars about its status. A proximity group around an ambulance, which informs nearby cars about its presence, requires in contrast a relative specification. The group membership management layer is supported by a partition prediction component. This component predicts the probability of a partition and is employed to guarantee consistency of group membership and message delivery. A failure predictor is recommended to predict link failures due to node crashes and battery status. A movement planner predicts mobility related link failures based on the notion of a safe distance (Roman et al. 2001). An environment evaluator is envisioned that infers knowledge from environmental conditions that have an influence on link quality.

While we like the idea of partition prediction, Killijian et al.'s (2001) work provides no details on how such partition prediction can be realized or how such a component integrates into the group membership management layer. Furthermore, only the prediction of link failures is envisioned. However, the prediction of a future network topology also requires a prediction of link recoveries since they also change the network topology. Moreover, Killijian et al. (2001) suggest LQP similar to Roman et al. (2001) for ensuring consistency. As discussed in previously, group communication systems provide consistency guarantees that cannot be realized solely by imperfect LQP.

9. Handoff algorithms

Handoff, which is also termed 'handover', is the process of transferring a connection with a mobile phone to a different base station than currently used in a cellular network. This transfer can be realized by switching the frequency band, time slot or codeword used or any combination of these. Handoff algorithms are specific to the structure of the infrastructure networks for which they are designed. For example, an algorithm that is designed to work in a macrocellular environment is unlikely to perform well in an overlay network of macro-

and microcells. If the infrastructure network controls the handoff process, the handoff is centralized; otherwise it is decentralized. The goal of the handoff process is to maximize communication quality while minimizing the number of handoffs. Handoff, admission control, channel allocation and power control are all resource management tasks that should be integrated to obtain better overall system performance. Most handoff algorithms are based on signal-power and noise-power measurements, which indicate that these criteria are well suited in determining link quality. An excellent survey article for handoff algorithms is given by Tripathi et al. (1998).

The handoff process is related to LQP in MANETs since both topics are concerned with link quality. However, these topics differ greatly in other respects. The handoff process switches between base stations in an infrastructure network. The infrastructure network is static with respect to mobility and therefore has well-defined properties such as certain coverage and fixed positions of the base stations. Furthermore, base stations are elevated, have more powerful transmitters and more sensitive receivers than mobile stations such that up- and down links have different properties. Moreover, base stations are connected by wired links and therefore have global knowledge about all mobile users and other base stations. All these properties are absent in MANETs that consists of mobile nodes, which act as peers only. Furthermore, the aim differs for both topics. The handoff process focuses on selecting the base station with the best communication quality and the accurate prediction of single link failures is of no particular concern. The mobile station can also control the communication quality to a certain extent via the base station due to a feedback loop. The handoff problem is a global optimization problem with global knowledge that aims to simultaneously maximize the communication quality and the capacity of the network. LQP in MANETs, however, aims to predict the probability of link failures without any global knowledge and without the possibility of influencing the communication quality. Furthermore, MANETs require multi-hop communication while cellular networks rely on single-hop wireless communication.

10. Real-world measurement studies related to link quality prediction

In recent years an increased number of measurement studies can be found in the literature due to the deployment of wireless networks, especially in universities. However, most of these measurements are concerned with higher-layer metrics in infrastructure networks and indoor studies are predominant. A comprehensive list of measurement studies is given by the 'Community Resource for Archiving Wireless Data At Dartmouth' project that intends to provide a wireless network data resource for the research community (Dartmouth College 2006). In this section we focus on some studies that measure link quality using low-layer metrics in outdoor 802.11 networks and that are related to our own measurements.

10.1 Anastasi, Borgia, Conti & Gregori (2004)

Anastasi et al. (2004) measured link quality in terms of the higher-level metrics of TCP and UDP throughput for an 802.11b network in ad hoc mode. All measurements were done outdoors with static nodes. The authors observed that the maximum transmission range is lower for higher data rates than for lower ones. This result confirms theory since with constant transmission power more energy per symbol is available for lower than for higher data rates. The maximum transmission range was estimated to be 30 m, 70 m, 90–100 m and 110–130 m for the data rates of 11 Mbps, 5.5 Mbps, 2 Mbps and 1 Mbps respectively. Therefore, Anastasi et al. (2004) conclude (erroneously) that the maximum transmission range of 250 m, which is assumed in most MANET simulations, is 2-3 times higher than in a real-world, obstacle-free environment. Furthermore, it was demonstrated that the throughput is seriously degraded if a node height is below 1.20 m. This effect depends on the First Fresnel Zone that contains most of the radio energy. If nodes are too low, the First Fresnel Zone touches the ground such that a significant amount of radio energy is lost.

Moreover, Anastasi et al. (2004) showed that the UDP and TCP throughput between two nodes was significantly reduced, if other nodes, which were placed further away than the maximum transmission range, transmitted on the same channel. However, the throughput increased in two steps with when the distance between the interfering nodes was increased. Therefore, the authors concluded that the maximum transmission range is smaller than the maximum physical carrier sensing range, which in turn is smaller than the maximum interference range. Consequently, Anastasi et al. (2004) suggest using different values for each of these ranges in simulations that traditionally use only one value for all ranges.

10.2 Aguayo, Bicket, Biswas, Judd & Morris (2004)

Aguayo et al. (2004) measured packet loss for a static mesh network of 38 nodes that were distributed over roofs in Cambridge, Massachusetts. Each node consisted of a PC with an 802.11b card with Intersil Prism 2.5 chipset in ad hoc mode and an omni-directional antenna. All measurements of signal power, noise power and packet loss were made with 1500-byte long packets at different data rates. The accuracy of the signal power and noise power measurements was shown to be within 4 dB for the type of 802.11 card used.

The 'neighbour' abstraction is commonly used in the design of MANET protocols and it expresses the idea that nodes can be grouped into partitions in which nodes can either communicate with 100% or 0% probability. However, Aguayo et al. (2004) demonstrated that a high number of nodes had packet loss ratios within the range of 10% to 90%. Therefore, the classical neighbour abstraction did not hold for this wireless test bed. Furthermore, it was shown that the observed packet loss was only partly depending on the signal-to-noise ratio (SNR) or distance. In an experiment with a hardware channel emulator it was determined that a delay spread higher than a few hundred nanoseconds affected the packet loss negatively. This observation agrees with the 802.11 cards' specification which states that the RAKE receiver, which is the main receiver structure in an 802.11 card, and the channel equalizer only support delay spreads up to 250 ns. While this delay spread is appropriate for indoor environments, urban microcells exhibit delay spread often exceeding 1 ms (Aguayo et al. 2004). Moreover, it was shown that the packet loss typically varies by only a few per cent within a time frame of one second.

While Aguayo et al. (2004) present interesting results, it is not clear how general these are. All their measurements have been made with only one card type. The propagation environment is over urban roofs in which diffraction is especially high. If the same observations would be made on the ground for moving nodes is open to question. Additionally, all measurements were made with very long packets of only one size and therefore it is open if the same results could be observed with different packet lengths.

10.3 Sridhara, Shin & Bohacek (2006)

Sridhara et al. (2006) examined signal-power variations due to pedestrians for an 802.11 network in infrastructure mode. Two nodes were placed at heights of 1.2 m and 4.3 m respectively and the signal power was simultaneously recorded for both transmitting nodes. It was observed that the signal power varied due to moving pedestrians albeit all nodes were static. Sridhara et al. (2006) noticed further that with increasing density of pedestrians the variance of the signal power increased. Different node heights, however, only slightly affected the variance of the signal power. Moreover, it was demonstrated that the variations can be characterized well in the time domain by a one-dimensional diffusion process with four parameters. This statistical description allows a computationally inexpensive simulation of the observed variations.

The proposed statistical description of signal-power variations, which are caused by pedestrians, is valid for stationary nodes only. Whether these results extend to mobile scenarios has not been explored yet. Furthermore, the authors do not investigate to what degree these signal-power fluctuations are caused by small-scale fading. Small-scale fading can also be experienced by static nodes if other objects move in the radio channel (Rappaport 2002, p. 178).

11. Small-scale wideband signal envelope fading

In general the wideband signal envelope shows less severe small-scale fading than the narrowband signal envelope (Lee 1991). The reason for this is that wideband signals inherently exploit frequency diversity (Lee 1991, Kozono 1994, Yamaguchi et al. 1995). While the probability distributions of small-scale narrowband signal envelope fading, for which closed-form expressions exist, have been extensively studied (see, for example, Young 1952, Nakagami 1960, Jakes & Reudink 1967, Clarke 1968, Suzuki 1977), comparatively few such studies are reported for wideband systems (Kozono 1994, Yamaguchi et al. 1995, Yan & Kozono 1999, Oh et al. 2001) and closed-form expressions for the CDF and PDF of the small-scale fading envelope are unavailable.

11.1 Kozono (1994) and Yan & Kozono (1999)

Kozono (1994) and Yan & Kozono (1999) have proposed a wideband signal propagation model with which they investigated the small-scale fading of the received signal envelope of a mobile receiver. They verified their model with extensive measurements performed in the Tokyo region (Kozono 1994, Nakabayashi & Kozono 1998, Yan & Kozono 1999, Nakabayashi et al. 2001) and observed that the fading depth is strongly dependent not only on the ratio of the direct to indirect power of the wideband signal but also on what they term the 'Equivalent Received Bandwidth'. They also observed that the distribution of the signal envelope is almost independent of the carrier frequency. Their results also demonstrate that wideband signals in general exhibit shallower fades than narrowband signals and that small-scale wideband signal envelope fading in non-line-of sight (NLOS) conditions cannot in general be characterized using the Rayleigh distribution. They did not investigate however whether other known distributions can do this. Similar results to those given above were reported by Yamaguchi et al. (1995).

Yan & Kozono's (1999) model assumes that N_{wv} multipath waves arrive at the receiver under the following conditions: each wave has an amplitude A_i, a path length L_i and an angle of arrival ϑ_i. A_i and L_i are independent of each other and are distributed uniformly over a given range. ϑ_i is distributed uniformly over 2π in the horizontal plane. A_0 and L_0 denote the amplitude and path length of the LOS wave. The ratio of the direct to indirect power is defined as:

$$a = A_0^2 / \sum_{i=1}^{N_{Wv}-1} A_i^2 \tag{27}$$

The bandwidth of each arriving wave (assumed to have a flat power spectral density) is taken to be greater than the receiver bandwidth $2\Delta f$. The received signal power R in watts is then expressed by (Yan & Kozono 1999, (2)):

$$R(2\Delta f) = 2\Delta f \left(\sum_{i=0}^{N_{Wv}-1} A_i^2 + \underbrace{\sum_{i=0}^{N_{Wv}-1} \sum_{j=0}^{N_{Wv}-1} \frac{c}{2\pi} \frac{A_i A_j}{\Delta f \Delta L_{ij}} \cos\left(\frac{2\pi f_c \Delta L_{ij}}{c} \right) \sin\left(\frac{2\pi f_c \Delta L_{ij}}{c} \right)}_{i \neq j} \right) \tag{28}$$

where ΔL_{ij} is the difference in the path lengths of the ith and jth arriving waves. As mentioned earlier Yan & Kozono (1999) also propose a propagation parameter termed the 'equivalent received bandwidth' which can be used to estimate the fading depth (Cardoso & Correia 2003). The equivalent received bandwidth is the product of the receiver bandwidth $2\Delta f$ and the maximum difference in path lengths ΔL_{max} of the arriving waves (i.e. $\Delta L_{max} = \max|L_i - L_j|$). In summary, the equivalent bandwidth and hence observed fading depth are dependent on the receiver bandwidth and the propagation environment.

11.2 Nakabayashi & Kozono (1998)

Two properties of the stochastic process of small-scale wideband signal envelope fading are important: the probability distribution of the process and its autocorrelation. The probability distribution is the summary of the amplitude structure of the process; the autocorrelation is the summary of the time structure (Yates & Goodman 2005, p. 353). Nakabayashi & Kozono (1998) investigated the autocorrelation properties of small-scale wideband signal envelope fading based on Kozono's (1994) wideband signal propagation model. It was shown that the normalized autocorrelation of the small-scale fading signal envelope r is independent of the receiver bandwidth and is described for both, narrowband and wideband signals, by (Nakabayashi & Kozono 1998, (2)), (Goldsmith 2005, pp. 73,74):

$$\rho_{rr}(\Delta_t) = J_0(2\pi f_D \Delta_t) \tag{29}$$

where f_D is the Doppler frequency, Δ_t is the time lag of the normalized autocorrelation, and J_0 is the zeroth-order Bessel function of the first kind. This result was derived analytically and also verified by measurements (Nakabayashi & Kozono 1998).

11.3 Oh, Lee, Choi & Kim (2001)

In their work on a statistical model of a W-CDMA receiver, Oh et al. (2001) showed that the distribution of the small-scale fading signal envelope can be described using the Nakagami distribution for a CDMA system. Though this simulation study was limited to only two receiver bandwidths deployed in two types of urban environment, namely, urban high-rise and urban residential, it suggested to us the possibility that the Nakagami and the closely related Rice distribution could be used as a general means with which to describe the distribution of the small-scale fading wideband signal envelope, albeit to a certain degree of approximation.

12. Accuracy and methodology of MANET simulations

Simulation is still the most popular means of evaluation in the MANET community (Kurkowski et al. 2005) since real-world measurements are expensive in terms of labour and equipment. Simulations also allow the influence of parameters on evaluation metrics to be isolated and simulation experiments are easy to repeat. Nevertheless, the accuracy of current MANET simulators is debatable as some studies suggest. Furthermore, the methodology of MANET studies is often questionable and seems to be poorer than in more established fields (Kurkowski et al. 2005). Therefore, we discuss here some key studies that enabled credible evaluations in this work.

12.1 Cavin, Sasson & Schiper (2002)

Cavin et al. (2002) implemented a simple broadcast algorithm to compare the basic metrics of this protocol in three widely used MANET simulators, namely GloMoSim, NS-2 and OPNET without further assessment of these results in real-world experiments. The used broadcast algorithm floods the network by forwarding every message received for the first time to neighbouring nodes. This algorithm is simple enough to be implemented equally with the same parameters in all three simulators while also being relevant since broadcasting is an important building block in MANET protocols. Cavin et al. (2002) demonstrated that the evaluation results were quantitatively and qualitatively divergent. Not only did the absolute values of the evaluation metrics differ but also in some cases the general behavior varied. They explained these observations by the different choice of models and different level of detail in the simulators used, especially in the physical layer. Similar results were obtained by Takai et al. (2001) who showed that the AODV and DSR routing protocols lead to different evaluation results in NS-2 than in GloMoSim, mainly due to differences in the physical layer. Both studies, as well as our own real-world measurements, lead us to the conclusion that the standard NS-2 simulator is unsuitable for credible evaluations.

12.2 Gray, Kotz, Newport, Dubrovsky, Fiske, Liu, Masone, McGrath & Yuan (2004)

Gray et al. (2004) conducted a study in which they compared the performance of four ad hoc routing protocols in an outdoor trial, an indoor trial and simulations. The network consisted of 33 laptops that were equipped with an 802.11 card and a GPS receiver each. In the outdoor trial, people moved randomly, similar to the Random Waypoint Mobility Model with no pauses, and the nodes would broadcast their GPS positions every three seconds. The locations were saved in files and a connectivity trace was derived that represents the network topology. Two nodes were modeled to be connected within the beacon interval, if they received a beacon; otherwise they were modeled to be disconnected. The experiment occurred on an athletic field that can roughly be divided into four flat, equally sized sections of which one was approximately 4 m to 6 m lower. This difference in height caused blockage in LOS paths. In the indoor trials, all laptops were located on the same tabletop and the GPS and connectivity traces were used to emulate mobility and connectivity. In the simulation experiment, six different combinations of radio propagation models and the connectivity trace were compared. The Free Space Propagation Model, the Two-Ray Ground Propagation Model and a single-slope path-loss model with an empirically derived path-loss exponent and large-scale fading were used either with or without the connectivity trace. If the connectivity trace was enabled, the simulator would first check that nodes were in range according to the trace file before the radio propagation model calculated the received power. If the power was higher than a threshold, a packet is received; otherwise it is lost. SWAN (Perrone & Nicol 2002) was the simulator of choice and the mobility was given by the GPS trace file.

Gray et al. (2004) demonstrated that the indoor experiments did not represent the outdoor behavior despite relying on the outdoor connectivity trace. The performance metrics differed in values and even the relative ranking between the routing protocols varied. This observation was attributed mainly to interference that was caused by the 33 nodes all being located on the same tabletop. The comparison of simulations with outdoor measurements showed that the single-slope path-loss model with large-scale fading was fairly accurately predicting the outdoor performance while the other model combinations led to poorer representations of the real-world environment. This result indicated to us that stochastic radio propagation models with empirical constants for path loss and large-scale fading can lead to accurate outdoor simulations of MANETs, especially if these models are extended with simulation of wideband small-scale fading.

12.3 Stepanov, Herrscher & Rothermel (2005)

Stepanov et al. (2005) compared the NS-2 simulator with a single-slope path-loss model (without any fading simulation) to an augmented NS-2 version that calculates the received power at nodes via a ray tracing based radio propagation model. The latter type of radio propagation model is known to describe the radio channel accurately when detailed geographical maps with building materials are used (Neskovic et al. 2000). Stepanov et al. (2005) demonstrated that the network topology differed significantly between simulations with the single-slope path-loss model with different constants and the ray tracing model. Additionally, it was shown that the packet delivery ratio, the routing packet overhead and the packet latency differed for simulations of the AODV protocol. Unfortunately, no comparison was made between the ray tracing model and a single-slope path-loss model with additional simulation of large-scale and small-scale fading.

For their study, Stepanov et al. (2005) precomputed the signal power for all possible transmitter-receiver positions in the city centre of Stuttgart with a commercial ray tracing based radio propagation simulator. This precomputed data was then fetched during simulations in NS-2. The signal power was precomputed for a 5 m by 5 m grid that leads to 32 billion position pairs for the city centre of Stuttgart. A finer sampling would have lead to significantly more position pairs. The precalculation step produced 120 GB of data and required the computing power of three days of a PC cluster of 50 machines. These required resources indicated clearly to us that ray traced based methods do not scale to ad hoc networks. While in infrastructure networks the position of base stations is fixed, every node in ad hoc networks is potentially mobile such that the problem of precomputing signal-power values increases by a whole dimension. Consequently, an improved stochastic simulation of the radio channel is better suited to the simulation of MANETs due to significantly lower resource requirements than using ray tracing based methods.

12.4 Kurkowski, Camp & Colagrosso (2005)

Kurkowski et al. (2005) investigated the methodology on which the MANET community performs evaluations. Their findings are derived by surveying all papers that have been published in the premier conference MobiHoc from 2000 to 2005. Four different criteria were identified for credible research. First, results should be repeatable. Secondly, results should be unbiased and not specific to an unrepresentative scenario. Thirdly, results should be rigorous and therefore scenarios and conditions of simulations should truly represent the aspects of the MANET protocol being studied. Fourthly, the results must be statistically sound. Based on these criteria, Kurkowski et al. (2005) concluded that the state of art in MANET simulations is poor and that most studies lack believability.

75.5% of all surveyed studies were based on simulations. Nevertheless, in 29.8% of all studies the simulator was not identified. NS-2, followed by GloMoSim, was by far the most popular MANET simulator and NS-2/GloMoSim were used in 43.8%/10.0% of all studies in which the simulator was identified. 57.9% of all studies did not state the type of simulation (e.g. termination or steady-state). Steady-state simulations represent the long-term behavior of MANET protocols after the initialization phase and they are the most common type of simulations since researchers are normally interested in the long-term behavior of protocols. However, only 7.0% of all studies addressed the initialization bias albeit using the unreliable method of arbitrarily discarding data from the beginning of the simulation. The number of simulation iterations was only published in 35.8% of all surveyed studies and confidence intervals were provided only in 12.5%. However, it is unclear how well results represent the population under study when the number of simulation iterations and confidence intervals are unknown.

12.5 PalChaudhuri, Boudec & Vojnovic (2005)

Steady-state simulations of MANETs require that the mobility distribution is in steady state within the whole simulation run since protocol performance varies significantly between the transient phase and the steady state as, for example, PalChaudhuri et al. (2005) showed for the packet reception ratio of the DSR protocol. Steady-state simulation of mobility, which is also called perfect simulation, is provided for the random trip model by a tool that was developed by PalChaudhuri et al. (2005). The random trip model is a generic model that includes the widespread Random Waypoint Mobility Model (Johnson & Maltz 1996) and also the Random

Waypoint Mobility Model City Section (Saha & Johnson 2004). In the latter model the nodes move according to a geographical map that is often derived from the real world such that this mobility model is often considered to be realistic (Saha & Johnson 2004).

13. Summary

All the surveyed link quality prediction (LQP) algorithms are based on radio propagation models that only characterize path loss to various degrees of accuracy but fail to also consider large-scale and wideband small-scale fading effects. Consequently, neither these LQP algorithms themselves nor their performance evaluations are applicable to real-world environments in which these two propagation effects are pronounced (Neskovic et al. 2000, Kotz et al. 2004). Hence, novel LQP algorithms are required that incorporate these propagation effects and make the promise of accurate LQP in real-world urban environments a reality. Furthermore, the need for credible evaluations by real-world studies or advanced simulations, which account for the propagation effects of path loss, large-scale and small-scale fading, is evident.

A number of LQP algorithms rely on location measurements for their prediction. However, typical GPS measurements have a user-equivalent range error of 19.2 m and location information is often unobtainable in dense urban environments where a connection with at least three satellites cannot be continuously maintained. Therefore, location measurements are currently impractical for these environments.

In the literature indirect evaluations of LQP algorithms prevail over direct ones. In these indirect evaluations the prediction accuracy of LQP algorithms is assessed by comparing the performance of communication protocols with and without LQP, typically by comparing metrics such as packet latency and data throughput. Since various studies use different communication protocols to evaluate their LQP algorithms, a direct comparison of LQP accuracy for different studies is difficult.

Link quality prediction is currently in its nascent stage. It is fundamentally a propagation problem which applies not just to ad-hoc networks *per se* but to cognitive radio in general where it is necessary to be able to calculate the electromagnetic footprint of a cognitive radio in order to guarantee non-interference. In conclusion an effort, similar to that seen forty years ago with the dawn of cellular radio in propagation modeling for network planning, must now take place in real-time radio propagation prediction.

14. References

Agarwal, S., Ahuja, A., Singh, J. P. & Shorey, R. (2000), Route-Lifetime Assessment Based Routing (RABR) Protocol for Mobile Ad-Hoc Networks, *in* 'IEEE International Conference on Communications', Vol. 3, New Orleans, LA, USA, pp. 1697–1701.

Aguayo, D., Bicket, J., Biswas, S., Judd, G. & Morris, R. (2004), Link-Level Measurements from an 802.11b Mesh Network, *in* 'ACM SIGCOMM Conference on

Communications Architectures, Protocols and Applications', Vol. 4, Portland, OR, USA.

Anastasi, G., Borgia, E., Conti, M. & Gregori, E. (2003), IEEE 802.11 Ad Hoc Networks: Performance Measurements, *in* 'Workshop on Mobile and Wireless Networks', Providence, RI, USA.

Bajaj, L., Takai, M., Ahuja, R., Tang, K., Bagrodia, R. & Gerla., M. (1999), Glomosim: A Scalable Network Simulation Environment., Technical report, UCLA Computer Science Department.

Cavin, D., Sasson, Y. & Schiper, A. (2002), On the Accuracy of MANET Simulators, *in* 'Workshop on Principles of Mobile Computing', Toulouse, France, pp. 38–43.

Clarke, R. H. (1968), 'Statistical Theory of Mobile-Radio Reception', *Bell System Technical Journal* 47(6), 957–1000.

Cardoso, F. D. & Correia, L. M. (2003), 'Fading Depth Dependence on System Bandwidth in Mobile Communications - an Analytical Approximation', *IEEE Transactions on Vehicular Technology* 52(3).

Dartmouth College (2006), 'Community Resource for Archiving Wireless Data at Dartmouth'. http://crawdad.cs.dartmouth.edu/

De Couto, D. S. J., Aguayo, D., Bicket, J. & Morris, R. (2003), A High-Throughput Path Metric for Multi-Hop Wireless Routing, *in* 'International Conference on Mobile Computing and Networking', San Diego, CA, USA, pp. 134–146.

Dowling, J., Curran, E., Cunningham, R. & Cahill, V. (2005), 'Using Feedback in Collaborative Reinforcement Learning to Adaptively Optimize MANET Routing', *IEEE Transactions on Systems, Man and Cybernetics* 35(3), 360–372.

Dube, R., Rais, C. D., Wang, K.-Y. & Tripathi, S. K. (1997), 'Signal Stability-Based Adaptive Routing (SSA) for Ad Hoc Mobile Networks', *IEEE Personal Communications* 4(1), 36–45.

Fischer, M. J., Lynch, N. A. & Paterson, M. S. (1985), 'Impossibility of Distributed Consensus with One Faulty Process', *Journal of the ACM* 32(2), 374–382.

Gerharz, M., Waal, C. d., Frank, M. & Martini, P. (2002), Link Stability in Mobile Wireless Ad Hoc Networks, *in* 'IEEE Conference on Local Computer Networks', Tampa, FL, USA, pp. 30–42.

Gerla, M., Pei, G., Lee, S.-J. & Chiang, C.-C. (1998), 'On-Demand Multicast Routing Protocol (ODMRP) for Ad Hoc Networks'. Internet Draft, draft-ietf-manet-odmrp-00.txt, Nov. 1998, work in progress.

Gerla, M., Tang, K. & Bagrodia, R. (1999), TCP Performance in Wireless Multi-Hop Networks, *in* 'Workshop on Mobile Computer Systems and Applications', New Orleans, LA, USA, pp. 41–50.

Goff, T., Abu-Ghazaleh, N. B., Phatak, D. S. & Kahvecioglu, R. (2001), Preemptive Routing in Ad Hoc Networks, *in* 'International Conference on Mobile Computing and Networking', Rome, Italy, pp. 43–52.

Goldsmith, A. (2005), *Wireless Communications*, 1st edn, Cambridge University Press.

Gray, R. S., Kotz, D., Newport, C., Dubrovsky, N., Fiske, A., Liu, J., Masone, C., McGrath, S. & Yuan, Y. (2004), Outdoor Experimental Comparison of Four Ad Hoc Routing

Algorithms, *in* 'International Workshop on Modeling Analysis and Simulation of Wireless and Mobile Systems', Venice, Italy, pp. 220–229.

Grewal, M. S., Weill, L. R. & Andrews, A. P. (2001), *Global Positioning Systems, Inertial Navigation and Integration*, John Wiley and Sons.

Harvey, A. C. (1993), *Time Series Models*, 2nd edn, The MIT Press.

Hines, W. W., Montgomery, D. C., Goldsman, D. M. & Borror, C. M. (2002), *Probablity and Statistics in Engineering*, 4th edn, John Wiley and Sons.

Holland, G. & Vaidya, N. (2002), 'Analysis of TCP Performance over Mobile Ad Hoc Networks', *Wireless Networks* 8(2/3), 275–288.

Jakes, W.C., J. & Reudink, D. (1967), 'Comparison of Mobile Radio Transmission at UHF and X-Band', *IEEE Transactions on Vehicular Technology* 16, 10–13.

Jiang, S., He, D. & Rao, J. (2005), 'A Prediction-Based Link Availability Estimation for Routing Metrics in MANETs', *IEEE/ACM Transactions on Networking* 13(6), 1302–1312.

Johnson, D. & Maltz, D. (1996), Dynamic Source Routing in Ad Hoc Wireless Networks, *in* Imelinsky & H. Korth, eds, 'Mobile Computing', Kluwer Academic Publishers, pp. 153–181.

Kay, D. A. (2001), *Trigonometry*, 1st edn, Cliffs Notes.

Killijian, M.-O., Cunningham, R., Meier, R., Mazare, L. & Cahill, V. (2001), Towards Group Communications for Mobile Participants, *in* 'ACM Workshop on Principles of Mobile Computing', Newport, RI, USA.

Kotz, D., Newport, C., Gray, R. S., Liu, J., Yuan, Y. & Elliott, C. (2004), Experimental Evaluation of Wireless Simulation Assumptions, *in* 'International Workshop on Modeling Analysis and Simulation of Wireless and Mobile Systems', Venice, Italy, pp. 78–82.

Kozono, S. (1994), 'Received Signal-Level Characteristics in a Wide-Band Mobile Radio Channel', *IEEE Transactions on Vehicular Technology* 43(3).

Kurkowski, S., Camp, T. & Colagrosso, M. (2005), 'MANET Simulation Studies: The Incredibles', *ACM SIGMOBILE Mobile Computing and Communications Review* 9(4), 50–61.

Lee, W. C. Y. (1991), Theory of Wideband Radio Propagation, *in* 'IEEE Vehicular Technology Conference', St. Louis, MO, USA.

Lee, S.-J., Su, W. & Gerla, M. (1999), Ad Hoc Wireless Multicast with Mobility Prediction, *in* 'IEEE International Conference on Computer Communications and Networks', Boston, MA, USA, pp. 4–9.

McDonald, A. B. & Znati, T. (1999*a*), 'A Mobility-Based Framework for Adaptive Clustering in Wireless Ad-Hoc Networks', *IEEE Journal on Selected Areas in Communication* 17(8).

McDonald, A. B. & Znati, T. (1999*b*), A Path Availability Model for Wireless Ad-Hoc Networks, *in* 'IEEE Wireless Communications and Networking Conference', New Orleans, LA, USA.

MIL 3 Inc. (2006), 'OPNET'. http://www.opnet.com.

Nakabayashi, H. & Kozono, S. (1998), 'Autocorrelation Characteristics for Received Signal-Level in a Wide-Band in a Wide-Band Mobile Radio Channel', *Transactions IEICE* J81-B-II(2).

Nakabayashi, H., Yan, J., Masui, H., Ishii, M., Sakawa, K., Shimizu, H., Kobayashi, T. & Kozono, S. (2001), 'Validation of Equivalent Received Bandwidth to Characterize Received Signal Level Distribution through Experiment and Simulation', *IEICE Transactions on Communications* E84-B(9), 2550–2559.

Nakagami, M. (1960), The M-Distribution; a General Formula of Intensity of Rapid Fading, *in* W. G. Hoffman, ed., 'Statistical Methods in Radio Wave Propagation: Proceedings of a Symposium held at the University of California', Permagon Press, pp. 3–36.

Neskovic, A., Neskovic, N. & Paunovic, G. (2000), 'Modern Approaches in Modeling of Mobile Radio Systems Propagation Environment', *IEEE Communications Surveys* 3(3), 2–12.

Oh, D., Lee, G., Choi, D. & Kim, C. (2001), Statistical Model of W-CDMA Signals over Realistic Multipath Channel, *in* 'Vehicular Technology Conference', Vol. 1, Rodes, Greece, pp. 347–351.

PalChaudhuri, S., Boudec, J.-Y. L. & Vojnovic, M. (2005), Perfect Simulations for Random Trip Mobility Models, *in* 'Annual Simulation Symposium', San Diego, CA, USA, pp. 72–79.

Paul, K., Bandyopadhyay, S., Mukherjee, A. & Saha, D. (1999), Communication-Aware Mobile Hosts in Ad-Hoc Wireless Network, *in* 'International Conference on Personal Wireless Communications', Jaipur, India, pp. 83–87.

Perkins, C. E. & Bhagwat, P. (1994), *Highly Dynamic Destination-Sequenced Distance-Vector Routing (DSDV) for Mobile Computers, in 'ACM SIGCOMM Conference on Communications Architectures, Protocols and Applications'*, London, UK, pp. 234–244.

Perkins, C., Royer, E. & Das, S. (2001), 'Ad Hoc on-Demand Distance Vector (AODV) Routing'. Internet Draft, Internet Engineering Task Force, Mar. 2001. http://www.ietf.org/internet-drafts/draft-ietf-manet-aodv-08.txt

Perrone, L. F. & Nicol, D. M. (2002), A Scalable Simulator for TinyOS Applications, *in* 'Winter Simulation Conference', San Diego, CA, USA.

Polya, G. (1921), 'Ueber eine Aufgabe der Wahrscheinlichkeitstheorie betreffend die Irrfahrt im Strassennetz', *Mathematische Annalen* 84(1-2), 149–160.

Punnoose, R. J., Nikitin, P. V., Broch, J. & Stancil, D. D. (1999), Optimizing Wireless Network Protocols Using Real-Time Predictive Propagation Modeling, *in* 'IEEE Radio and Wireless Conference', Denver, CO, USA.

Rappaport, T. S. (2002), *Wireless Communications - Principles and Practice*, Prentice Hall Communications Engineering and Emerging Technologies Series, 2nd edn, Prentice Hall.

Qin, L. & Kunz, T. (2002), 'Pro-Active Route Maintenance in DSR', *ACM SIGMOBILE Mobile Computing and Communications Review* 6(3), 79–89.

Qin, L. & Kunz, T. (2003), Increasing Packet Delivery Ratio in DSR by Link Prediction, *in* '36th Annual Hawaii International Conference on System Sciences', Vol. 9, Hawaii, USA.

Roman, G.-C., Huang, Q. & Hazemi, A. (2001), Consistent Group Membership in Ad Hoc Networks, *in* 'IEEE International Conference in Software Engineering', Toronto, Canada, pp. 381–388.

Saha, A. K. & Johnson, D. B. (2004), Modeling Mobility for Vehicular Ad Hoc Networks, *in* 'Workshop on Vehicular Ad Hoc Networks (VANET)', Philadelphia, PA, USA.

Schiller, J. (2000), *Mobile Communications*, 1st edn, Addison-Wesley Professional

Sridhara, V., Shin, H.-C. & Bohacek, S. (2006), Observations and Models of Time-Varying Channel Gain in Crowded Areas, *in* 'Workshop on Wireless Network Measurements', Boston, MA, USA.

Stepanov, I., Herrscher, D. & Rothermel, K. (2005), On the Impact of Radio Propagation Models on MANET Simulation Results, *in* 'IFIP International Conference on Mobile and Wireless Communication Networks', Marrakech, Morocco.

Suzuki, H. (1977), 'A Statistical Model for Urban Radio Propagation', *IEEE Transactions on Communications* 25(7), 673–680.

Takai, M., Martin, J. & Bagrodia, R. (2001), Effects of Wireless Physical Layer Modeling in Mobile Ad Hoc Networks, *in* 'International Conference on Mobile Computing and Networking', Long Beach, CA, USA, pp. 87–94.

Tobagi, F. & Kleinrock, L. (1975), 'Packet Switching in Radio Channels: Part II—the Hidden Terminal Problem in Carrier Sense Multiple-Access and the Busy-Tone Solution', *IEEE Transactions on Communications* 23(12).

Tripathi, N. D., Reed, J. H. & Van Landingham, H. F. (1998), 'Handoff in Cellular Systems', *IEEE Personal Communications* 5(6), 26–37.

Wang, Y., Martonosi, M. & Peh, L.-S. (2006), A Supervised Learning Approach for Routing Optimizations in Wireless Sensor Networks, *in* 'REALMAN', Florence, Italy, pp. 79–86.

Woo, A., Tong, T. & Culler, D. (2003), Taming the Underlying Challenges of Reliable Multihop Routing in Sensor Networks, *in* 'International Conference on Embedded Networked Sensor Systems', Los Angeles, CA, USA, pp. 14–27.

Xu, S. & Saadawi, T. (2001), 'Does the IEEE 802.11 MAC Protocol Work Well in Multihop Wireless Ad Hoc Networks?', *IEEE Communications Magazine* 39(6), 130–137.

Yamaguchi, A., Suwa, K. & Kawasaki, R. (1995), Received Signal Level Characteristics for Wideband Radio Channel in Microcells, *in* 'International Symposium on Personal, Indoor and Mobile Radio Communications', Toronto, Canada.

Yan, J. & Kozono, S. (1999), 'A Study of Received Signal-Level Distribution in Wideband Transmissions in Mobile Communications', *IEEE Transactions on Vehicular Technology* 48(5).

Yates, R. D. & Goodman, D. J. (2005), *Probability and Stochastic Processes*, 2nd edn, John Wiles and Sons.

Young, W.R., J. (1952), 'Comparison of Mobile Radio Transmission at 150, 450, 900, and 3700 MC', *Bell System Technical Journal* 31(6), 1068–1085.

Part 2

Cooperative Sensing

5

Collaborative Spectrum Sensing for Cognitive Radio Networks

Aminmohammad Roozgard[1], Yahia Tachwali[2], Nafise Barzigar[1] and Samuel Cheng[1]
[1]*School of Electrical and Computer Engineering, University of Oklahoma*
[2]*Agilent Technologies*
USA

1. Introduction

The fast user growth in wireless communications has created significant demands for new wireless services in both the licensed and unlicensed frequency spectra[1]. Recent studies show that a fixed spectrum assignment policy cannot achieve good spectrum utilization (FCC, 2005) . To address this problem, cognitive radio (CR) has been developed as a technology to facilitate the utilization of temporally idle frequency bands (commonly known as white space or spectrum holes) to increase the spectral efficiency.

A key component of cognitive radio is spectrum sensing. The basic tasks of spectrum sensing are to detect the PUs and to identify available spectrum bands. Spectrum sensing is responsible for a CR system to satisfy two fundamental requirements. First, a CR system should be able to identify white spaces or spectrum holes to satisfy the throughput and quality of service requests of SUs. Second, the CR system needs to ensure that any SU will not cause harmful interference to the other users' frequency bands.

Currently, shadowing, multi-path fading, and receiver uncertainty problems (Cabric et al., 2004; Ghasemi & Sousa, 2005) at receivers decreased significantly the performance of spectrum sensing. However, partial diversity of CR locations can potentially improve the accuracy of spectrum sensing. SUs at different locations can observe different portions of the wireless spectrum, if SUs can cooperate and share the sensing results, the collaborative decision from the collected observations can be better than the local decision at each SU. This is why cooperative spectrum sensing is an attractive and effective approach to overcome different fading, shadowing, and receiver uncertainty problems (Mishra et al., 2006).

The primary idea of cooperative spectrum sensing is to improve the sensing performance by using observations from diversely located SUs. Not only that SUs can improve their decisions through collaboration, the sensing time can also be reduced by cooperation among SUs and this allows each SU to have more time for data transmission over white spaces. Furthermore, with cooperative spectrum sensing, the hardware requirement to maintain the same accurate sensing can be reduced.

[1] In the literature, the users of licensed and unlicensed frequency spectra are known as Primary User (PU) and Secondary User (SU) respectively.

The cooperative sensing for finding the spectrum holes starts with individual or local sensing performed by each SU, which needs to determine the presence ($\mathcal{H}_1$) and absence ($\mathcal{H}_0$) of PU in one frequency band. According to different hypothesis, the received signal at SU $x(t)$ can be modeled by (Akyildiz et al., 2009)

$$x(t) = \begin{cases} n(t) & ,\mathcal{H}_0 \\ h(t) \times s(t) + n(t) & ,\mathcal{H}_1 \end{cases} \tag{1}$$

where $h(t)$ denotes the channel gain of sensing channel, $s(t)$ is the transmitted signal of PU, and $n(t)$ is zero mean additive white Gaussian noise. In addition, we denote the probability of presence ($\mathcal{H}_1$) and absence ($\mathcal{H}_0$) of PU as $P(\mathcal{H}_0)$ and $P(\mathcal{H}_1)$ respectively. We will use three common metrics to measure the performance of a spectrum sensing algorithm: the probability of detection P_d, the probability of false alarm P_f, and the probability of missed detection P_f, which are defined by

$$\begin{aligned} P_d &= Pr\{decision = \mathcal{H}_1|\mathcal{H}_1\} = Pr\{Y > \lambda|\mathcal{H}_1\}, \\ P_f &= Pr\{decision = \mathcal{H}_1|\mathcal{H}_0\} = Pr\{Y > \lambda|\mathcal{H}_0\}, \\ P_m &= Pr\{decision = \mathcal{H}_0|\mathcal{H}_1\} = 1 - P_d, \end{aligned} \tag{2}$$

where Y is the decision statistic and λ is the decision threshold.

One important concern for the cooperative spectrum sensing design is the overhead for exchange sensing results. This overhead directly depends on the structure of the SUs' cooperation which can be divided into three categories: "centralized" (Unnikrishnan & Veeravalli, 2008), "distributed" (Li et al., 2010) and "relay-assisted" (Zhang & Letaief, 2008). In the centralized structure, the fusion center (FC) or base station (BS), which is an identified SU to manage the communication and decision making, selects one of the channels and asks the other SUs to send their local sensing results through the selected channel (known as the common control channel). Then, the FC receives the SUs' sensing results via a reporting channel. Lastly, the FC combines the SUs' information[2] and determine the existence of a PU.

Unlike centralized approach, we do not have a FC in the distributed spectrum sensing. All SUs communicate among themselves and make decisions by a distributed algorithm. The distributed algorithm is applied iteratively until all SUs converge to a decision. For example, one particular SU senses and sends its information to others. Then, the SU combines its data with received data of others and makes a decision. If its decision does not satisfy the local criterion, the SU must perform the following three steps iteratively until converge to a decision that satisfies the criterion. The first step is to send combined data through the report channel. The second step is to combine the received data with the local sensing data. And the last step is to make a decision and validate it with the local criterion. Also, in a distributed structure, for evaluation and comparison of the spectrum sensing algorithms, the rate and speed of convergence and convergence guarantee are the essential metrics.

[2] Based on received data of SUs, the FC can combine the information in three ways such as "soft combining", "quantized soft combining" and "hard combining". If SU sends complete local sensing data, the fusion process is classified as soft combining. If SU quantizes the local sensing information, before sending to the FC, the fusion process is classified as quantized soft combining. For hard combining fusion, a SU makes decision after sensing and sends one bit as its decision to the FC.

The third type of cooperative spectrum sensing is relay-assisted. As we mentioned before, the channels are not perfect, especially the report channel and the sensing channel. The relay-assisted schema can improve the performance. For example, a SU can find out that its report channel is weak and its sensing channel is strong; yet another SU can have a strong report channel and a weak sensing channel. These two SU can complement and cooperate with each other in a relay-assisted schema. This structure can be used in distributed or centralized schema. Here, a SU with a strong reporting channels will help carry the sensing information of a SU that has a weak reporting channel. In the literature, the centralized and distributed schema are considered as one-hop cooperative sensing and relay-assisted approach is considered as multi-hop cooperative sensing.

In this chapter, we review the recent studies about cooperative spectrum sensing in cognitive radio. In addition, the open research challenges will be identified. The chapter goes through the theoretical studies of the performance bounds and limits of cooperative spectrum sensing in Section (2). In this section, we will review and compare the research for optimizing the parameters which affect the throughput of cooperative spectrum sensing such as frame size, sensing time, fusion rules and etc. Also, a brief review of the sensing scheduling techniques is included in Section (2). Section (3) addresses the game theoretical approaches in cooperative cognitive radio. In this section, two main categories of game theory applications are described. The first category is coalitional games for formation of cooperative sets, which are used to reduce the overhead of communication in distributed cooperative spectrum sensing. The second category of game theoretical approach is evolutionary games which are useful for analyzing and managing the behavior of SUs. Also, the strategy selection for large networks will be described. Section (4) reviews the compressed sensing studies for cooperative cognitive radio. This section contains a brief introduction about basic concepts in the area of compressed sensing, then the usage of compressed sensing in cognitive radio will be described. Section (5) explores the evaluation studies of cooperative sensing performance in various types of wireless environments such as Radar, UAV, VANET, WiFi Networks. Finally, the chapter is concluded in section (6).

2. Theoretical bounds and limits

In cognitive radio networks, the opportunities for transmission are limited. All SUs would like to increase their chances for data transmission and save more power without causing harmful interference. Theoretical studies for optimizing the cooperation spectrum sensing parameters, which affect the data transmission opportunities, will help SU to improve their throughput and decrease the energy consumption. Also, to efficiently find spectrum holes, one needs to model the PU behavior, carefully distribute sensing tasks among SUs, and optimize the sensing time for each SU. In following Section (2.1), the trade-off between sensing time and throughput of system will be described. Then, in Section (2.2), dynamic spectrum sensing will be reviewed.

2.1 Sensing vs throughput Trade-off

In the literature, a frame-based structure (or periodic) spectrum sensing is broadly studied for cognitive radios (Fan & Jiang, 2009; Jiang et al., 2009; Liang et al., 2008; Lifeng et al., 2011). We refer to a frame as the duration of time composed of ***spectrum sensing slot*** and ***data***

transmission slot. During the spectrum sensing slot, SUs sense the spectrum to check for the peresence of any PU using the spectrum or not. Then, during the following transmission slot, they will initiate communication if no PU is detected. In general, the duration of sensing slot affects the accuracy of spectrum sensing. A longer sensing slot leads to a higher probability of detection P_d and lower probability of false alarm P_f. However, it reduces the time available for data transmission and thus decrease the throughput of SUs. The optimal trade-off between sensing time and the other parameters such as resource allocation[3] (Fan et al., 2011), fusion center properties (Peh et al., 2009) and cooperation time (Zhang et al., 2010) is still under investigation.

The optimal trade-off between sensing and throughput is formulated by (Liang et al., 2008) to maximize a SU's throughput without affecting the quality of service (QoS) of PU. (Liang et al., 2008) considered two scenarios. The first scenario is when SU detects an idle slot and indeed there is no PU using the channel. In this case, the SU throughput is $C_0 = \log_2(1 + \frac{P_s}{P_n})$, where P_s and P_n are the received SU power and the noise power, respectively. The second scenario happens when there is a detection error that a SU thinks that the slot is idle but is actually occupied by an undetected active PU. The throughput in this case will be $C_1 = \log_2(1 + \frac{P_s}{P_p+P_n})$, where P_p is the interference power due to the undetected PU. Then, they combined the throughput of each scenario and obtained the average throughput of the SU as

$$\begin{aligned} R(\tau,\varepsilon) &= C_0 \frac{T-\tau}{T}(1 - P_f(\tau,\varepsilon))P(\mathcal{H}_0) + C_1 \frac{T-\tau}{T}(1 - P_d(\tau,\varepsilon))P(\mathcal{H}_1) \\ &\triangleq R_0(\tau,\varepsilon) + R_1(\tau,\varepsilon), \end{aligned} \tag{3}$$

where T, τ and ε are the frame duration and the sensing slot duration and threshold for energy detectors, respectively. The goal of (Liang et al., 2008) is then to maximize the expected throughput without sacrificing the QoS of PUs through varying the sensing slot duration τ. That is,

$$\max_{\tau} R(\tau,\varepsilon) = R_0(\tau,\varepsilon) + R_1(\tau,\varepsilon) \quad s.t. \quad P_d(\tau,\varepsilon) \geq \bar{P}_d \tag{4}$$

where $\bar{P}_d$ is the target probability of detection for sufficient protection of PUs. With assumption $C_0 > C_1$, the value of $R_0(\tau,\varepsilon)$ would dominate the value of $R_1(\tau,\varepsilon)$ and the problem (4) is solved in simplified form where

$$\max_{\tau} \tilde{R}(\tau,\varepsilon) = R_0(\tau,\varepsilon) \quad s.t. \quad P_d(\tau,\varepsilon) \geq \bar{P}_d. \tag{5}$$

The simulation result confirmed the accuracy of formulation and estimated solution. The work of (Liang et al., 2008) was the foundation of later research on sensing throughput trade-off.

Beside the studies about the effect of sensing time on the throughput of cognitive radio networks, the k value of "k out of N" fusion rule (Peh et al., 2009) and the frame size (Zhang et al., 2008) are other important parameters in cognitive radio networks that can affect the throughput. In (Peh et al., 2009), they extended equations (3) and (4) for general case with N

[3] In resource allocation, each secondary user is assigned one channel or a portion of channel at a certain transmission power.

nodes[4] and formulated the sensing-throughput trade-off problem to determine the optimal k and sensing time subject to sufficient protection of PU. They achieve this by rewriting (3) as

$$R(\tau,k,\varepsilon) = R_0(\tau,k,\varepsilon) + R_1(\tau,k,\varepsilon) \tag{6}$$

$$= C_0\frac{T-\tau}{T}(1-\mathbb{P}_f(\tau,\varepsilon))P(\mathcal{H}_0) + C_1\frac{T-\tau}{T}(1-\mathbb{P}_d(\tau,\varepsilon))P(\mathcal{H}_1) \tag{7}$$

where

$$\mathbb{P}_d(\tau,k,\varepsilon) = \sum_{i=k}^{N}\binom{N}{i}P_d(\tau,\varepsilon)^i(1-P_d(\tau,\varepsilon))^{N-1} \tag{8}$$

and

$$\mathbb{P}_f(\tau,k,\varepsilon) = \sum_{i=k}^{N}\binom{N}{i}P_f(\tau,\varepsilon)^i(1-P_f(\tau,\varepsilon))^{N-1} \tag{9}$$

are the corresponding detection and false alarm probabilities given τ and ϵ. Here, a slot is considered idle only when k or more SUs cannot detect any activities during the sensing slot.

They proposed an iterative algorithm to find the maximum $R(\tau,k,\varepsilon)$ with two steps. In the first step, it finds the optimum k^* value for an initial given τ by exhaustively trying all possible values in $[1,N]$. The second step of the algorithm uses the k^* value of the first step and solves equation (10) to find the optimum length of sensing slot τ^*.

$$\max_{\tau} R(\tau,k,\varepsilon) \quad s.t. \quad \mathbb{P}_d(\tau,\varepsilon) \geq \bar{\mathbb{P}}_d \quad and \quad 0 \leq \tau \leq T \quad and \quad 1 \leq k \leq N \tag{10}$$

τ^* obtained in step two is used in step one again for the next iteration of the algorithm τ. These two steps repeat alternatively until k^* and τ^* converge. Their results showed that values for k and τ are very sensitive to the level of noise. In addition, the throughput of their iterative algorithm consistently performs better than the OR-fusion-rule and the AND-fusion-rule.

Prior research works for finding optimum sensing duration usually assumed that the transmission power and rate are fixed and dynamic resource allocation was not considered (Fan & Jiang, 2009; 2010; Kang et al., 2009; Liang et al., 2008). In addition, there are some research works on wide-band spectrum sensing (Quan et al., 2008; 2009) which detect the PU over multiple band of frequency rather than over one band at a time. These studies have motivated the investigation on techniques of optimizing spectrum sensing and resource allocation jointly. As an example for dynamic resource allocation application, in (Fan et al., 2011), the authors considered a network with M frequency bands and three slots for each frame to be responsible for sensing, reporting, and transmission. They used wide-band spectrum sensing technique (Bletsas et al., 2010) for sensing the M channels and estimated the average throughput for secondary user n as

$$R_n = (1-\frac{\tau}{T})\sum_{m=1}^{M}[(1-P_m^f(\tau,\varepsilon_m))P(H_m^0).r_{m,n}^0 + (1-P_m^d(\tau,\varepsilon_m))P(H_m^1).r_{m,n}^1] \tag{11}$$

[4] N nodes include one secondary base station and $N-1$ SUs, where the secondary base station acts as a sensor node to cooperatively detect the presence of a PU.

where $r^0_{m,n}$ and $r^1_{m,n}$ are the achievable transmission rates of SU when the channel is free and when the channel is busy (missed detection was happened), respectively. Also, the subscript $_m$ under parameters shows those are belong to the mth channel. They found the maximum of R_n for all channels by changing the polyblock algorithm[5]. The numerical simulations show that the result of modified polyblock algorithm is close to global maximum of R_n.

2.2 Sensing schedule

According to Section (2.1), the increase in sensing cooperation for a specific channel will increase the probability of detection P_d and reduce the probability of false alarm P_f. Also, as mentioned in Section (2.1), there have been some investigations for finding optimum sensing time to reach the maximum throughput. On the other hand, long sensing duration could decrease the overall performance. For example, some SUs can sense cooperatively to reach higher sensing accuracy, but they will lose opportunity to exploit the other channels. Therefore, there is a trade-off between sensing accuracy on one channel and exploring all channels to seek opportunities for data transmission. Research works such as (Chen et al., 2009; Fan & Jiang, 2010; Wu & Tsang, 2009) show that the dynamic spectrum sensing could be an efficient way to seize the opportunities of transmission without sacrificing sensing accuracy. The dynamic spectrum sensing is used where the SUs can not sense multiple frequency bands and there is more than one PU. In this situation, if there is a schedule for sensing different bands by one particular SU, it can give more opportunities to SUs to find hole of spectrum and improve their average throughput.

In addition, as PUs start and stop their communications, the dynamic spectral activities can only be partially observed by SUs because of imperfect spectrum sensing techniques. Thus, the dynamic spectrum sensing methods used for time-varying environments need to be able to model system with partially observed variables. Markov modeling is one of the well-known and widely studied techniques for modeling the behavior of PUs. For example, (Zhao, Tong, Swami & Chen, 2007) considered the optimal distribution MAC protocols for opportunistic spectrum access in the partially observable Markov decision process (POMDP) framework. Also, Zhao et al. investigated the structure of myopic policy in their POMDP problem and compare it with the optimal policy (Zhao, Krishnamachari & Liu, 2007). (Zhao & Krishnamachari, 2007) worked on similar problem and considered the imperfect sensing performance. (Chen et al., 2009) studied the threshold structure of optimal sensing and access policies for reducing the complexity of optimal policy searching.

Recently, (Zhang et al., 2010) worked on dynamic scheduling for cooperative sensing under time-varying spectrum environments. They formulate the sensing schedule problem with POMDP. They derived an optimal policy that determined which SUs must sense which channels with what sensing accuracy. Also, they studied the solution structure for the myopic and optimal policy under a simplified model with only two SUs analytically.

The time-domain combining spectrum sensing (TDC–SS) based on the Bayesian method and the Neyman–Pearson theorem was proposed in (Lee & Kim, 2011). (Lee & Kim, 2011) assumed that the state of the PU evolves according to the Markov ON–OFF process (Sung et al.,

[5] The polyblock algorithm is a monotonic programming method which can solve monotonic optimization problems. The solution of polyblock algorithm is ϵ-optimal. For more information, see (Floudas & Pardalos, 2009)

2006). By using Bayesian method and considering rate of the ON–OFF, the TDC–SS algorithm sequentially updated the likelihood ratio of PU state and decided the current PU state from Neyman–Pearson criterion. Also, the TDC–SS algorithm can adapt itself with transition rate of PU state.

2.3 Challenges

In Section (2.1), the studies of optimal trade-off between sensing time and transmission time are reviewed. In these studies, the report channel is assumed to be an error-free channel, but it may not be true in practice. Optimizing the cooperative spectrum sensing parameters with noisy reports channel is one of the research challenges in cognitive radio. It seems that applying coding techniques on the report channel and studying its effects can be one way to mitigate the problem of noisy report channels (Cheng et al., 2009). Also, in most prior works, finding the theoretical bound of spectrum sensing parameters is limited to the centralized structure. The distributed cooperative spectrum sensing needs more theoretical investigations to optimize its key parameters such as sensing time. Comparing to one-hop structure, the multi-hop structured systems is more difficult to analyze because of their complex structures in contrast to one-hop systems.

3. Game theoretical approaches

In cognitive radio networks, SUs are able to sense, learn and act to optimize their throughput. They can also cooperate with each other to improve their performance with the same or less power consumption. In addition, the wireless environments change continuously due to the movement of SU and/or PU, the change of background noise and the variation of users' traffics. In traditional spectrum sensing, each change will force the network controller to re-allocate the spectrum resources. The re-allocation of the spectrum resource through all SUs will create a lot of communication overhead and more time for convergence.

Some studies such as (Saad et al., 2011; 2009b; Wang, Liu & Clancy, 2010; Wang, Wu & Liu, 2010) tried to use the the game theoretical techniques to tackle these challenges. By using game theory framework in cognitive radio, the SUs and PUs behaviors can be studied and analyzed as a game where each user can choose a strategy to maximize its own utility function. Also, as we briefly described in Sections (2.1) and (2.2), the optimization of spectrum usage is a multi-objective optimization problem.

There are two major approaches based on game theory to model cognitive radio systems, there are "Evolutionary games" (bargaining games) and "Coalitional games". The example of evolutionary game can be seen in the work of (Wang, Liu & Clancy, 2010) who studied the behavior of selfish SUs in cooperative and non-cooperative spectrum sensing games. The other type of cooperative spectrum sensing, called coalitional game, was studied in (Saad et al., 2011), where SUs self-organize into coalition groups and select coalition heads[6] by a distributed algorithm. Then, they only share their sensing information through the coalition group. In Sections (3.1) and (3.2), the details of coalitional and evolutionary spectrum sensing games will be described.

[6] In a coalition set S, the SU with the lowest non-cooperative probability of missed detection will be chosen as a coalition head.

3.1 Coalition game for spectrum sensing

In the literature, the centralized approach for cooperation has been most widely studied for cooperative spectrum sensing. In this approach, the SUs send their sensnig data, or decisions, to a fusion center and then the fusion center will make a decision about the presence or absence of PU and will inform the SUs. Also, it is possible to have some SUs, which are working for different service providers, collaborate with each other without a unique centralized fusion center. The solution using a centralized fusion center will significantly increase the complexity and overhead when the number of SUs increases.

To tackle the problem of complexity and communication overhead, in some papers such as (Kim & Shin, 2008; Sun et al., 2007), the idea of making clusters for cooperative spectrum sensing was used to improve the spectrum utilization of SUs. For example, in (Sun et al., 2007), the centralized fusion center employed clustering to reduce the error over a reporting channel. However, it did not propose any specific algorithm for creating clusters. Further, (Kim & Shin, 2008) focused on analyzing the performance of cooperative spectrum sensing based on different detection algorithms with respect to geographical restriction on clustering of the users. The research shows composing clusters and making decisions for the SUs belonging to the same cluster will reduce the complexity of decision making and overhead of communication.

Followed the aforementioned cluster-based cognitive spectrum sensing, (Saad et al., 2011) recently employed game theoretical techniques for forming optimal clusters and conducting distributed decision making. The major contribution of (Saad et al., 2011) is the derived distributed strategy for cooperative spectrum sensing between the SUs. Also, they studied the effect of network topology on the probability of detection $Q_{m,S}$ and false alarm $Q_{f,S}$ for each coalition group S, which are given by

$$Q_{m,S} = \prod_{i \in S} [P_{m,i}(1 - P_{e,i}) + (1 - P_{m,i})P_{e,i}], \tag{12}$$

$$Q_{f,S} = 1 - \prod_{i \in S} [(1 - P_{f,i})(1 - P_{e,i}) + P_{f,i}P_{e,i}], \tag{13}$$

where $P_{m,i}$ and $P_{f,i}$ are the probability of missed detection and false alarm for each SU i in the coalition set S, respectively. Also, $P_{e,i}$ is the probability of error on the report channel[7]. In (Saad et al., 2009b), based on importance of false alarm, the logarithmic barrier function (Boyd & Vandenberghe, 2004) was used for penalizing the false alarm in the cost function. (Saad et al., 2009b) defined the cost function $C(Q_{f,S}, \alpha)$ and the utility function $v(S)$ by

$$C(Q_{f,S}, \alpha) = \begin{cases} -\alpha^2.log\left(1 - \left(\frac{Q_{f,s}}{\alpha}\right)^2\right) & ,if\ Q_{f,S} < \alpha \\ +\infty & ,if\ Q_{f,S} \geq \alpha \end{cases} \tag{14}$$

$$v(S) = Q_{d,S} - C(Q_{d,S}, \alpha_S) = (1 - Q_{m,S}) - C(Q_{d,S}, \alpha_S) \tag{15}$$

[7] According to distributed structure for the fusion center, the report channels are defined between the SUs and the coalition head of same coalition set S.

They modeled the problem as a non-transferable[8] coalition game (Saad et al., 2009a), and they applied a distributed algorithm to collaborative sensing. This distributed algorithm has three phases: "neighborhood discovery", "coalition group formation", and "coalition sensing" respectively.

In the first phase, the distributed algorithm needs to acquire the distances between neighbors, i.e., the distance between the SU and the PU or the distances between the SUs. In addition, the probability of missed detection and false alarm of each SU in the neighbors and the existence of other coalitions are necessary for the coalition group formation. This type of information can be obtained:

- directly via control channel such as Cognitive Pilot Channel (CPC) (Filo et al., 2009; Sallent et al., 2009);
- estimated via signal processing techniques (Cardoso et al., 2008; Damljanovic, 2008; Hossain et al., 2009);
- estimated via ad-hoc routing neighborhood discovery (Arachchige et al., 2008; Damljanovic, 2008; Krishnamurthy et al., 2009);
- estimated via geo-location techniques (Shellhammer et al., 2009).

The second phase is the coalition group formation. This phase is a distributed iterative algorithm based on information about the neighborhood discovery performed in the first phase. The main operations of the second phase are merge, split, and adjust operations that find the suitable topology of the network. The merge operation tries to group some SUs to increase the utility function $v(S)$. The split operation tries to separate some SUs for the coalition group S if the splitting will increase the $v(S)$. The coalitions decide to merge or split if at least one of the SUs can improve its utility function without decreasing the others' utility function. The adjust operation excludes some SUs from coalition group S that do not affect the $v(S)$. All operations in the coalition group formation must repeat every period θ to adapt the topology of the network with the dynamic changes of PUs and SUs.

The complexity of the second phase of algorithm depends on the number of merge and split operations in each iteration. The worst case scenario for N number of SUs before finding the suitable topology needs to $\sum_{i=1}^{N-1} i = \frac{N(N-1)}{2}$ merge operation attempts. In practice, the number of attempts is much less than the worst case. Also, for the split operations, the number of possible partitions is proportional to the *Bell number*[9], which grows exponentially with the number of the SUs in each coalition set. However, a coalition does not need to search all possible cases. It can find a way of splitting in the middle of a search. In addition, the number of SUs in the coalition sets is very small; therefore, the possible splitting options are restricted.

In the last phase, coalition sensing, each SU reports its sensing bit to the coalition head about the presence or absence of the PU through the wireless channel, and the coalition head makes

[8] A coalition game is non-transferable if the value of the utility function of a coalition can not be arbitrarily apportioned between coalition's players (for more information see (Ray, 2007; Saad et al., 2009a))

[9] The number of ways a set of n elements can be partitioned into nonempty subsets is called a Bell number and is denoted as B_n. For example, there are five ways the numbers $\{1,2,3\}$ can be partitioned: $\{\{1\},\{2\},\{3\}\}$,$\{\{1,2\},\{3\}\}$,$\{\{1,3\},\{2\}\}$, $\{\{1\},\{2,3\}\}$ and $\{\{1,2,3\}\}$, so $B_3 = 5$ and similarly $B_4 = 15$.

the final decision by using the decision fusion OR-rule. Clearly, the required bandwidth for communication in each coalition set to reach the final decision is significantly less than the common fusion center, because it needs to send a bit to each PU only through its group. Also, this strategy has an advantage in reducing the probability of missed detection and power consumption by sending the sensing bit of each SU to its coalition head only.

3.2 Evolutionary game for spectrum sensing

In most of the studies about cooperative spectrum sensing, a fully cooperative spectrum sensing is considered. It means in all frames, all SUs take part in sensing and send their own sensing results to the other SUs for fusion and make final decision about presence or absence of PUs. According to discussions in Sections (2.1) and (2.2), sensing the PU band consumes energy and needs certain amount of time. Depend on QoS, performance and throughput of cognitive radio systems, a SU needs to optimize the length of sensing time in each sensing slot. Furthermore, with the emergence of new wireless applications and networks, there may be some SUs which are selfish and attend to take advantage of other sensing result, save their energy and time and have more time for data transmission. Therefore, the study of dynamic cooperative behavior of the selfish SUs in the competitive environments is essential to improve the performance of system and fairness between the SUs.

Modeling the behaviors of SUs as an evolutionary game, where the payoff is defined as SU's throughput, has attracted lots of attention in the research community recently. In (Wang, Liu & Clancy, 2010), an evolutionary game model is studied where selfish SUs tend to overhear the others' sensing results and contribute less to the common task. For example, the SUs like to spent less time in sensing and more time for data transmission. (Wang, Liu & Clancy, 2010) incorporated practical multi-user effect and constraints into the spectrum sensing game. They analyzed the strategy of updating SU's profile using the replicator dynamics equations (Fudenberg & Levine, 1998). They derived the evolutionary stable strategy (ESS) of game and proved its convergence by analyzing the dynamic of SU's behavior. Also, they proposed a distributed learning algorithm that SUs update their strategies by exploring different actions at each time, adaptively learning during the strategic interaction, and approaching the best response strategy.

Similar to equation (3), they defined the average throughput function by

$$\begin{aligned} R(N) &= R_{\mathcal{H}_0}(N)P(\mathcal{H}_0) + R_{\mathcal{H}_1}(N)P(\mathcal{H}_1) \\ &= \frac{T-\delta(N)}{T}(1-P_f)C_{\mathcal{H}_0}P(\mathcal{H}_0) + \frac{T-\delta(N)}{T}(1-P_d)C_{\mathcal{H}_1}P(\mathcal{H}_1) \end{aligned} \tag{16}$$

where N is the number of collected samples and $C_{\mathcal{H}_0}$ and $C_{\mathcal{H}_1}$ are the data rate of SU under $\mathcal{H}_0$ and $\mathcal{H}_1$ respectively. Also, $\delta(N) = N/f_s$ where f_s is the frequency of sampling. They assumed that there was only one PU and its licensed band was divided into K sub-bands. When the PU was idle, each SU operated exclusively in one of K sub-bands. They defined a narrow-band signaling channel for exchanging the sensing results. By this setup, some of SUs can contribute in the sensing part, called, "C" (Contributer) and some of them can refuse to contribute in the sensing part, in the hope that other will do it for them, called, "D" (Denier). If no SU contributes in the sensing part, the average throughput or utility of SUs will be zero.

(Wang, Liu & Clancy, 2010) defined payoff functions for homogeneous[10] and heterogeneous[11] players separately. For Homogeneous players, the payoff function was defined as

$$\mathcal{U}_C(J) = \mathcal{U}_0(1 - \frac{\tau}{J}),\ if\ J \in [1, K] \tag{17}$$

and

$$\mathcal{U}_D(J) = \begin{cases} \mathcal{U}_0\ ,\ if\ J \in [1, K-1] \\ 0\ \ ,\ if\ \quad J = 0 \end{cases} \tag{18}$$

where J is the number of contributer SUs and $\mathcal{U}_C$ and $\mathcal{U}_D$ are the payoff for contributers and deniers respectively. $\mathcal{U}_0$ denotes the throughput of a denier and $\tau = \delta(N)/T$. Here, the denier can use all of the its time to transmit the data, but the contributer will spend τ/J of its time for the spectrum sensing and the rest of time for the data transmission. Also, equations (17) and (18) show that the payoff of players, who do not contribute in the sensing part, will be more than the payoff of contributer players except no one perform the spectrum sensing.

Consider a simple heterogeneous environment with only two players "1" and "2" with different false alams probabilities $P_{f,1}$ and $P_{f,2}$. We also assume that a contributing player will spend τ of its time for spectrum sensing and $1 - \tau$ of time for the data transmission and a user will transmit at the data rate C_i when no PU is observed. Similar to a homogeneous environment, at least one of the heterogeneous players has to be a contributer to avoid zero payoff. The payoff function for the players can be easily found and is shown in Table (1).

player 1 \ player 2	**Contributing**	**Denying**
Contributing	$C_1P(\mathcal{H}_0)(1-P_f)(1-\frac{\tau}{2})$, $C_2P(\mathcal{H}_0)(1-P_f)(1-\frac{\tau}{2})$	$C_1P(\mathcal{H}_0)(1-P_{f,1})(1-\tau)$, $C_2P(\mathcal{H}_0)(1-P_{f,1})$
Denying	$C_1P(\mathcal{H}_0)(1-P_{f,2})$, $C_2P(\mathcal{H}_0)(1-P_{f,2})(1-\tau)$	0,0

Table 1. Payoff table for heterogeneous players

(Wang, Liu & Clancy, 2010) derived the replicator dynamic equations (Fudenberg & Levine, 1998) and found an ESS for this problem. In addition, they proposed a distributed learning algorithm for ESS and showed its convergence by computer simulations.

3.3 Challenges

As we discussed in Section (3.1), the rate of changes in the environment can affect the performance of coalition group formation. To avoid performance reduction, the operations in coalition group formation should not be repeated periodically to adapt the dynamic changes of networks. How to tune the length of period and how to learn it based on the changes of environment are some of meaningful research challenges. A potential direction is to use modeling techniques to analyze environmental changes so as to reduce the overhead of periodic coalition group formation.

[10] The individual players with identical data rate and received primary SNR and the same set of pure strategies will be called homogeneous players.

[11] Two players are heterogeneous if they do not have the identical set of parameters (data rates, received primary, the set of strategies, etc.).

4. Compressed sensing

Wide-band spectrum sensing is essential for efficient dynamic spectrum sharing in cognitive radio networks. The wider the range of shared spectral resources the higher the cognitive network capacity. However, this gain comes at the cost of increasing the complexity of network spectral resource management at several levels. In particular, the cost of spectrum sensing is proportional to the range of the sensible spectrum.

While cooperative techniques are used to relax the complexity constraints of wide-band spectrum sensing, this relaxation is not sufficient or effective in large-scale cognitive radio networks. In such networks, the range of shared spectrum and/or the number of cooperative nodes can be too large rendering high cost of sensing data acquisition unit and wide bandwidth for control channel to exchange sensing information among cooperative nodes.

Compressed sampling is one solution used to address scalability problems in signal processing by reducing the cost of acquisition at the sensor (Donoho, 2006; Polo et al., 2009; Romberg, n.d.). Using compressive sensing framework, it is possible to recover the sampled signals on a probabilistic basis from a few number of measurements, which are less than the number imposed by the Nyquist rate for band-limited signals. However, successful signal recovery is conditioned by two factors: signal sparsity and incoherency (Candès & Wakin, 2008; Donoho, 2006). Signal sparsity means that the rate of information to be captured from the acquired signal is small relatively compared to the signal dimension. In spectrum sensing applications, this condition implies that the occupancy of the spectrum should be relatively small compared to the complete range of the spectrum, i.e., the sensed spectrum is sparse. The incoherency condition means that representation domain of the signal should be incoherent with the sampling domain. This condition is satisfied in spectrum sensing applications since the representation domain of the signal is the frequency domain while the sampling domain is the time domain. The time domain is incoherent with the frequency domain as sudden changes in one of these domains spreads out in the other one.

Motivated by the fact that the wireless spectrum in rural areas is highly underutilized, i.e., sparse (Rysavy, 2009), recent studies have examined the application of these compressive techniques in spectrum sensing (Bazerque & Giannakis, 2010; Candès et al., 2006; Laska et al., 2011; Liang et al., 2010; Tian, 2008; Tian & Giannakis, 2007; Zeng et al., 2010). The research efforts attempt to exploit the efficiency of compressive sensing technique to reduce the cooperation overhead or sampling complexity in cooperative spectrum sensing.

The compressive sensing framework is comprised of two stages: signal sampling and signal recovery. The compressive sensing requires a random access to the sensed signal samples or a linear combination of them. There are several signal sampling systems that have been proposed in literature for that purpose such as the *Random Demodulator* (Tropp et al., 2010), the *Modulated Wide-band Converter* (Mishali & Eldar, 2010) and the *Analog to Information Converter* (Kirolos et al., 2006). Also, a frequency selective filter has been recently suggested as means of obtaining linear combinations of multiple channel information at the sensing circuit (Meng et al., 2011).

In addition to the signal sampling systems, several signal recovery technique based on compressive samples have been proposed in literature. They can be grouped into two main categories: based on solving linear convex optimization problem such as *Basis Pursuit technique* (BP) (Chen et al., 1999) and greedy algorithms such as *Matching Pursuit* (MP) (Mallat & Zhang, 1993) and Orthogonal Matching Pursuit (OMP) (Tropp & Gilbert, 2007).

Orthogonal matching pursuit technique is proposed in a distributive spectrum sensing framework when the received signals among cooperative nodes are highly correlated (Liang et al., 2010). In (Tian, 2008; Zeng et al., 2010), a distributive compressive sensing technique based on BP signal recovery technique is presented. The dissemination of sensing information in this distributive technique enforces consensus among local spectral estimates and shown to converge with reasonable computational cost and power overhead. The lowest SNR at which the collaborative compressed sensing in (Zeng et al., 2010) can perform with is probability of detection around 90% and compression ratio of 50% is approximately 5 dB. However, this sensing technique assumes identical spectral observations among cooperative nodes. This can be a limiting factor for deploying it in large scale networks where the observed spectrum might vary among widely scattered cooperative nodes.

Using similar signal recovery technique, (Bazerque & Giannakis, 2010) has proposed a cooperative compressive sensing technique that exploits two forms of signal sparsity, i.e. in the frequency domain and spatial domain. The sparsity in the frequency domain is a result of the narrow-band nature of transmitted signal compared to the overall range of shared spectrum. The second form of sparsity emerges from scattered active primary users' radios that co-exist with the cooperative cognitive radio. The join spectral-spatial analysis featured in this technique allows it to be used in large-scale networks since it takes into consideration the variation of spectral observation due to spatial diversity among cooperative nodes.

The work of (Meng et al., 2011) has addressed the sensing problem in small and large scale networks jointly using compressive sensing techniques. As a result, two spectrum recovery approaches were proposed. The first one is based on matrix completion and the second spectrum recovery approach is based on joint sparsity recovery. The matrix completion approach is appropriate for small scale cooperative networks as it recovers the channel occupancy information from a small number of cooperative sensing. The joint sparsity approach, on the other hand, scales better to large networks because it skips recovering the reports and directly reconstructs channel occupancy information by exploiting the fact that each occupied channel is observable by multiple cooperative sensing nodes.

4.1 Challenges

A number of challenges arise from using compressive sensing techniques for spectrum sensing in general, and in cooperative sensing in particular. One challenge resulted from sensing densely occupied spectrum where spectrum sparsity condition does not hold. Furthermore, cognitive radio networks attempt to utilize the available spectral resources in sparse spectrum through dynamic spectrum access. This can negatively impact the efficiency of compressive sensing techniques as it increases the occupancy (reduces sparsity) of the sensed spectrum (Laska et al., 2011). Sensing dense spectrum using compressive sensing technique is still an open problem. However, (Zhang et al., 2011) have addressed this problem by proposing a detection mechanism of compressive sensing failure due to sparsity reduction. This is achieved by modeling the compressed spectrum reconstruction as a Gaussian process so that the correlation of model parameters can be exploited to detect the failure of compressed reconstruction due to non-sparse spectrum.

Sensing heterogeneous sets of active radios poses another challenge because the detection thresholds for each one might be different. Hence, the ability to classify signals using compressive samples can be useful in this context and there are a number of recent studies that have addressed this matter (Davenport et al., 2010). Also, certain signal features can be exploited to distinguish between legitimate signals and spurious interference from adjacent cognitive radios. In (Zeng et al., 2010), the orthogonality between the spectrum of primary users and that of cognitive radio users was utilized as a constraint for consensus optimization during distributed collaborative sensing.

Another intriguing direction for future research is shadowing effects in cooperative compressive sensing techniques. (Tian & Giannakis, 2007) suggested complementing the distance-only dependent propagation functions used in the compressive sensing design with non-parametric models that can be learned from the data (Mateos et al., 2009).

5. Applications

The characteristics of the wireless environment have considerable impacts on the design and development of cooperative sensing. Among these characteristics are the mobility of sensing nodes, the spatial density, power and bandwidth of the co-existing wireless networks. The literature has reported several performance evaluation studies of cooperative sensing in various types of wireless environments such as Radar (Wang et al., 2008), UAV (Gu et al., 2006), VANET (Wang & Ho, 2010), WiFi Networks (You et al., 2011), Relaying-based network (Roshid et al., 2010), Positioning Services in ad-hoc Networks (Gellersen et al., 2010), Small scale wireless devices (Min et al., 2011).

5.1 Radar

Cooperative sensing is used to enable sharing different parts of the spectrum with various types of primary user networks. S-band radar spectrum features great spectral opportunities for band sharing compared with those at the TV spectrum frequencies (Research, 2005; SYSTEM, 2006). Studies have examined the feasibility of using cooperative sensing for managing the coexistence between radar and other communication systems. One study has concluded that sharing the radar band with another communication system is feasible on the basis of causing "***acceptable interference***" to the radar. Acceptable interference is determined by calculating the safe regions where radar and communication systems can coexist without causing unacceptable interference to the radar systems. Priori knowledge about the radar systems, such as swept radar rotation rate and main beam to side-lobe ratio, and the radar system maximum interference-to-noise ratio (INR) threshold, can assist in determining the borders of the safe region. It turns out that the number of cooperative nodes, their channel correlation levels, channel conditions and the detection probability threshold have considerable impact on the cooperation gain obtained in such wireless environments (Wang et al., 2008). Therefore sensing node grouping strategies are needed in order to maximize the cooperative gain with minimum number of sensors.

5.2 UAV

High mobility in wireless environments imposes considerable challenges for cooperative sensing. Unmanned Air Vehicle networks and Vehicular ad-hoc Network are examples of such wireless environments. Cooperative sensing using unmanned aerial vehicles (UAVs) is a fundamental element in combat intelligence, surveillance, and reconnaissance (ISR) to estimate the position and velocity of the ground moving target (GMT). In (Gu et al., 2006), the performance of cooperation sensing among a set of pulse Doppler radar sensors mounted on a group of UAVs is studied. Specifically, the number and geometry of sensor distributions impact on the quality of the position and velocity estimates is analyzed. The quality of the estimation is characterized by its statistical variance and the theoretical minimum variance is derived based on Cramer-Rao bounds. However, the complexity of such analysis increases as more types of sensor measurements are involved. In such application and wireless environment, various type of measurements are obtained; namely: azimuth, range and range rate measurements. One common assumption to maintain the tractability of the mathematical analysis is the variance is identical for each measurement type at each UAV. In case that this assumption is not held, numerical methods are used instead for estimation error minimizations (Gu et al., 2006). Also, it was found that optimizing the configuration of the UAV team is essential for efficient "***ground moving target identification***" (GMTI) and "***moving surface target engagement***" (MSTE) operations in ISR systems. One possible representation of the optimal geometrical set up can be defined by the bearing angle formed by each UAV and the GMT (Gu et al., 2006).

5.3 VANET

Another type of highly mobile wireless environment is VANET networks. The accuracy and efficiency of spectrum sensing is essential to maintain reliable information exchange and data transmission between nearby On-Board Units (OBUs) and Roadside Units (RSUs) for applications such as route planning, traffic management, as well as other wireless Internet service in the unlicensed spectrum resources (Wang & Ho, 2010). One fundamental challenge of cooperative sensing in such networks is the sparsity and variability of spatial density of vehicles equipped with CR technologies. Another challenge is the heterogeneity of primary user signals. Nevertheless, cooperative sensing is used in such environments to resolve some problems that can-not be mitigated by individual sensing such as coordination and fairness among nodes for efficient resource utilization. In this context, a sensing coordination node is introduced to establish an adaptive sensing coordination framework (Wang & Ho, 2010) that combines the advantages of both stand alone and cooperative sensing between CR-VANETs. Rather than having complete control of spectrum sensing and access, the coordinating nodes assist and coordinate the VANETs for efficient spectral utilization and minimal communication interruptions. This framework relies on some priori knowledge such as feature carriers, channel spacing of frequency bands, proactive fast sensing information, as well as user-based class information, to perform intelligent channel selection of spectrum.

5.4 WiFi

Recent explosive data increase (mostly due to smart phones) for wireless internet access results in a spectrum shortage and increasing demand for additional spectrum to support the dramatic increase in data traffic on WiFi networks. One good solution for the spectrum shortage is to use the underutilized TV spectrum based on Cognitive radio techniques. Cooperative sensing is proposed in WiFi networks coordinate the coexistence between WiFi and wireless devices in TV spectrum (You et al., 2011). The study has analyzed the performance "$k\ out\ of\ N$" rule for cooperative sensing in WiFi networks (You et al., 2011). The performance of sensing is controlled by sensing time and number of sensors. Both factors are optimized to maximize data transmission throughput and minimize detection error respectively.

5.5 Integration of cooperative communication and cooperative sensing

Cooperation strategies are not limited to spectrum sensing application. Other form of cooperation appears in wireless communication such as relaying. An interesting extension of cooperative strategies is to integrate cooperative sensing and cooperative communication in wireless networks with relaying support. This integration was suggested in (Roshid et al., 2010). Cooperative communication is performed through relaying with different resource allocation strategies (power allocation for relaying PU traffic and bandwidth allocation for relaying CR traffic). Such integration of cooperation strategies necessitates an efficient spectrum management system e.g., spectrum broker in centralized systems and multi-agent and cognitive pilot channels (CPC) in distributed systems (Roshid et al., 2010). However, there is no clear description to this date on how the integration can be done. Also, another open problem is the signaling that enables a joint cooperation of communication and sensing that can impact the architecture of current networks leading to higher transmission capacity and transparent coexistence among heterogeneous set of wireless networks.

5.6 Positioning

Cooperative sensing technique is also applied for positioning application in ad-hoc wireless networks. Unlike most infrastructure based positioning systems that are based on heterogeneous architecture (fixed/reference and mobile nodes), cooperative-relative-positioning systems relies on a homogeneous set of nodes that can measure the relative distance between each pair of nodes. Performing cooperative positioning in such network setup pertains the following challenges: additional measurements such as angle measurements may be required, scalable coordination and synchronization system that can operate in ad-hoc environment, and increasing computational load on sensing node due to the absence of a centralized entity at which most computational load can be outsourced. The communication among cooperative nodes is performed in time division multiple access basis. Each node broadcasts to the rest of nodes its measurements along with the received measurements from other nodes. Synchronous distributed jam signaling (SDJS) for data fusion on the physical layer is used to manage a systematic measurement error removal by estimating statistical measurement parameters from a coordinated transmission of jamming signals.

5.7 Small scale primary user

As more wireless standards continues to evolve to fulfill the requirements of the new wireless services, the coexistence among heterogeneous set of wireless networks becomes a critical challenge. This is particularly true when heterogeneity is not only a result of difference in the waveform but also in the coverage range. For example, the WiFi network coverage is around 20-30m in 2.4 GHz while Bluetooth coverage in that portion of the spectrum is around 5 m (for Bluetooth class 3). Another example is the digital TV signal (DTV) signal and wireless microphone signals. Most of the cooperative sensing research has focused on detecting large scale networks. However, detecting small scale networks bears additional challenges due to the unpredictability of their transmission schedule and high mobility profile i.e., highly dynamic temporal and spatial transmission pattern. In such cases, the knowledge of the transmitter location is more critical than in the case for detecting large scale networks. Recent studies have examined the problems of detecting small-scale networks. One solution is to use a disabling beacon additional information such as the signature/authentication and geo-location of the small scale transmitter as is the case for wireless microphones in the 802.22 Task Group 1 (TG 1) proposal (Lei & Chin, 2008). More recently, a solution proposed in (Min et al., 2011) identifies the optimal range of sensing fusion in cooperative sensing strategies so that the average detection delay is minimized for a given sensing performance requirements. This solution, however, is sensitive to power and location estimation errors. Hence, a join power-location estimation technique (DeLOC) is proposed (Min et al., 2011) to minimize the estimation errors and improve the overall detection performance with optimal fusion range.

Limited work has been done in this area, further improvements can be made in the performance of detection by incorporating further communication layers in addition to the MAC layer (Min et al., 2011). Also, the fusion of different sensing measurements such as time, location, orientation and wireless signal power using different types of sensors can further improve the detection performance.

6. Conclusion

In this chapter, we reviewed the recent studies about cooperative spectrum sensing in cognitive radio and identified the research challenges and unsolved issues. The chapter introduced some basic concepts of cooperative spectrum sensing and reviewed theoretical studies of known bounds and limits. Then, we surveyed the game theoretical approaches in cooperative cognitive radio briefly. The chapter followed by reviewing compressed sensing studies for cooperative cognitive radio. In the end, we explored the application of cooperative spectrum sensing in various types of wireless environments such as Radar, UAV, VANET, WiFi Networks and etc.

7. References

Akyildiz, I., Lee, W. & Chowdhury, K. (2009). Crahns: Cognitive radio ad hoc networks, *Ad Hoc Networks* 7(5): 810–836.

Arachchige, C., Venkatesan, S. & Mittal, N. (2008). An asynchronous neighbor discovery algorithm for cognitive radio networks, *New Frontiers in Dynamic Spectrum Access Networks, 2008. DySPAN 2008. 3rd IEEE Symposium on*, IEEE, pp. 1–5.

Bazerque, J. & Giannakis, G. (2010). Distributed spectrum sensing for cognitive radio networks by exploiting sparsity, *Signal Processing, IEEE Transactions on* 58(3): 1847–1862.

Bletsas, A., Dimitriou, A. & Sahalos, J. (2010). Interference-limited opportunistic relaying with reactive sensing, *Wireless Communications, IEEE Transactions on* 9(1): 14–20.

Boyd, S. & Vandenberghe, L. (2004). *Convex optimization*, Cambridge Univ Pr.

Cabric, D., Mishra, S. & Brodersen, R. (2004). Implementation issues in spectrum sensing for cognitive radios, *Signals, Systems and Computers, 2004. Conference Record of the Thirty-Eighth Asilomar Conference on*, Vol. 1, IEEE, pp. 772–776.

Candès, E., Romberg, J. & Tao, T. (2006). Robust uncertainty principles: Exact signal reconstruction from highly incomplete frequency information, *Information Theory, IEEE Transactions on* 52(2): 489–509.

Candès, E. & Wakin, M. (2008). An introduction to compressive sampling, *Signal Processing Magazine, IEEE* 25(2): 21–30.

Cardoso, L., Debbah, M., Bianchi, P. & Najim, J. (2008). Cooperative spectrum sensing using random matrix theory, *Wireless Pervasive Computing, 2008. ISWPC 2008. 3rd International Symposium on*, IEEE, pp. 334–338.

Chen, S., Donoho, D. & Saunders, M. (1999). Atomic decomposition by basis pursuit, *SIAM journal on scientific computing* 20(1): 33–61.

Chen, Y., Zhao, Q. & Swami, A. (2009). Distributed spectrum sensing and access in cognitive radio networks with energy constraint, *Signal Processing, IEEE Transactions on* 57(2): 783–797.

Cheng, S., Stankovic, V. & Stankovic, L. (2009). An efficient spectrum sensing scheme for cognitive radio, *Signal Processing Letters, IEEE* 16(6): 501–504.

Damljanovic, Z. (2008). Cognitive radio access discovery strategies, *Communication Systems, Networks and Digital Signal Processing, 2008. CNSDSP 2008. 6th International Symposium on*, IEEE, pp. 251–255.

Davenport, M., Boufounos, P., Wakin, M. & Baraniuk, R. (2010). Signal processing with compressive measurements, *Selected Topics in Signal Processing, IEEE Journal of* 4(2): 445–460.

Donoho, D. (2006). Compressed sensing, *Information Theory, IEEE Transactions on* 52(4): 1289–1306.

Fan, R. & Jiang, H. (2009). Channel sensing-order setting in cognitive radio networks: A two-user case, *Vehicular Technology, IEEE Transactions on* 58(9): 4997–5008.

Fan, R. & Jiang, H. (2010). Optimal multi-channel cooperative sensing in cognitive radio networks, *Wireless Communications, IEEE Transactions on* 9(3): 1128–1138.

Fan, R., Jiang, H., Guo, Q. & Zhang, Z. (2011). Joint optimal cooperative sensing and resource allocation in multi-channel cognitive radio networks, *Vehicular Technology, IEEE Transactions on* 60(99): 722–729.

Federal Communications Commission (2005). Notice of proposed rule making and order: Facilitating opportunities for flexible, efficient, and reliable spectrum use employing cognitive radio technologies, *ET, Docket No. 03-108*.

Filo, M., Hossain, A., Biswas, A. & Piesiewicz, R. (2009). Cognitive pilot channel: Enabler for radio systems coexistence, *Cognitive Radio and Advanced Spectrum Management, 2009. CogART 2009. Second International Workshop on*, IEEE, pp. 17–23.

Floudas, C. & Pardalos, P. (2009). *Encyclopedia of optimization, Second edition*, Springer-Verlag.

Fudenberg, D. & Levine, D. (1998). *The theory of learning in games*, Vol. 2, The MIT press.

Gellersen, H., Lukowicz, P., Beigl, M. & Riedel, T. (2010). Cooperative relative positioning, *Pervasive Computing, IEEE* 9(99): 78–89.

Ghasemi, A. & Sousa, E. (2005). Collaborative spectrum sensing for opportunistic access in fading environments, *First IEEE International Symposium on New Frontiers in Dynamic Spectrum Access Networks(DySPAN 2005)*, IEEE, pp. 131–136.

Gu, G., Chandler, P., Schumacher, C., Sparks, A. & Pachter, M. (2006). Optimal cooperative sensing using a team of uavs, *Aerospace and Electronic Systems, IEEE Transactions on* 42(4): 1446–1458.

Hossain, E., Niyato, D. & Han, Z. (2009). Dynamic spectrum access and management in cognitive radio networks.

Jiang, H., Lai, L., Fan, R. & Poor, H. (2009). Optimal selection of channel sensing order in cognitive radio, *Wireless Communications, IEEE Transactions on* 8(1): 297–307.

Kang, X., Liang, Y., Garg, H. & Zhang, L. (2009). Sensing-based spectrum sharing in cognitive radio networks, *Vehicular Technology, IEEE Transactions on* 58(8): 4649–4654.

Kim, H. & Shin, K. (2008). In-band spectrum sensing in cognitive radio networks: energy detection or feature detection?, *Proceedings of the 14th ACM international conference on Mobile computing and networking*, ACM, pp. 14–25.

Kirolos, S., Laska, J., Wakin, M., Duarte, M., Baron, D., Ragheb, T., Massoud, Y. & Baraniuk, R. (2006). Analog-to-information conversion via random demodulation, *Design, Applications, Integration and Software, 2006 IEEE Dallas/CAS Workshop on*, IEEE, pp. 71–74.

Krishnamurthy, S., Mittal, N., Chandrasekaran, R. & Venkatesan, S. (2009). Neighbour discovery in multi-receiver cognitive radio networks, *International Journal of Computers & Applications* 31(1): 50–57.

Laska, J., Bradley, W., Rondeau, T., Nolan, K. & Vigoda, B. (2011). Compressive sensing for dynamic spectrum access networks: Techniques and tradeoffs, *New Frontiers in Dynamic Spectrum Access Networks (DySPAN), 2011 IEEE Symposium on*, IEEE, pp. 156–163.

Lee, S. & Kim, S. (2011). Optimization of time-domain spectrum sensing for cognitive radio systems, *Vehicular Technology, IEEE Transactions on* 60(4): 1937–1943.

Lei, Z. & Chin, F. (2008). A reliable and power efficient beacon structure for cognitive radio systems, *Broadcasting, IEEE Transactions on* 54(2): 182–187.

Li, Z., Yu, F. & Huang, M. (2010). A distributed consensus-based cooperative spectrum-sensing scheme in cognitive radios, *Vehicular Technology, IEEE Transactions on* 59(1): 383–393.

Liang, J., Liu, Y., Zhang, W., Xu, Y., Gan, X. & Wang, X. (2010). Joint compressive sensing in wideband cognitive networks, *Wireless Communications and Networking Conference (WCNC), 2010 IEEE*, IEEE, pp. 1–5.

Liang, Y., Zeng, Y., Peh, E. & Hoang, A. (2008). Sensing-throughput tradeoff for cognitive radio networks, *Wireless Communications, IEEE Transactions on* 7(4): 1326–1337.

Lifeng, L., El Gamal, H., Jiang, H. & Poor, H. . (2011). Cognitive medium access: Exploration, exploitation, and competition, *Mobile Computing, IEEE Transactions on* 10(2): 239–253.

Mallat, S. & Zhang, Z. (1993). Matching pursuits with time-frequency dictionaries, *Signal Processing, IEEE Transactions on* 41(12): 3397–3415.

Mateos, G., Bazerque, J. & Giannakis, G. (2009). Spline-based spectrum cartography for cognitive radios, *Signals, Systems and Computers, 2009 Conference Record of the Forty-Third Asilomar Conference on*, IEEE, pp. 1025–1029.

Meng, J., Yin, W., Li, H., Hossain, E. & Han, Z. (2011). Collaborative spectrum sensing from sparse observations in cognitive radio networks, *Selected Areas in Communications, IEEE Journal on* 29(2): 327–337.

Min, A., Zhang, X. & Shin, K. (2011). Detection of small-scale primary users in cognitive radio networks, *Selected Areas in Communications, IEEE Journal on* 29(2): 349–361.

Mishali, M. & Eldar, Y. (2010). From theory to practice: Sub-nyquist sampling of sparse wideband analog signals, *Selected Topics in Signal Processing, IEEE Journal of* 4(2): 375–391.

Mishra, S., Sahai, A. & Brodersen, R. (2006). Cooperative sensing among cognitive radios, *IEEE International Conference on Communications(ICC'06)*, Vol. 4, IEEE, pp. 1658–1663.

Peh, E., Liang, Y., Guan, Y. & Zeng, Y. (2009). Optimization of cooperative sensing in cognitive radio networks: a sensing-throughput tradeoff view, *Vehicular Technology, IEEE Transactions on* 58(9): 5294–5299.

Polo, Y., Wang, Y., Pandharipande, A. & Leus, G. (2009). Compressive wide-band spectrum sensing, *Acoustics, Speech and Signal Processing, 2009. ICASSP 2009. IEEE International Conference on*, IEEE, pp. 2337–2340.

Quan, Z., Cui, S., Poor, H. & Sayed, A. (2008). Collaborative wideband sensing for cognitive radios, *Signal Processing Magazine, IEEE* 25(6): 60–73.

Quan, Z., Cui, S., Sayed, A. & Poor, H. (2009). Optimal multiband joint detection for spectrum sensing in cognitive radio networks, *Signal Processing, IEEE Transactions on* 57(3): 1128–1140.

Ray, D. (2007). *A game-theoretic perspective on coalition formation*, Oxford University Press, USA.

Research, R. M. (2005). Review of bandsharing solutions-final report, *the cave independent audit of spectrum* Report No. 72/05/R/281/R(1).

Romberg, J. (n.d.). L1 magic, a collection of matlab routines for solving the convex optimization programs central to compressive sampling.
URL: *http://www.acm.caltech.edu/l1magic/*

Roshid, R., Aripin, N., Fisal, N., Ariffin, S. & Yusof, S. (2010). Integration of cooperative sensing and transmission, *Vehicular Technology Magazine, IEEE* 5(3): 46–53.

Rysavy, P. (2009). Spectrum crisis?, *InformationWeek Magazine* pp. 23–30.

Saad, W., Han, Z., Basar, T., Debbah, M. & Hjorungnes, A. (2011). Coalition formation games for collaborative spectrum sensing, *Vehicular Technology, IEEE Transactions on* 60(99): 276–297.

Saad, W., Han, Z., Debbah, M., Hjorungnes, A. & Basar, T. (2009a). Coalitional game theory for communication networks: A tutorial, *Signal Processing Magazine, IEEE* 26(5): 77–97.

Saad, W., Han, Z., Debbah, M., Hjorungnes, A. & Basar, T. (2009b). Coalitional games for distributed collaborative spectrum sensing in cognitive radio networks, *INFOCOM 2009, IEEE*, IEEE, pp. 2114–2122.

Sallent, O., Pérez-Romero, J., Agustí, R. & Cordier, P. (2009). Cognitive pilot channel enabling spectrum awareness, *Communications Workshops, 2009. ICC Workshops 2009. IEEE International Conference on*, IEEE, pp. 1–6.

Shellhammer, S., Sadek, A. & Zhang, W. (2009). Technical challenges for cognitive radio in the tv white space spectrum, *Proc. of Information Theory and Applications Workshop (ITA)*, pp. 323–333.

Sun, C., Zhang, W. & Ben, K. (2007). Cluster-based cooperative spectrum sensing in cognitive radio systems, *Communications, 2007. ICC'07. IEEE International Conference on*, IEEE, pp. 2511–2515.

Sung, Y., Tong, L. & Poor, H. (2006). Neyman-pearson detection of gauss-markov signals in noise: closed-form error exponentand properties, *Information Theory, IEEE Transactions on* 52(4): 1354–1365.

SYSTEM, B. (2006). Study into spectrally efficient radar systems in the L and S bands, *Ofcom spectral efficiency scheme 2004–2005* SES-2004-2.

Tian, Z. (2008). Compressed wideband sensing in cooperative cognitive radio networks, *Global Telecommunications Conference, 2008. IEEE GLOBECOM 2008. IEEE*, IEEE, pp. 1–5.

Tian, Z. & Giannakis, G. (2007). Compressed sensing for wideband cognitive radios, *Acoustics, Speech and Signal Processing, 2007. ICASSP 2007. IEEE International Conference on*, Vol. 4, IEEE, pp. IV–1357–IV–1360.

Tropp, J. & Gilbert, A. (2007). Signal recovery from random measurements via orthogonal matching pursuit, *Information Theory, IEEE Transactions on* 53(12): 4655–4666.

Tropp, J., Laska, J., Duarte, M., Romberg, J. & Baraniuk, R. (2010). Beyond nyquist: Efficient sampling of sparse bandlimited signals, *Information Theory, IEEE Transactions on* 56(1): 520–544.

Unnikrishnan, J. & Veeravalli, V. (2008). Cooperative sensing for primary detection in cognitive radio, *Selected Topics in Signal Processing, IEEE Journal of* 2(1): 18–27.

Wang, B., Liu, R. & Clancy, T. (2010). Evolutionary cooperative spectrum sensing game: how to collaborate?, *Communications, IEEE Transactions on* 58(3): 890–900.

Wang, B., Wu, Y. & Liu, K. (2010). Game theory for cognitive radio networks: An overview, *Computer Networks* 54(14): 2537–2561.

Wang, L., McGeehan, J., Williams, C. & Doufexi, A. (2008). Application of cooperative sensing in radar-communications coexistence, *Communications, IET* 2(6): 856–868.

Wang, X. & Ho, P. (2010). A novel sensing coordination framework for cr-vanets, *Vehicular Technology, IEEE Transactions on* 59(4): 1936–1948.

Wu, Y. & Tsang, D. (2009). Dynamic rate allocation, routing and spectrum sharing for multi-hop cognitive radio networks, *Communications Workshops, 2009. ICC Workshops 2009. IEEE International Conference on*, IEEE, pp. 1–6.

You, C., Kwon, H. & Heo, J. (2011). Cooperative tv spectrum sensing in cognitive radio for wi-fi networks, *Consumer Electronics, IEEE Transactions on* 57(1): 62–67.

Zeng, F., Tian, Z. & Li, C. (2010). Distributed compressive wideband spectrum sensing in cooperative multi-hop cognitive networks, *Communications (ICC), 2010 IEEE International Conference on*, IEEE, pp. 1–5.

Zhang, T., Wu, Y., Lang, K. & Tsang, D. (2010). Optimal scheduling of cooperative spectrum sensing in cognitive radio networks, *Systems Journal, IEEE* 4(4): 535–549.

Zhang, W. & Letaief, K. (2008). Cooperative spectrum sensing with transmit and relay diversity in cognitive radio networks-[transaction letters], *Wireless Communications, IEEE Transactions on* 7(12): 4761–4766.

Zhang, W., Mallik, R., Letaief, B. et al. (2008). Cooperative spectrum sensing optimization in cognitive radio networks, *Communications, 2008. ICC'08. IEEE International Conference on*, IEEE, pp. 3411–3415.

Zhang, Z., Li, H., Yang, D. & Pei, C. (2011). Collaborative compressed spectrum sensing: what if spectrum is not sparse?, *Electronics Letters* 47: 519.

Zhao, Q. & Krishnamachari, B. (2007). Structure and optimality of myopic sensing for opportunistic spectrum access, *Communications, 2007. ICC'07. IEEE International Conference on*, IEEE, pp. 6476–6481.

Zhao, Q., Krishnamachari, B. & Liu, K. (2007). Low-complexity approaches to spectrum opportunity tracking, *Proc. of the 2nd International Conference on Cognitive Radio Oriented Wireless Networks and Communications (CrownCom)*, Citeseer.

Zhao, Q., Tong, L., Swami, A. & Chen, Y. (2007). Decentralized cognitive mac for opportunistic spectrum access in ad hoc networks: A pomdp framework, *Selected Areas in Communications, IEEE Journal on* 25(3): 589–600.

6

Improving Spectrum Sensing Performance by Exploiting Multiuser Diversity

Tuan Do[1] and Brian L. Mark[2]
[1]*Global Wireless Solutions, Inc., Dulles, Virginia*
[2]*Dept. of Electrical and Computer Eng. George Mason University, Fairfax, Virginia*
USA

1. Introduction

In traditional wireless systems, spectrum or frequency is allocated to licensed users over a geographic area. Within these constraints, spectrum is considered a scarce resource due to static spectrum allocation. Recent empirical studies of radio spectrum usage have shown that licensed spectrum is typically highly underutilized (Broderson et al., 2004; McHenry, 2003). To recapture the so-called "spectrum holes," various schemes for allowing unlicensed or secondary users to opportunistically access unused spectrum have been proposed. Opportunistic or dynamic spectrum access is achieved by cognitive radios that are capable of sensing the radio environment for spectrum holes and dynamically tuning to different frequency channels to access them. Such radios are often called *frequency-agile* or *spectrum-agile*.

On a given frequency channel, a spectrum hole can be characterized as spatial or temporal. A *spatial* spectrum hole can be specified in terms of the maximum transmission power that a secondary user can employ without causing harmful interference to primary users that are receiving transmissions from another primary user that is transmitting on the given channel. Spectrum reuse in this context is similar to frequency reuse among cochannel cells in a cellular network. A *temporal* spectrum hole is a period of time for which the primary transmitter is idle. During such idle periods, a secondary user may opportunistically transmit on the given channel without causing harmful interference.

Various technologies have been proposed for spectrum sensing: matched filter, cyclostationary feature detector, and energy detector (Cabric et al., 2004). The matched filter maximizes the received signal-to-noise ratio, but requires demodulation of the primary user signal. To demodulate the primary signal, secondary users require prior knowledge of the primary signal at both PHY and MAC layers, e.g., modulation type and order and packet format. Moreover, demodulation requires timing and carrier synchronization with the primary signal. Cyclostationary feature detectors exploit the cyclostationary characteristics of the modulated signal to estimate signal parameters such as the carrier phase and pulse timing at the receiver. Assuming the modulation type is known, a cyclostationary feature detector can be used for detection of a random signal in noise and interference. In this chapter, we shall focus on spectrum sensing based on an energy detector, which uses noncoherent detection, and is applicable in a wide range of scenarios.

In a wireless network with fading, multiuser diversity is the phenomenon whereby different users experience different channel fading conditions during the same observation period. Multiuser diversity can be leveraged to achieve higher throughput by scheduling users to transmit when their channel conditions are favorable (Knopp & Humblet, 1995). Multiuser diversity systems can be centralized or distributed. In centralized systems, a central processor maintains channel state information for all users and always schedules the user with the best channel for transmission. In distributed multiuser diversity systems, each user has knowledge of its own channel state, but does not has knowledge of the fading levels of other users.

In this chapter, we propose a distributed approach to temporal spectrum sensing that exploits multiuser diversity among secondary users to improve sensing performance in a cognitive radio networks. Since our focus is on temporal spectrum sensing, we shall use the term spectrum sensing to mean *temporal* spectrum sensing, unless otherwise specified. Nevertheless, the proposed scheme can be integrated with spatial spectrum sensing to achieve a joint spatial-temporal spectrum sensing system (Do & Mark, 2009; 2010). The proposed multiuser diversity spectrum sensing scheme uses a cooperative sensing framework to overcome low signal-to-noise ratio (SNR) and shadowing. Unlike traditional multiuser diversity schemes for wireless networks, fairness and delay issues can be ignored in the spectrum sensing scenario because the only performance metric of interest is the detection probability given the false positive is the same.

We also propose a MAC protocol bases on carrier sense multiple access (CSMA) protocol to facilitate the transmission of observation from secondary users to fusion center. In this chapter, our proposed MAC protocol uses different backoff window to exploit the multiuser diversity inherent in secondary networks. We name our MAC protocol as cognitive CSMA MAC protocol which controls the communication between secondary users and fusion center. Our numerical results show that the proposed spectrum sensing scheme significantly outperforms schemes that do not exploit multiuser diversity. Furthermore, we show by simulation the benefit of using multiple antennas for spectrum sensing.

For our numerical and simulation results, we consider two main scenarios: 1) secondary users are equipped with single antenna and multiple antennas, and 2) channel between primary user and secondary users are Rayleigh with and without shadowing. We also study the performance of our multiuser diversity scheme in the context of IEEE 802.22 WRAN (Wireless Regional Area Network) standard (IEEE 802.22 Working Group, 2011) since it is the leading standard on the cognitive radio systems. The intention of the WRAN system is to provide internet services in rural area by utilizing unused TV white spaces.

The remainder of the chapter is organized as follows. In Section 2, we provide some background on spectrum sensing and discuss related work on multiuser diversity and IEEE 802.22. The system model considered in this chapter is detailed in Section 3. In Section 4, we develop a distributed scheme for exploiting multiuser diversity to improve spectrum sensing capability. In Section 5, we describe a practical MAC protocol to coordinate transmissions between the secondary users and fusion center. Simulation results are presented in Section 6. In Section 6, we also study the performance of our scheme in scenarios based on the IEEE 802.22 standard. Finally, the chapter is concluded with a summary and a discussion of future work in Section 7.

2. Background and related work

In this section, we provide a brief overview of spectrum sensing and the IEEE 802.22 standard. We then discuss related work on multiuser diversity and spectrum sensing techniques applied to IEEE 802.22.

2.1 Spectrum sensing

Spatial spectrum sensing is investigated (Mark & Nasif, 2009; Nasif & Mark, 2009), wherein the maximum interference-free transmit power (MIFTP) of a given secondary user is estimated based on signal strengths received by a group of secondary nodes. To calculate the MIFTP for a secondary node, estimates of both the location and transmit power of the primary transmitter are estimated collaboratively by a group of secondary nodes. Using these estimates, each secondary node determines its approximate MIFTP, which bounds the size of its spatial spectrum hole. The problem of detecting when the primary is ON or OFF is called *temporal* spectrum sensing (Unnikrishnan & Veeravalli, 2008). In (Do & Mark, 2009; 2010), a joint spatial-temporal spectrum sensing scheme is performed wherein the secondary node performs spatial sensing to determine its MIFTP when the primary transmitter is ON and uses localization information obtained in the process of spatial sensing to improve the performance of temporal sensing, which estimates the ON/OFF state of the primary transmitter. Joint spatial-temporal sensing has higher achievable capacity compared to pure spatial or pure temporal sensing (Do & Mark, 2010).

Cooperative sensing has been studied in a number of papers (Mishra et al., 2006; Unnikrishnan & Veeravalli, 2008; Visotsky et al., 2005). Cooperation between secondary nodes can mitigate the effects of low signal to noise ratio (SNR), shadowing, and hidden terminals Unnikrishnan & Veeravalli (2008). In cooperative sensing, secondary users at different locations sense the channel independently and send their observation to a fusion center. They can communicate either the soft information about the channel or a one-bit hard decision to the fusion center (Ma & Li, 2007). The optimum soft combination rule is derived in (Ma & Li, 2007), wherein the optimal weight coefficients are shown to be identical to those for *maximal ratio combining* (MRC).

2.2 IEEE 802.22

In Section 6.3, we study the performance of our proposed multiuser diversity spectrum sensing scheme in the context of the IEEE 802.22 WRAN (Wireless Regional Area Network) standard (Cordeiro et al., 2005; IEEE 802.22 Working Group, 2011). The 802.22 standard is currently the leading international standard for cognitive radio systems. The intention of the WRAN system is to provide broadband internet services in rural and remote areas by utilizing unused TV white spaces. While availability of broadband access may not be so critical in urban and suburban areas, this certainly is an issue in rural and remote areas. Therefore, this has triggered the FCC (Federal Communications Commission) to stimulate the development of new technologies based on cognitive radio that increase broadband availability in these markets (Challapali et al., 2004; Federal Communications Commission, 2004; 2005). In fact, broadband access in rural areas was one of the reasons why the FCC selected the TV bands for providing such service, as this lower spectrum of frequencies facilitates the propagation of wireless service to rural and remote areas. Moreover, studies of wireless spectrum occupancy have shown that many TV channels are largely unoccupied in many parts of the United States (McHenry, 2003).

Another motivation for the WRAN standard is that IEEE 802.22 devices in the TV band will be unlicensed, which will further lower cost, thus providing affordable wireless service. The commercial markets of 802.22 may include single-family residential, small office/home office (SOHO), small businesses, multi-tenant buildings, and public and private campuses. In the U.S., TV channels are in the VHF (Very High Frequency) and UHF (Ultra High Frequency) regions, i.e., from channel 2 to 69. All of these channels are 6 MHz wide and span from 54-72 MHz, 76-88 MHz, 174-216 MHz, and 470-806 MHz.

The 802.22 system specifies a fixed point-to-multipoint topology whereby a base station (BS) manages its own cell and all associated Consumer Premise Equipments (CPEs). The medium access control of all CPEs in a cell is controlled by the BS. In order to ensure the protection of primary user services, the 802.22 system follows a masters/slave relationship, wherein the BS performs the role of the master and the CPEs are the slaves. No CPE is allowed to transmit before receiving authorization from a BS. The BS also controls all the RF characteristics (e.g., modulation, coding) used by the CPEs. The IEEE 802.22 also manages distributed spectrum sensing which is needed to protect primary user services.

The 802.22 system specifies spectral efficiencies in the range of 0.5 bps/Hz up to 5 bps/Hz or an average of 3 bps/Hz. For 6 MHz TV channel, this would correspond to a total PHY data rate of 18 Mbps. In order to obtain the minimum data rate per CPE, a total of 12 simultaneous users have been considered which leads to a required minimum peak throughput rate at edge of coverage of 1.5 Mbps per CPE in the downstream direction. In the upstream direction, a peak throughput of 384 kbps is proposed, which is also comparable to DSL (Digital Subscriber Line) services. The BS coverage range of IEEE 802.22 can go up to 100 Km if power is not an issue (current specified coverage range is 33 Km at 4 Watts CPE EIRP). Compared to the existing 802.22 standards, WRANs have a much larger coverage range, which is primarily due to its higher power and the favorable propagation characteristics of TV frequency bands. This enhanced coverage range offers unique technical challenges as well as opportunities.

2.3 Multiuser diversity

Opportunistic MAC protocols which exploit multiuser diversity have been investigated in literature. In (Zhao & Tong, 2005), opportunistic CSMA for energy-efficient information retrieval in sensor networks is investigated. The key idea is to exploit the channel state information (CSI) in the backoff strategy of carrier sensing in which the backoff time is a decreasing function of CSI. This scheme ensures that only sensor with the best channel transmit. In (Hwang et al., 2006), the authors incorporate multiuser diversity into p-persistent CSMA. Each user sends a packet if the CSI is above threshold which is determined such that the probability of accessing the medium is p. The proposed opportunistic p-persistent CSMA has a significant capacity increase compare to traditional p-persistent CSMA. The paper by (Hwang & Cioffi, 2007) investigates opportunistic CSMA/CA to achieve multi-user diversity in a wireless LAN.

In (Qin & Berry, 2006), a distributed approach for exploiting multiuser diversity is proposed, based on a protocol called channel-aware slotted ALOHA wherein each user decides, based on the channel state, in which slot to transmit and how much power to use. The design of a multiuser diversity system should consider two important issues: fairness and delay

(Viswanath et al., 2002). In the ideal situation when users fading statistics are the same, the multiuser diversity maximizes not only the total capacity of the system but also the throughput of individual users. However, in reality, users that are closer to the base station have a better average SNR. Some users are stationary, while others are moving. A pure multiuser diversity strategy maximizes long-term average throughput, without regard to delay requirement.

2.4 Spectrum sensing for IEEE 802.22

Spectrum sensing with application to the IEEE 802.22 standard has been studied in (Cordeiro et al., 2007; Kim & Andrews, 2010; Lim et al., 2009; Shellhammer et al., 2006). The spectrum sensing approach proposed in (Kim & Andrews, 2010) uses a spectral covariance sensing algorithm which exploits different statistical correlations of the signal and noise in frequency domain. The spectral covariance sensing algorithm is studied in the context of IEEE 802.22 systems. The algorithm is shown to be very robust to noise uncertainty, which is one of the critical performance measures of spectrum sensing.

An overview of blind sensing techniques in IEEE 802.22 WRANs is provided in (Sai Shanka, 2008). Cooperative spectrum sensing for IEEE 802.22 is studied in (Lim et al., 2009), in which the authors proposed two data fusion schemes for cooperative spectrum sensing. The data fusion structure is appropriate for the WRAN system due to its centralized structure and it can provide reliable sensing performance. The proposed data fusion schemes improve sensing reliability by utilizing a confidence vector of the sensing results.

The performance of power detector sensors for digital TV signals in IEEE 802.22 is studied in (Shellhammer et al., 2006). The authors studied the performance of a power detector in various IEEE 802.22 scenarios such as the "keep out region," also referred as the noise-limited contour. The use of multiple independent sensors is recommended to address the issue of shadow fading. Even with the use of multiple sensors, however, the power detector performance is strongly degraded by the effects of noise uncertainty at the sensors.

3. System model

We consider a discrete-time system model with a single primary transmitter and S secondary users equipped with frequency-agile cognitive radios. Each user makes local decisions about the presence of the primary user and communicate a one-bit hard or soft decision to the fusion center, which makes the final decision. Alternatively, the system can operate in a distributed manner wherein secondary users exchange their local decisions with each user. Without loss of generality, we shall assume a fusion center in this chapter.

Due to communication constraints between secondary users and the fusion center, not all the secondary users are able to communicate their decisions to the fusion center. We assume that N out of S secondary users are able to communicate with the fusion center. Because of multiuser diversity, each of the S secondary users has different fading channel parameters during a given observation time period.

We adopt a spectrum sensing model similar to that in (Ma & Li, 2007). Each secondary user uses M samples for energy detection. We define two hypotheses: H_1 is the hypothesis that the primary is ON and located close to the secondary nodes and H_0 is the hypothesis that the primary is OFF or far away. In other words, H_0 is the hypothesis that the spectrum hole exists and the frequency channel is available for reuse by secondary users. The observed energy value at the jth user is given by

$$Y_j = \begin{cases} \sum_{i=1}^{M} n_{ji}^2, & \text{under } H_0, \\ \sum_{i=1}^{M} (s_{ji} + n_{ji})^2, & \text{under } H_1, \end{cases} \tag{1}$$

where n_{ji} is the white noise signal in the ith sample of the jth user and s_{ji} denotes the received primary signal at each secondary user, $1 \leq j \leq N$, $1 \leq i \leq M$. The noise samples n_{ji} are assumed to be independently and identically distributed (i.i.d.) Gaussian random variables with zero mean and unit variance.

The instantaneous SNR of the jth secondary user is defined as

$$\gamma_j \triangleq \frac{1}{M} \sum_{i=1}^{M} s_{ji}^2.$$

Following(Ma & Li, 2007), we assume that the total energy of the transmitted primary signal is constant within each observation blocks. Thus, the γ_j's represent the power of the instantaneous channel gain and can be modeled by a Rayleigh or Nakagami distribution (Digham et al., 2003) and are i.i.d. over different secondary users j and observation blocks. Within a given observation block, multiuser diversity exists because of the differences in γ_j across users $j = 1, \ldots, N$.

If the primary user is absent or in the OFF state, Y_j can be modeled as a central chi-square random variable with M degrees of freedom. Otherwise, if the primary user is in the ON state, Y_j follows a non-central chi-square distribution with M degree of freedom and a non-centrality parameter $\lambda_j = M\gamma_j$ (Ma & Li, 2007):

$$\begin{aligned} H_0 &: Y_j = \chi_M^2, \\ H_1 &: Y_j = \chi_M^2(\lambda_j). \end{aligned}$$

For large M, Y_j can be approximated by a Gaussian distribution (Ma & Li, 2007):

$$\begin{aligned} H_0 &: Y_j \sim \mathcal{N}(M, 2M), \\ H_1 &: Y_j \sim \mathcal{N}(M(1+\gamma_j), 2M(1+\gamma_j)). \end{aligned} \tag{2}$$

In (Ma & Li, 2007), a Gaussian approximation of the received energy distribution is used to derived the optimal soft combination weights. The weighted summation at fusion center is given by

$$Y = \sum_{j=1}^{N} \omega_j Y_j. \tag{3}$$

where the optimal weight coefficients are given by(Ma & Li, 2007)

$$\omega_j = \frac{\gamma_j}{\sqrt{\sum_{k=1}^{N} \gamma_k^2}}. \tag{4}$$

where γ_j is the instantaneous SNR for node j. The distribution of Y can be approximated by a Gaussian distribution as follows: Under H_0,

$$H_0 : Y \sim \mathcal{N}\left(M\sum_{j=1}^{N}\omega_j, 2M\sum_{j=1}^{N}\omega_j^2\right),$$

$$H_1 : Y \sim \mathcal{N}\left(M\sum_{j=1}^{N}(1+\gamma_j), 2M\sum_{j=1}^{N}\omega_j^2(1+\gamma_j)\right). \tag{5}$$

The fusion center chooses hypothesis H_1 if $Y > \tau_f$ and H_0 otherwise, where τ_f is the decision threshold at the fusion center. The performance metrics of interest are the false alarm probability and the detection probability, given respectively by

$$P_f \triangleq \Pr\{Y > \tau_f | H_0\},\ P_d \triangleq \Pr\{Y > \tau_f | H_1\}. \tag{6}$$

For a given false alarm probability, the objective is to maximize the probability of (correct) detection. The performance of different spectrum sensing schemes can be evaluated by comparing P_d at a predetermined P_f value.

4. Multiuser diversity spectrum sensing

In this Section, we develop a multiuser diversity spectrum sensing scheme for cognitive radio networks. We assume that there are S secondary nodes, which are equipped with identical energy detectors. The received signal powers between pairs of secondary nodes are i.i.d. with a Rayleigh or Nakagami fading distribution. We first consider the case of secondary users equipped with a single antenna and then address the case of multiple antennas.

4.1 Soft combination

Let τ_l and τ_u be predefined lower and upper thresholds, respectively, where $\tau_l < \tau_u$. In the proposed scheme, a node j ($j = 1, \ldots, S$) with received energy level satisfying

$$Y_j > \tau_u \quad \text{or} \quad Y_j < \tau_l \tag{7}$$

is given priority to send its observation to the fusion center. As stated earlier, we assume that the communication capacity of the channel between the secondary nodes and the fusion center is limited such that only N out of S nodes can communicate with the fusion center. We assume that there exists a dedicated control channel for enabling the communication between secondary users and fusion center.

In a *perfect* Medium Access Control (MAC) protocol, a centralized scheduler selects N nodes to communicate their observations to the fusion center. Let $\tilde{N}$ denote the number of nodes with received energy levels that satisfy (7). If $\tilde{N} \geq N$, the centralized scheduler (randomly) selects N out of the $\tilde{N}$ nodes to send their observations to the fusion center. Otherwise, if $\tilde{N} < N$, the N nodes selected by the scheduler consists of the $\tilde{N}$ nodes plus an additional $N - \tilde{N}$ nodes randomly selected from among the remaining nodes. Thus, the total number of observations sent to the fusion center is always equal to N. A practical, distributed MAC protocol for coordinating communications between the secondary nodes and the fusion center, based on CSMA, is proposed in Section 5.

To understand the benefit of exploiting multiuser diversity, we consider a simple soft information equal gain combining (EGC) strategy at the fusion center:

$$Y = \sum_{j=1}^{N} Y_j.$$

The distribution of Y can be approximated by a Gaussian distribution as given in (5) with $\omega_j = 1, j = 1, 2 \ldots, N$ for EGC. For $S \gg N$, the thresholds τ_l and τ_u can be chosen such that

$$\Pr(Y_j < \tau_l | H_1) \approx 0, \; \Pr(Y_j > \tau_u | H_0) \approx 0, \; j = 1, 2, \ldots S.$$

Suppose that $\tilde{N} > 0$ nodes satisfy (7) and let their received energy levels be denoted by $\tilde{Y}_j$, $j = 1, \ldots \tilde{N}$. Under hypothesis H_1, the following inequality holds with probability one:

$$\sum_{j=1}^{\tilde{N}} \tilde{Y}_j \geq \sum_{j=1}^{\tilde{N}} Y_j,$$

where $\{Y_j\}_{j=1}^{N}$ denotes a set of observations that does not exploit multiuser diversity; i.e., a set of N out of S nodes is randomly selected to send their observations to the fusion center. Hence,

$$\tilde{Y} = \sum_{j=1}^{\tilde{N}} \tilde{Y}_j + \sum_{j=\tilde{N}+1}^{N} Y_j \geq Y = \sum_{j=1}^{N} Y_j, \tag{8}$$

where the inequality is understood to hold almost surely. Thus,

$$P_{\text{mud}} \triangleq \Pr\{\tilde{Y} > \tau_f\} \geq \Pr\{Y > \tau_f\} \triangleq P_c \tag{9}$$

where P_{mud} and P_c denote the detection probability of the multiuser diversity spectrum sensing scheme and a conventional scheme, respectively. Therefore, the multiuser diversity spectrum sensing results in a superior detection probability compared to conventional spectrum sensing. A similar approach can be applied for hypothesis H_0. In this case, the false alarm probability of the multiuser diversity spectrum sensing scheme can be shown to be smaller than that of the conventional scheme. Simulation results presented in Section 6 validate the benefit of exploiting multiuser diversity for spectrum sensing.

The optimal soft combination is derived in (Ma & Li, 2007), where the optimal weight coefficients are given by (4) and the soft combination rule is given by (3). Since ω_j, derived in (4), is similar to the weights used in *maximal ratio combining* (MRC), we refer to this approach as the MRC scheme. In this case, the fusion center compares the obtained soft combination metric Y in (3) with a predetermined threshold τ_f and decides on hypothesis H_1 if $Y > \tau_f$ and H_0 otherwise. The value of τ_f is determined by simulation (Ma & Li, 2007) such that the probability of interference is smaller than or equal to a threshold on the probability of false alarm, P_F.

4.2 Hard combination

The soft combination scheme may be impractical due to the overhead of sending the observation data to the fusion center. As an alternative, a hard combination scheme could be adopted at the fusion center. In this scheme, each node compares its observation Y_j with a given threshold τ_n. If Y_j satisfies (7), the node will send a hard decision $U_i = 1$ to the fusion center if $Y_j > \tau_n$ and $U_i = 0$ otherwise:

$$U_i \triangleq I_{\{Y_i > \tau_n\}},$$

where I_A denotes the indicator function on the event A. At the fusion center two fusion rules that could be applied are:

1. 1-out-of-N (OR) rule (Ghasemi & Sousa, 2005): The primary signal will be declared present if any one of the cooperative users decides locally that the primary signal exists.
2. Counting rule: The final decision is made by comparing the sum $\sum_{i=1}^{N} U_i$ to a decision threshold. The value of this threshold is obtained through simulation (Unnikrishnan & Veeravalli, 2008).

The threshold at each node τ_n for the OR rule is also determined by simulation such that the constraint on the probability of false alarm is satisfied at the fusion center. However, OR rule tends to have a high probability of false alarm (Ghasemi & Sousa, 2005). Moreover, the OR rule may not be used in case communication is not available between the secondary nodes and the fusion center. The counting rule ensures that the constraint on the probability of false alarm is met both at individual nodes and at the fusion center. However, *randomization* between two fusion thresholds may be required at the fusion center in order for the counting rule to achieve the false alarm probability constraint (Unnikrishnan & Veeravalli, 2008).

4.3 Multiple antenna case

We now extend the preceding discussion for the case when each secondary user has N_t antennas. As before, we assume that the primary transmitter has a single antenna. An energy detector is used at each antenna of a secondary user. We assume that the distance between the antennas is sufficiently far that the fading for different the antennas may be considered i.i.d. Assume that M samples are collected at each detector. The observed energy from the kth antenna at a node j is given by

$$Z_{j,k} = \begin{cases} \sum_{i=1}^{M} n_{ji}^2, & \text{under } H_0, \\ \sum_{i=1}^{M} (s_{ji} + n_{ji})^2, & \text{under } H_1. \end{cases} \tag{10}$$

For a multiple-receive antenna system, *equal gain combination* (EGC) is used (Ma & Li, 2007):

$$Y_j = \sum_{k=1}^{N_t} Z_{j,k}, \; j = 1, \ldots, S, \tag{11}$$

where Y_j is the combined total received energy at the output of secondary user j.

Similar to the single antenna scenario, Y_j is compared against two thresholds $\tilde{\tau}_u$ and $\tilde{\tau}_l$. If Y_j satisfies

$$Y_j > \tilde{\tau}_u \quad \text{or} \quad Y_j < \tilde{\tau}_l, \tag{12}$$

node j will be given priority to communicate its observation to the fusion center. In the multiple antenna case, we assume that the each node sends a one-bit hard decision (if any) to the fusion center for hard combination. If the observation Y_j satisfies $Y_j > \tilde{\tau}_n$, node sends the value 1 to the fusion center and 0 otherwise. The OR rule or the counting rule can be used at fusion center as the detection rule. The threshold $\tilde{\tau}_f$ at the fusion center and the thresholds $\tilde{\tau}_n$ are determined by simulation to meet the false alarm probability requirement.

The MAC protocol selects a subset of N secondary nodes to transmit their hard decisions to the fusion center, based on the condition (12). Let $\tilde{N}$ denote the number of nodes with received energy levels satisfying (12). Similar to the single antenna case, a perfect, centralized MAC scheduler makes a (possibly random) selection of N out of $\tilde{N}$ nodes, if $\tilde{N} \geq N$. If $\tilde{N} < N$, the N nodes selected by the scheduler consists of the $\tilde{N}$ nodes satisfying (12) plus a random selection of $N - \tilde{N}$ additional nodes. A more practical, distributed MAC protocol based on CSMA is discussed next.

5. CSMA-based MAC protocol

In this Section, we develop MAC protocol based on Carrier Sense Multiple Access (CSMA) for secondary users to transmit their observations to the fusion center. The proposed MAC is used to enable communications between secondary users and the fusion center during spectrum sensing period. Clearly, a different MAC protocol may be used for communications between secondary users during the spectrum hole period. As mentioned earlier, we assume that there exists a dedicated control channel for secondary users to exchange information with the fusion center. Also, the physical layer between fusion center and user is assumed to be perfect, i.e., the fusion center receives what the users send without error. Our proposed MAC protocol based on the IEEE 802.11 MAC (IEEE 802.11 Working Group, 1997). In our scenario, the MAC protocol is used to enable communications from the secondary users to the fusion center. Since there is only one receiver in this scenario, i.e., the fusion center, there is no hidden terminal issue, and Request to Send/Clear to Send (RTS/CTS) packets are not needed.

Time is divided into slots and each user is allowed to transmit only at the beginning of each time slot. If a secondary user wishes to communicate its observation to the fusion center, it monitors the channel activity. If the channel is idle for a specified time period, i.e., the distributed interframe space (DIFS) in the 802.11 standard, the secondary user transmits. Otherwise, if the channel is busy, the user continues to monitor the channel until it remains idle for an interval of DIFS. In this case, the user generates a random backoff interval before transmitting, in accordance with the 802.11 collision avoidance feature.

We adopt the exponential backoff scheme in 802.11 standard with a modification to exploit multiuser diversity. User i generates a random backoff time which is drawn from the interval $(0, w_i - 1)$ according to a uniform distribution, where $w_i > 1$ is an integer called the contention window of user i. At the first transmission attempt or after a successful transmission, $w_i = CW_1$ if the observation Y_i satisfies the condition (7); otherwise, $w_i = CW_2$. After each failed transmission, i.e., when there is more than one user transmitting at the beginning of a time slot, w_i is doubled until it reaches CW_{max} where $CW_{max} = 2^m CW_1$ if Y_i satisfies the condition (7); otherwise, $CW_{max} = 2^m CW_2$.

The backoff time counter is decremented as long as the channel is sensed idle and is frozen when the channel is sensed busy. When the backoff time counter reaches zero, the user transmits its observation to the fusion center. We choose $CW_1 \ll CW_2$. Hence, users satisfying

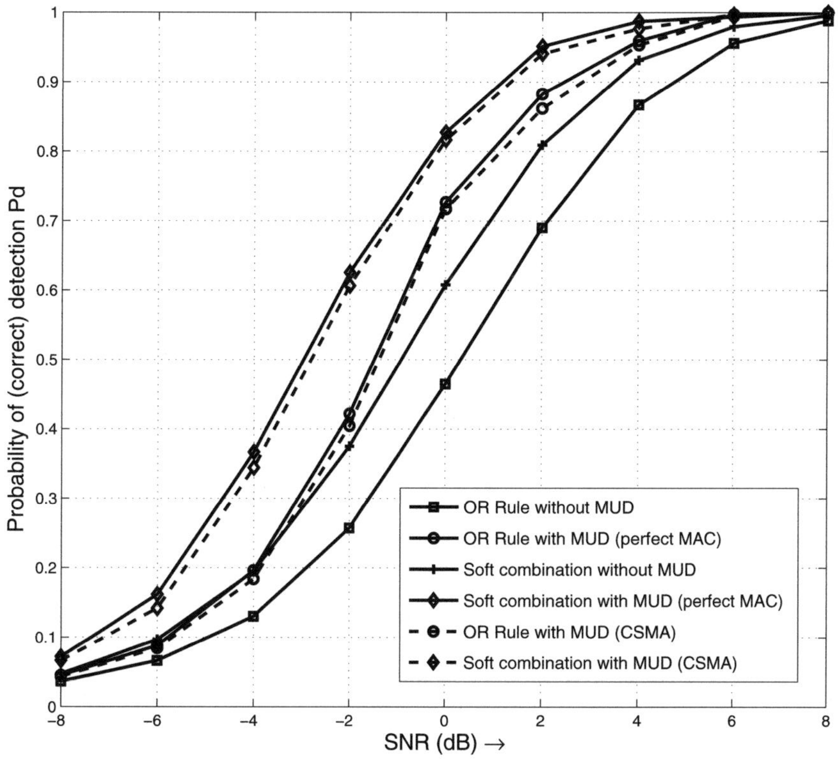

Fig. 1. Performance of OR rule rule and soft combination scheme with multiuser and conventional spectrum sensing.

condition (7) will likely have a smaller random backoff timer and will have channel access with higher probability. The fusion center will make the final decision whenever it receives the observations of N users. Once the fusion center receive observations from N nodes, it broadcasts a signal to stop the other nodes from transmitting further observations.

6. Numerical results

In this Section, we compare the performance of the proposed multiuser diversity spectrum sensing scheme with a conventional scheme that does not exploit multiuser diversity. The following parameters are used for all simulations discussed in this Section.

- False alarm probability requirement $\tau_{FA} = 0.01$;
- Number of secondary nodes $N = 4$.
- $CW_1 = 8$, $CW_2 = 64$, $m = 3$

In addition, the number of samples $M = 6$ for the simulations corresponding to Fig. 1 through Fig. 6. For all of the simulation results, 95% confidence intervals were computed, but they have been omitted from the figures for clarity of presentation. For each simulation result, the width of the 95% confidence interval is less than 0.2 of a unit on the vertical scale.

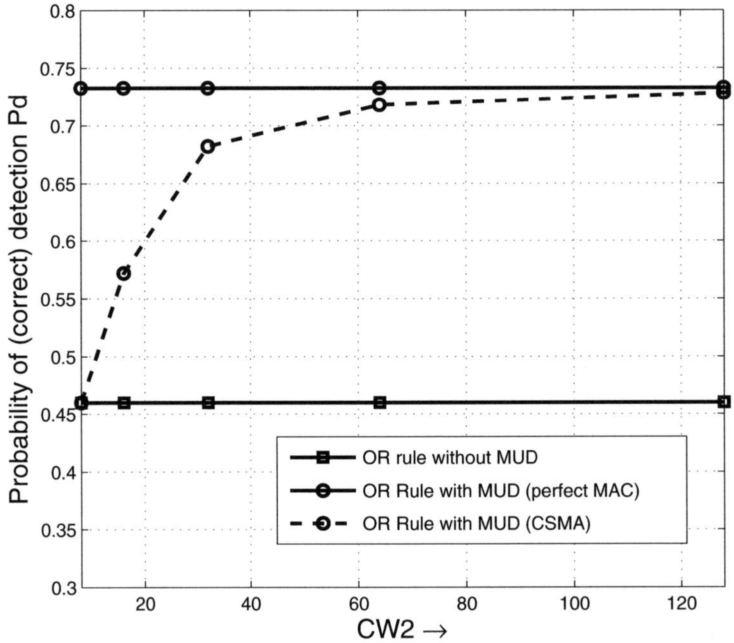

Fig. 2. Performance of OR Rule with perfect MAC and CSMA-based MAC vs. the contention window CW_2.

6.1 Spectrum sensing with Rayleigh fading only

In Fig. 1, we compare the performance of multiuser diversity spectrum sensing with conventional spectrum sensing when soft optimal combination and the OR rule are used at the fusion center. Here, the total number of users is $S = 12$. We reproduce the results for the conventional OR rule and optimal soft combination considered in (Ma & Li, 2007) and compare with the corresponding results from the multiuser diversity scheme. The thresholds τ_u and τ_l are chosen to satisfy

$$P(Y_i < \tau_l | H_0) = \frac{N}{S} \text{ and } P(Y_i > \tau_u | H_1) = \frac{N}{S}. \tag{13}$$

These thresholds can easily be calculated using (2). The threshold τ_n is set by simulation to meet the false alarm probability requirement. Fig. 1 shows that the detection probability of our proposed scheme is much better than that of the conventional spectrum sensing scheme scheme especially when the signal-to-noise ratio (SNR) is relatively small. For $S = 12$ and $N = 4$, the hard combination (OR rule) with multiuser diversity can outperform the optimal soft combination scheme.

In Fig. 2, we compare the performance of CSMA-based MAC protocol described in Section 5 for different values of CW_2. The performance of the OR rule with the CSMA-based MAC approaches the performance of the OR rule with perfect MAC protocol (i.e., a centralized scheduler) at $CW_2 = 64$. However, as CW_2 increases, the MAC delay also increases.

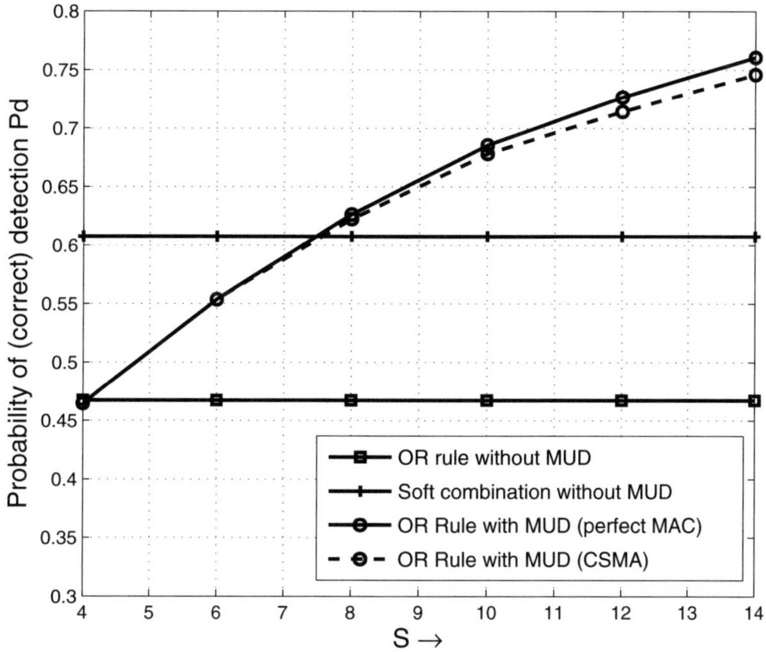

Fig. 3. Performance of conventional OR rule and soft combination scheme and OR rule with multiuser diversity vs. total number of users S.

In Fig. 3, we compare the performance of the multiuser diversity scheme with conventional optimum soft combination and the OR rule with different values of S with SNR $= 0$ dB. The thresholds τ_u and τ_l are similar to those used in the simulation of Fig. 1 with SNR $= 0$ dB. When the total number of users S increases, the detection probability of multiuser diversity spectrum sensing increases. When $S \geq 8$, the multiuser diversity hard combination based OR rule outperforms the conventional soft optimal combination studied in (Ma & Li, 2007).

In Fig. 4, we compare the performance of the conventional and multiuser diversity spectrum sensing schemes with the counting rule at the fusion center. All of the schemes meet the requirement of τ_{FA} at the fusion center, but only the counting rule satisfies the τ_{FA} requirement both at the nodes and at the fusion center. The detection probability for the OR rule rule is higher than that of the conventional counting rule, but the OR rule also has a higher false alarm probability at each node (Ghasemi & Sousa, 2005). At low SNR, single user detection can outperform the counting rule because of the effect of fading. At low SNR, nodes that experience severe fading can make wrong decisions about the presence of a primary user. This may in turn result in a wrong decision produced at the fusion center.

In Fig. 5, we compare the performance gain of secondary users equipped with multiple antennas over those equipped with only a single antenna. Hard combination with the OR rule is used. Each multi-antenna user is equipped with $N_t = 2$ antennas. We consider $S = 12$ nodes and set the thresholds as $\tilde{\tau}_u$=2τ_u and $\tilde{\tau}_l$=2τ_l, where τ_u and τ_l are obtained from the simulation corresponding to Fig. 1. Fig. 5 clearly shows the benefit of multiple antennas over single antennas and also the benefit of multiuser diversity in multi-antennas systems.

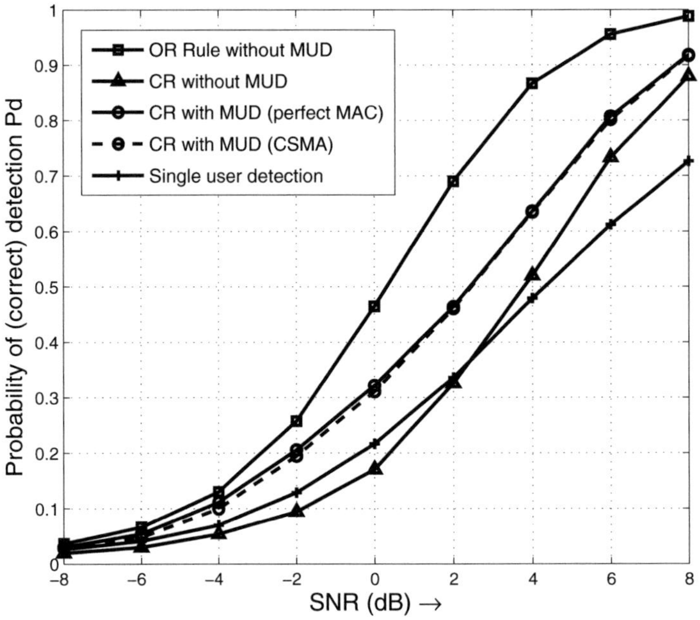

Fig. 4. Performance of Counting Rule (CR) with multiuser diversity and conventional spectrum sensing.

6.2 Spectrum sensing with Rayleigh fading and shadowing

In the previous simulation scenarios, we assumed that there was no shadowing in the received signal. In this Section, we consider the effect of both shadowing and small-scale fading on the received signal (Motamedi & Soleymani, 2007). In this case all users are affected by both large and small scale (Rayleigh) fading. The effect of shadowing causes random fluctuations in received power. These fluctuations can be modeled by multiplying the received power by a log-normal distributed random variable. The log-normal random variable w is specified by a mean μ_w (dB) and a standard deviation σ_w (dB). We have,

$$10 \log_{10} w \sim N(\mu_w, \sigma_w)$$

where $\mu_w = E[10 \log_{10} w]$ and $\sigma_w^2 = \mathrm{Var}[10 \log_{10} w]$. The probability density function (pdf) of w is given by (Motamedi & Soleymani, 2007)

$$P(w, \mu_w, \sigma_w) = \frac{1}{\sqrt{2\pi}\sigma_w} \exp[-(10 \log_{10} w - \mu_w)^2 / 2\sigma_w^2], \; w > 0. \tag{14}$$

In the presence of both shadowing and Rayleigh fading, the observed energy value at the jth user is given by

$$Y_j = \begin{cases} \sum_{i=1}^{M} n_{ji}^2, & \text{under } H_0, \\ \sum_{i=1}^{M} (s_{ji} w + n_{ji})^2, & \text{under } H_1, \end{cases} \tag{15}$$

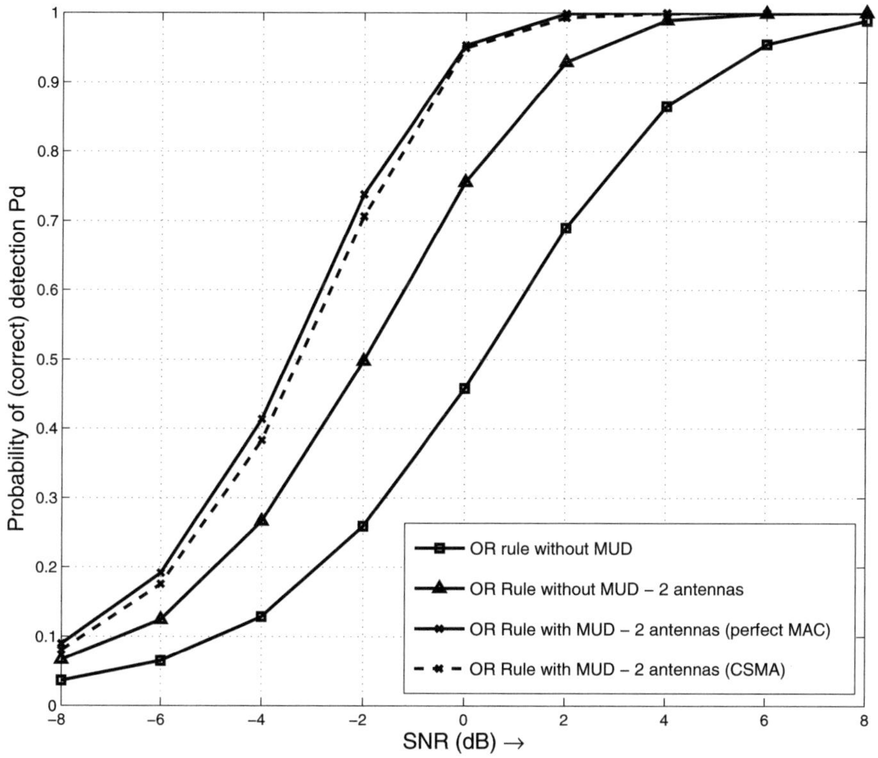

Fig. 5. Performance of OR rule with multiuser diversity and conventional spectrum sensing, 2 antennas used at secondary users.

where n_{ji} is the white noise signal in the ith sample of the jth user and s_{ji} denotes the received primary signal at each secondary user, $1 \le j \le N$, $1 \le i \le M$. The noise samples n_{ji} are assumed to be independently and identically distributed (i.i.d.) Gaussian random variables with zero mean and unit variance and w denotes the shadowing noise. In our numerical results, we assume that the shadowing noise is Gaussian distributed with zero mean and a variance of 5.5 dB. Figure 6 shows that spectrum sensing with multiuser diversity significantly outperforms spectrum sensing without multiuser diversity, even in the presence of shadowing. In this simulation experiment, the number of sample $M = 4$, the CSMA-based MAC protocol is used with $W_1 = 8$ and $W_2 = 64$, and the threshold values are same as in the previous simulations without shadowing.

6.3 IEEE 802.22 WRAN

IEEE 802.22 requires a detection probability of 0.9 with the false alarm probability of 0.1 at SNR $= -22$ dB (Kim & Andrews, 2010; Lim et al., 2009) In all simulations, we use: Number of secondary nodes $N = 4$, contention window for MAC protocol $CW_1 = 8$, $CW_2 = 64$, $m = 3$, total number of secondary user $S = 24$. The shadowing fading is modeled by a log-normal distribution with zero mean and a variance of 5.5 dB (Shellhammer et al., 2006). The total

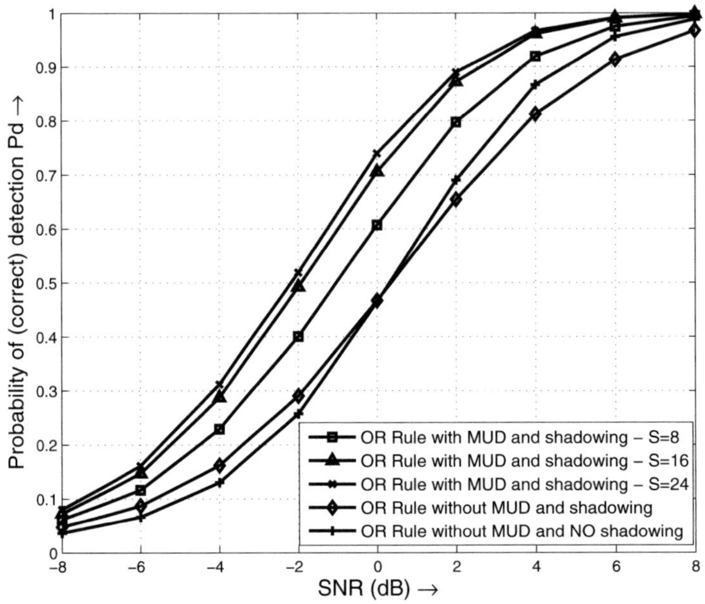

Fig. 6. Performance of OR rule with multiuser diversity and shadowing noise.

number of samples used by the energy detector is $M = 40$ except Fig. 10. The combination rule at fusion center is the OR rule.

In Fig. 7, the required probability of false alarm is $P_f = 0.01$. We compare the performance of our multiuser diversity spectrum sensing with spectrum sensing without multiuser diversity when only shadowing fading is assumed and no Rayleigh fading. As in Fig. 7, the multiuser spectrum sensing is outperformed the spectrum sensing without multiuser diversity. However, since it is assumed that there is no small-scale (Rayleigh) fading, the performance gain of multiuser diversity spectrum sensing is small. This result makes sense, since the degree of multiuser diversity is higher when small-scale fading is present among secondary users (Viswanath et al., 2002).

In Fig. 8, we compare the performance of multiuser diversity spectrum sensing with conventional spectrum sensing in the presence of both Rayleigh and shadowing fading. The simulation results show that multiuser diversity spectrum sensing significantly outperforms conventional spectrum sensing in this case. When SNR=-22 dB and $P_f = 0.1$, multiuser diversity spectrum sensing can achieve a detection probability close to 0.9, which is the spectrum sensing requirement of IEEE 802.22 when using only $M = 40$ samples. In this case, the detection probability for conventional spectrum sensing without multiuser diversity is only about 0.64,

In Fig. 9, we show the receiver operating characteristic (ROC) curve for energy detector with multiuser diversity spectrum sensing and conventional spectrum sensing without multiuser diversity. In Fig. 9, we use SNR=-21 dB, number of samples M = 40, both small scale fading (Rayleigh) and shadowing fading with zero mean and variance 5.5 dB. The ROC curve for spectrum sensing with multiuser diversity is much better than that of conventional spectrum

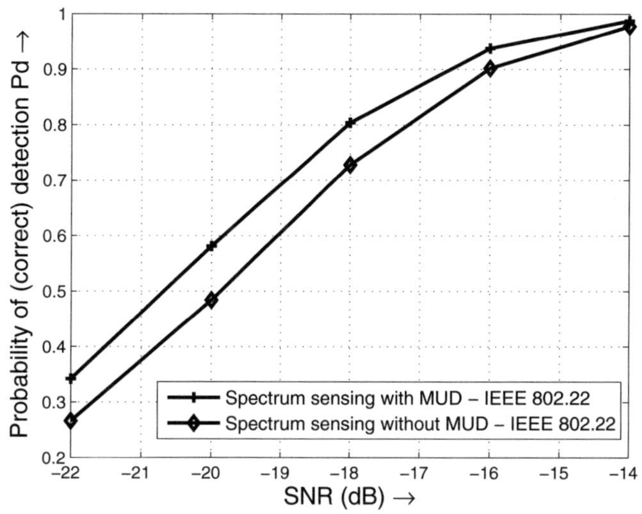

Fig. 7. Performance of OR rule without Rayleigh fading in IEEE 802.22: P_d vs. SNR.

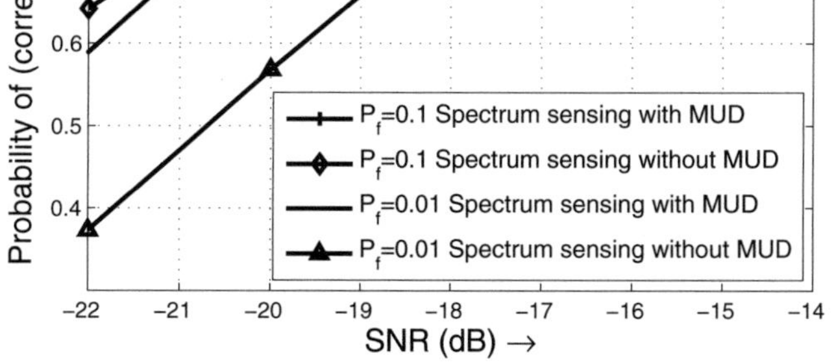

Fig. 8. Performance of OR rule with Rayleigh fading in IEEE 802.22: P_d vs. SNR.

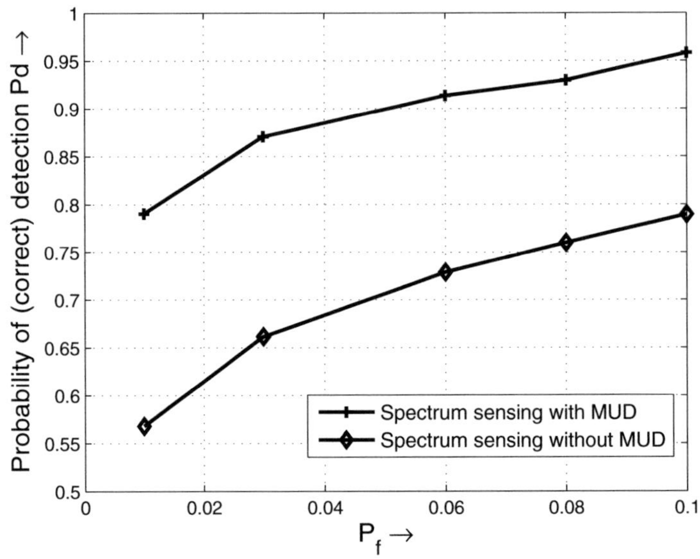

Fig. 9. IEEE 802.22 ROC curve with SNR=-21 dB

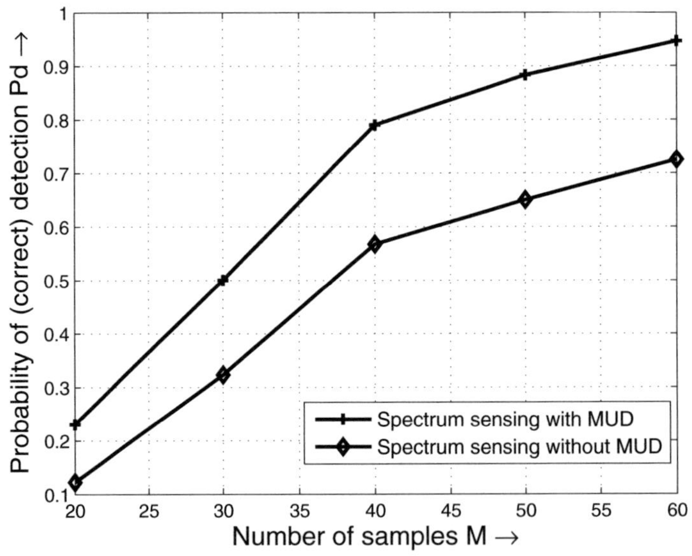

Fig. 10. Detection probability v.s. Number of sample M

sensing without multiuser diversity. In Fig. 10, the performance of the proposed multiuser diversity spectrum sensing scheme is compared with that of conventional spectrum sensing as the number of samples M is varied. We see that the detection probability for both schemes increase as M increases. For all values of M, the detection probability for multiuser diversity spectrum sensing is significantly higher than that of spectrum sensing without multiuser diversity.

7. Conclusion

We conclude this Chapter with a brief summary and a discussion of two topics for future work: sequential detection and multichannel systems.

7.1 Summary

We proposed a cooperative multiuser diversity spectrum sensing scheme that exploits the multiuser diversity inherent in the secondary network to improve the sensing capability of cognitive radio systems. We use a distributed approach in the sense that each secondary user only has local knowledge about its observed energy. We studied two detection rules for the fusion center, counting rule and the OR rule and also considered the case of users equipped with multiple antennas. We also proposed a CSMA-based MAC protocol for secondary users to communicate their observations to the fusion center.

We compared the performance of the proposed multiuser diversity spectrum sensing scheme and a conventional spectrum sensing scheme in a variety of scenarios. In particular, we studied the performance of the multiuser diversity scheme in the context of the IEEE 802.22 WRAN standard. Our simulation results showed that a substantial gain in spectrum sensing performance could be achieved by exploiting multiuser diversity, particularly in environments with both small-scale and large-scale fading.

7.2 Sequential detection

In spectrum sensing, detection delay is an important performance metric. When a primary user stops transmission, the secondary user should detect this event quickly in order to maximize its use of the newly created temporal spectrum hole. A small detection delay will allow secondary users to take advantage of short transmission opportunities. On the other hand, when the primary user starts transmission, the cognitive user should detect this event as quickly as possible in order to minimize interference to the primary user.

In this Chapter, the spectrum sensing decision is made after a fixed number of samples M. To improve detection delay performance, a *sequential* detection scheme could be incorporated into the multiuser diversity spectrum sensing scheme proposed in this Chapter. Sequential detection schemes exploit the fact that the number of samples required to achieve a given reliability level may well be dependent on the actual realization of the observed samples. For example, in a simple binary hypothesis testing context, Wald's sequential probability ratio test (SPRT) compares the likelihood ratio with two thresholds, and the decision is made as soon as the test statistic exceeds either one of the thresholds. It is known that SPRT minimizes the average sample number (ASN) among all tests with the same false alarm and detection probabilities. In (Kim & Giannakis, 2009), sequential sensing is proposed for orthogonal frequency division multiplexing (OFDM) cognitive radios. Lai *et al.* (Lai et al., 2008) develop a sequential sensing strategy based on a quickest detection framework.

7.3 Multichannel systems

In this Chapter, we focused on opportunistic spectrum access of a single channel. In multichannel cognitive radio networks, the licensed wireless spectrum consists of a set of N non-overlapping channels. Secondary users have the ability to access all of the available channels by switching frequencies. Multichannel cognitive radio networks have been studied in (Jiang et al., 2008; 2009; Kim & Giannakis, 2009; Min & Shin, 2008). For example, in (Jiang et al., 2008; 2009), a cognitive radio system with opportunistic transmissions is considered wherein the secondary user not only senses a channel to decide whether it is free, but also estimates the channel coefficients in order to determine the transmission rate. If a channel is sensed to be idle, but the channel quality between the secondary transceiver pair is not satisfactory, the secondary user may still skip this channel and keep sensing other channels.

A natural extension of the work presented in this Chapter would be to consider multiuser diversity spectrum sensing in a multichannel system. In such a system, the multiuser diversity principle could be exploited not only to improve spectrum sensing performance as studied in this Chapter, but also to improve the performance of opportunistic spectrum access when multiple channels are available.

8. References

Broderson, R. W., Wolisz, A., Cabric, D., Mishra, S. M. & Willkomm, D. (2004). Corvus: A cognitive radio approach for usage of virtual unlicensed spectrum, *Technical report*, Univ. California Berkeley.

Cabric, D., Mishra, S. & Brodersen, R. W. (2004). Implementation issues in spectrum sensing for cognitive radios, *Proc. 38th Asilomar Conf. on Signals, Systems and Computers '04*, pp. 772–776.

Challapali, K., Birru, D. & Mangold, S. (2004). Spectrum agile radio for broadband applications, *EE Times In-Focus* . URL: *http://www.eetimes.com/design/other/4003528/Spectrum-Agile-Radio-for-Broadband -Applications*

Cordeiro, C., Challapali, K., Birru, D. & Shankar, N. S. (2005). IEEE 802.22: the first worldwide wireless standard based on cognitive radios, *First IEEE International Symposium on New Frontiers in Dynamic Spectrum Access Networks (DySPAN)*, New Orleans, pp. 328 – 337.

Cordeiro, C., Ghosh, M., Cavalcanti, D. & Challapali, K. (2007). Spectrum sensing for dynamic spectrum access of TV bands, *Cognitive Radio Oriented Wireless Networks and Communications (CROWNCom)*, pp. 225 – 233.

Digham, F. F., Alouini, M. S. & Simon, M. K. (2003). On the energy detection of unknown signals over fading channels, *Proc. IEEE Int. Conf. on Comm. (ICC'03)*, pp. 3575–3579.

Do, T. & Mark, B. L. (2009). Joint spatial temporal spectrum sensing for cognitive radio networks, *Conference on Information Sciences and Systems - CISS 2009, submitted*, Baltimore, MD.

Do, T. & Mark, B. L. (2010). Joint spatial-temporal spectrum sensing for cognitive radio networks, *IEEE Trans. Veh. Technol.* 59(7): 3480–3490.

Federal Communications Commission (2004). Notice of proposed rule making, FCC - ET Docket no. 04-113.

Federal Communications Commission (2005). Report and order and memorandum opinion and order, FCC - ET Docket no. 05-56.

Ghasemi, A. & Sousa, E. S. (2005). Collaborative spectrum sensing for opportunitic access in fading enviroments, *Proc. IEEE DySPAN'05*, pp. 131–136.

Hwang, C.-S. & Cioffi, J. M. (2007). Using opportunistic CSMA/CA to achieve multi-user diversity in wireless LAN, *IEEE Globecom*, pp. 4952–4956.

Hwang, C.-S., Seong, K. & Cioffi, J. M. (2006). Opportunistic p-persistent CSMA in wireless networks, *IEEE Int. Conf. on Commun. (ICC)*, pp. 183–188.

IEEE 802.11 Working Group (1997). IEEE Standard for Wireless LAN Medium Access Control (MAC) and Physical Layer (PHY) Specifications.

IEEE 802.22 Working Group (2011). IEEE 802.22 standard.

Jiang, H., Lai, L., Fan, R. & Poor, H. V. (2008). Cognitive Radio: How to Maximally Utilize Spectrum Opportunities in Sequential Sensing, *Proc. IEEE Globecom 2008*, pp. 4851–4855.

Jiang, H., Lai, L., Fan, R. & Poor, H. V. (2009). Optimal selection of channel sensing order in cognitive radio, *IEEE Trans. Wireless Commun.* 8(1): 297–307.

Kim, J. & Andrews, J. G. (2010). Spectral covariance for spectrum sensing, with application to IEEE 808.22, *IEEE International Conference on Acoustics Speech and Signal Processing (ICASSP)*.

Kim, S.-J. & Giannakis, G. B. (2009). Rate-optimal and reduced-complexity sequential sensing algorithms for cognitive OFDM radios, *EURASIP Journal on Advances in Signal Processing, Special Issue on Dynamic Spectrum Access for Wireless Networking* .

Knopp, R. & Humblet, P. (1995). Information capacity and power control in single-cell multiuser communications, *Proc. IEEE Int. Conf. Comm. (ICC'95)*, Seattle, WA, pp. 331–335.

Lai, L., Fan, Y., & Poor, H. V. (2008). Quickest detection in cognitive radio: A sequential change detection framework, *Proc. IEEE Global Telecommunications Conference (GLOBECOM)*, New Orleans, LA, pp. 1–5.

Lim, S., Jung, H. & Song, M. S. (2009). Cooperative spectrum sensing for IEEE 802.22 WRAN system, *International Conference of Computer Communications and Networks (ICCCN)*.

Ma, J. & Li, Y. (2007). Soft combination and detection for cooperative spectrum sensing in cognitive radio networks, *Proc. IEEE Global Telecommun. Conf. (Globecom'07)*, pp. 3139–3143.

Mark, B. L. & Nasif, A. O. (2009). Estimation of maximum interference-free transmit power level for opportunistic spectrum access, *IEEE Trans. Wireless Commun.* 8(5): 2505–2513.

McHenry, M. (2003). Frequency agile spectrum access technologies, *FCC Workshop on Cognitive Radio*.

Min, A. W. & Shin, K. G. (2008). Exploiting multi-channel diversity in spectrum-agile networks, *Proc. IEEE Infocom 2008*, pp. 1921 – 1929.

Mishra, S., Sahai, A. & Brodersen, R. W. (2006). Cooperative sensing among cognitive radios, *Proc. IEEE Int. Conf. Communications*, Vol. 4, Istanbul, pp. 1658–1663.

Motamedi, Z. & Soleymani, M. R. (2007). For better or worse: The impact of shadow fading on the capacity of large MIMO networks, *IEEE Global Telecommunication Conference (Globecom 2007)*, pp. 3200 – 3204.

Nasif, A. O. & Mark, B. L. (2009). Opportunistic spectrum sharing with multiple cochannel primary transmitters, *IEEE Trans. Wireless Commun.* 8(11): 5702–5710.

Qin, X. & Berry, R. A. (2006). Distributed approaches for exploiting multiuser diversity in wireless networks, *IEEE Trans. Inf. Theory* 52(2): 392–413.

Sai Shanka, N. (2008). Overview of blind sensing techniques considered in IEEE 802.22 WRANs, *Sensor, Mesh and Ad Hoc Communications and Networks Workshops (SECON'08)*, pp. 1–4.

Shellhammer, S. J., Shankar, S., Tandra, R. & Tomcik, J. (2006). Performance of power detector sensors of dtv signals in IEEE 802.22 WRANs, *First International Workshop on Technology and Policy for Accessing Spectrum*.

Unnikrishnan, J. & Veeravalli, V. (2008). Cooperative sensing for primary detection in cognitive radio, *IEEE J. Sel. Topics Signal Process.* 2(1): 18–27.

Visotsky, E., Kuffner, S. & Peterson, R. (2005). On collaborative detection of TV transmissions in support of dynamic spectrum sharing, *Proc. DySPAN'05*, pp. 338–345.

Viswanath, P., Tse, D. N. C. & Laroia, R. (2002). Opportunistic beamforming using dumb antennas, *IEEE Trans. Inf. Theory* 48(6): 1277–1294.

Zhao, Q. & Tong, L. (2005). Opportunistic carrier sensing for energy-effcicient information retrieval in sensor networks, *EURASIP Journal on Wireless Communications and Networking* 2005(2): 231–241.

Part 3

Interference Management

7

Partial Response Signaling: A Powerful Tool for Spectral Shaping in Cognitive Radio Systems

Mohammad Mahdi Naghsh and Mohammad Javad Omidi
ECE Department, Isfahan University of Technology, Iran

1. Introduction

The demand for wireless services has increased rapidly in the past years and this trend is expected to continue faster in the future. So far most of the available spectrum resources have already been licensed and there is not much bandwidth left to set up new services [Farhang-Boroujeni et al. 2008]. On the other hand, studies reveal that a large percent of licensed spectra is rarely used. The basic idea of Cognitive Radio (CR) is to allow unlicensed users to use the licensed bandwidth under certain conditions [Weiss & Jondral, 2004; Mitola & Maguire, 1999]. In other words, CR is a new method to satisfy ubiquitous demand for wireless services while there is not enough unlicensed spectrum for use in the developing systems.

In order to utilize licensed bandwidth by unlicensed or secondary users (SUs), certain conditions should be considered. One of the most important issues is that in a CR system, SUs are allowed to transmit and receive data over portions of the licensed spectra when licensed users or primary users (PUs) are inactive. Communication among SUs should not interfere with the primary system. Secondary users have to detect portions of free spectrum to avoid PUs. This can be done through spectrum sensing and then careful considerations are required to limit any transmission in the purposed spectrum [Haykin 2005].

The physical layer (PHY) of a CR system should be able to utilize unoccupied spectra for the desired communication. In fact, it often occurs that pieces of spectra that are not used by PUs have discontinuous nature. The PHY layer of a CR system should be flexible enough to utilize such spectra as efficiently as possible. In essence, the PHY of a CR network should be able to shape the output spectrum dynamically and simply. Therefore, Orthogonal Frequency Division Multiplexing (OFDM) is a main candidate to implement the PHY layer of a CR system due to its ability to bring the required flexibility to the system inherently by simple implementation [Weiss et al. 2004a, b].

OFDM is a Multi-Carrier Modulation (MCM) technique that has been used in many conventional systems such as Wireless Local Area Networks (WLANs), Long Term Evolution systems (LTE), Digital Video Broadcast (DVB), and Digital Audio Broadcast (DAB). This is a robust method for transmitting and receiving data over frequency selective channels. Because of its excellent performance in multipath fading channel and simple implementation using Fast Fourier Transfom (FFT) algorithm, it has been the most popular MCM method.

However, one of the major drawbacks of the OFDM is relatively large Out-of-Band (OOB) components spectrum that may be inconsistent with Power Spectral Density (PSD) specification of WLANs, LTE, DVB, and other OFDM based standards as well as CR networks requirements. In other words, the most important drawback of OFDM based CR systems is the large OOB radiation that originates from the high level side-lobes of Inverse FFT (IFFT) modulated subcarriers. These side-lobes cause unwanted interference among SUs and also between SUs and PUs. Therefore, the OOB radiation of OFDM has been a considerable issue either in conventional applications or in CR networks.

The importance of the OOB components of the OFDM spectrum is high enough for CR application to challenge the application of OFDM based CR systems. This has resulted in more attention towards other MCM methods such as Filter-Bank. These are alternative approaches that have been proposed and considered to be used for CR networks within recent years [Farhang-Boroujeni et al. 2008 & Amini et al. 2005]. Filter-Bank based MCM methods are able to shape the spectrum and reduce the OOB spectrum components of the output signal significantly by imposing large computational load to the system.

Consequently, shaping the spectrum and reduction of the OOB components are the elementary steps for OFDM based CR networks. In this chapter, the use of Partial Response Signaling (PRS), also known as correlative coding, will be investigated in order to shape the spectrum of the OFDM based PHYs. In section 2, we provide a review of the subject and cite the most important works in the field with emphasis on more recent literature. The main focus of this section is the problem of OOB reduction in OFDM PHYs. Section 3 is devoted to description of the PRS including history, definition, theoretical background, properties, and various types of the PRS. In section 4, the analytical PSD of the OFDM signal is calculated for the case where modulated symbols are correlated. Investigation of the spectral shaping of the OFDM signal using PRS is addressed in section 5. This section also discusses block-diagram of the required receiver in the proposed scenario. Simulation results are provided in section 6. The effect of the proposed method on the Peak-to-Average-Power-Ratio (PAPR), a significant issue for OFDM PHYs, and word error rate (WER) is also investigated. In addition this section contains application of the proposed method for spectral shaping in CR networks. Finally, summary and conclusions along with proposed topics for future research are presented in section 7.

2. State-of-the-art

It is inferred from the nature of CR systems that their PHY system should possess not only high flexibility in spectral shaping but also they need to provide a spectrum with very low OOB components. On the other hand, as it is stated in the previous section, existence of the relatively large OOB components in the OFDM spectrum is one the most serious shortcomings of this MCM technique especially for CR networks. In this section a brief review of existing methods for spectral shaping and reduction of OOB components in OFDM spectrum will be introduced.

The problem is illustrated in figure 1. This figure shows the available bandwidth and OOB components for an OFDM PHY with 3 modulated subcarriers. The OOB components are fairly large and cannot be tolerated in most applications.

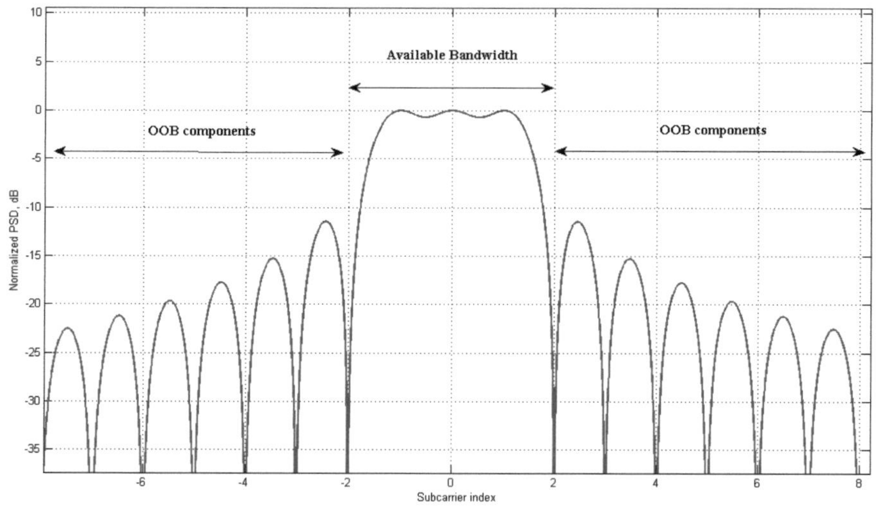

Fig. 1. Illustration of the OOB components in the OFDM spectrum, 3 modulated subcarriers

In the following, an overview of the previously suggested methods for spectral shaping and OOB reduction in OFDM systems is given. In [Weiss et al. 2004] and several standard documents, the insertion of guard bands at the border of OFDM spectrum has been proposed [IEEE 802.16, 2004; IEEE 802.11, 1999]. The drawback of this approach is the less effective use of the bandwidth. Another common method is time windowing at the transmitter. In this method, the time domain signal is multiplied by a proper window to smooth transition between consecutive OFDM blocks [Nee & Prasad, 2000]. Various window functions such as raised-cosine (RC), hamming, hanning, and so on can be used for this purpose. In this approach the effective time duration of transmitted signal is extended and inter-block interference is introduced [Nee & Prasad, 2000; EN 301 958, 2001]. In [Xu et al. 2009], a new class of window functions has been derived for OOB reduction but there is no investigation on the effect of the proposed method on the system performance.

In [Yu et al. 2010], an alternative windowing method has been suggested for OFDM based CRs where time windowing is applied to the entire OFDM symbol before addition of the guard interval. Although this method introduces a controlled amount of Inter-carrier-interference in the system, it does not waste time-frequency resource as opposed to conventional windowing methods.

A different technique is dual to time windowing and uses pulse shaping filter. In fact, each subcarrier is multiplied by a pulse shaping function which is equal to the convolution of time domain transmitted signal with the impulse response of pulse shaping filter. This approach suffers from high complexity and lack of guard interval [Naghsh & Omidi, 2010]. It should be noted that guard interval plays an important role in avoiding inter-block interference between OFDM blocks. Besides, guard interval not only extends the time duration of transmitted signal and reduces side-lobe levels but also increases robustness of system against synchronization errors [Phillip, 2001].

In [Brandes, 2006], insertion of the cancellation carrier at the edge of available bandwidth has been proposed. Proper values for cancellation carriers in each OFDM block have been obtained by solving a convex optimization problem. This method reduces OOB components considerably but has high complexity, and increases PAPR and WER. In [Cosovic, 2006], the modulated symbols in each OFDM block have been multiplied by proper sequence to reduce OOB components. Proper sequence can be found by solving a convex optimization problem. In addition to high complexity, this method results in WER loss.

In [Noreen & Azimi, 2010], a technique for OOB reduction in OFDM-based CRs has been suggested that maps groups of two input symbols onto extended constellation such that subcarriers in each group become 180 degrees out of phase. The mentioned technique does not require side information but degrades both PAPR and WER.

In the adaptive symbol transition method, the transition between consecutive OFDM blocks is smoothed adaptively. Multiple choice sequences method performs a mapping of each transmission sequence into a specific set of sequences. From this set, the sequence which offers the maximum reduction of out-of-band radiation is chosen for the actual transmission. These two approaches are discussed thoroughly in [Mahmoud & Arsalan, 2008a; Cosovic & Mazzoni, 2006] respectively. Note that in [Ghassemi et al. 2010], a generalization of the multiple choice sequences method has been proposed that reduces OOB and PAPR jointly.

Some references propose combination of the existing methods. For example, in [Mahmoud & Arsalan, 2008b] combination of the cancellation carrier insertion and RC windowing has been suggested. It has been shown that this method works well for small gaps in the spectrum in the cost of lower spectral efficiency and computational load. This reference also contains a review of the most important techniques for spectrum shaping in OFDM based CRs.

In [Yuan & Wyglinski, 2009, 2010], combination of the cancellation carrier insertion and frequency filtering has been proposed for side-lobe reduction of OFDM spectrum in CR PHYs. This method can reduce OOB components considerably in the cost of lower spectral efficiency, higher PAPR level, and higher WER.

In [Sokhandan & Safavi, 2010], a new method has been suggested that may be thoughtas a generalization of the combination of the cancellation carrier insertion and adaptive symbol transition. In essence, this method optimizes the value of cancellation carriers and time extension of the OFDM symbol jointly. The mentioned technique requires high computational load since it should solve an optimization problem for each OFDM symbol. Also, it does not affect the PAPR level and there is no report about its effect on the WER.

Finally we cite the work of [Zhou et al. 2011] that suggests mapping of the antipodal symbol pairs onto adjacent subcarriers to achieve a faster decay of OOB components in OFDM-based CR systems. It also proposes the use of power control scheme and channel coding with different rate for the balance between further reduction of the side-lobes level and system performance.

In this chapter, we propose a novel method to achieve side-lobe suppression. Our proposed method is to introduce proper carrier-by-carrier PRS for spectral shaping of the OFDM signal [Naghsh & Omidi, 2010]. In this approach a controlled amount of correlation is introduced among modulated symbols on each subcarrier in consecutive OFDM blocks. In this method the effective time duration and bandwidth of transmitted signal will remain unchanged. Also, guard interval can be used as before [Naghsh & Omidi, 2010].

It should be noted that the term "partial response OFDM" that has been used in [Vadde, 2001; Vadde & Gray, 2001, Kim & Km, 2005; Syed-yusof et al. 2006] refers to the introduced PRS between modulated symbols on the subcarriers in an OFDM block in order to decrease Inter-carrier-interference due to the receiver phase noise, PAPR reduction, etc. Therefore, stated methods are fundamentally different from our approach since our proposed method is developed for spectral shaping and OOB reduction by introducing correlation between consecutive OFDM blocks across the time (refer to section 5).

3. Partial response signalling

Partial response signalling (PRS), also known as correlative coding, was introduced for the first time in 1960s for high data rate communication [lender, 1960]. From a practical point of view, the background of this technique is related to the Nyquist criterion. Therefore, we present a brief review of this concept.

Assume a Pulse Amplitude Modulation (PAM), according to the Nyquist criterion, the highest possible transmission rate without Inter-symbol-interference (ISI) at the receiver over a channel with a bandwidth of W (Hz) is 2W symbols/sec. To achieve this rate, the transfer function of the overall system, i.e., $H(f)$, should satisfy $\sum_{m=-\infty}^{\infty} H(f-2mW) = cte$. The only transfer function that satisfies the Nyquist criterion over a W (Hz) bandwidth is

$$H(f) = \begin{cases} 1, & |f| \le W \\ 0, & \text{otherwise} \end{cases}$$

H(f) is an ideal Low Pass Filter (LPF) and is not practical for implementation. Therefore, in practice, other H(f) functions, i.e., $\hat{H}(f)$, that satisfy Nyquist criterion over a wider bandwidth have to be used. It can be shown that a wider bandwidth for $\hat{H}(f)$ in comparison with H(f) results in smoother transition of frequency response in cut off regions. For example, the required bandwidth for $\hat{H}(f)$ with RC shape[1] is $(1+\alpha)W$ where α is the roll-of-factor that controls the shape and bandwidth of the used transfer function. Note that we have $0 < \alpha \le 1$ and larger α means more smoothed frequency shaping.

The underlying assumption here is that the modulated data over consecutive pulses are independent while PRS relaxes this assumption and allows for a controlled amount of correlation in the system. Although this controlled correlation leads to ISI, it can be easily compensated for at the receiver side as it has a known pattern. Figure 2 shows the block diagram of Doubinary system which is commonly used for PRS.

Assume that $\{d_k\}$ denotes the sequence of modulated symbols with rate of 2W symbols/sec. The controlled correlation is introduced to the system by a digital filter adding two independent successive symbols, i.e., $a_k = d_k + d_{k-1}$ (This scheme is usually denoted by

[1] Raised Cosine Shape.

1+D where *D* is the T second delay operator[2]). Then, the sequence of $\{a_k\}$ is passed through an ideal Nyquist transfer function which guarantees no additional ISI at the receiver. Finally, at the receiver, removing of the controlled correlation and symbol detection is performed by means of a maximum likelihood sequence detector (MLSD) algorithm such as the Viterbi algorithm getting $\{z_k\}$ as its input. In some special cases, symbol detection can be implemented very easily as will be shown in section 6.

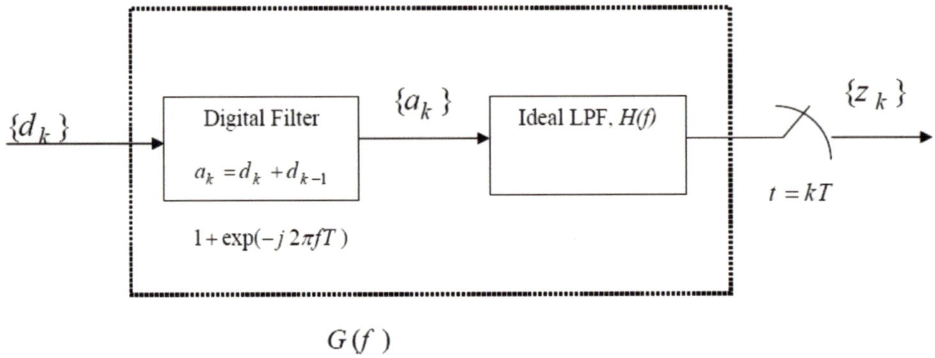

Fig. 2. Block diagram of Doubinary system

It is worth noting that the overall transfer function due to the cascade of the digital filter and H(f) is $G(f) = 2H(f)\cos(\pi fT)\exp(j\pi fT)$

As it is observable in figure 3, G(f) has a gradual roll-off to the band edge in comparison with H(f) and hence is practical for implementation.

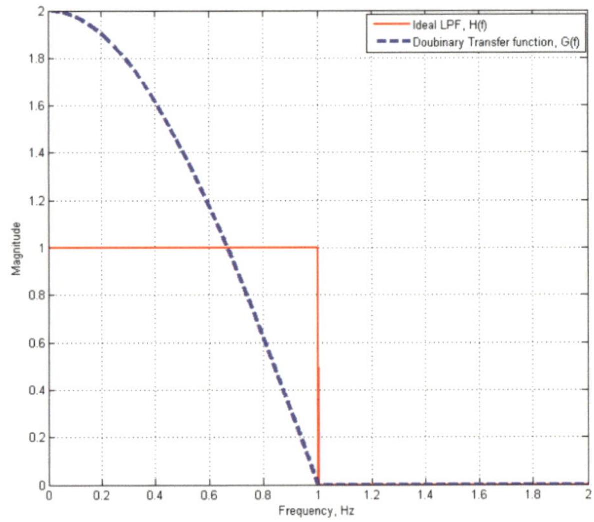

Fig. 3. Magnitude of G(f) and H(f), W=1(Hz)

---

[2] Note that $T \triangleq 1/2W$

The effect of the PRS on the error rate of the system is investigated in section 6. It should be noted that nine patterns of more common PRS schemes are introduced in [Passupathy, 1977; Kabal & Passupathy 1975]. It has been shown that all of those schemes require 2M-1 or 4M-1 level slicer in the receiver for M-array PAM. In addition, it has been shown that using PRS brings out more synchronization robustness for the system.

4. Analytical PSD of the OFDM signal

In this section, the PSD of OFDM signal will be drawn analytically. In essence, the OOB components are considered as PSD components that are out of the permitted bandwidth as it is illustrated in figure 1. Basically the finite symbol duration of OFDM symbol results in these OOB components, however, other reasons such as phase transition between successive blocks may contribute to this phenomenon.

In the OFDM, the baseband signal can be represented as

$$x(t) = \sum_{l=1}^{N} \sum_{k=-\infty}^{\infty} d_{k,\,l}\, w(t - kT_s)\exp(j2\pi f_l(t - kT_s - T_{GI})) \tag{1}$$

where l and k are the subcarrier and time indexes, complex number $d_{k,\,l}$ is the modulated symbol on l-th subcarrier at k-th time interval, $w(t)$ is the window function, e.g., rectangular, RC, hamming, etc., f_l is the l-th subcarrier frequency that equals to $1/T_{FFT}$, T_{FFT} is the pure OFDM block time duration, T_s is the total time duration of an OFDM block, T_{GI} is the guard interval (GI) duration, and finally N is the number of subcarriers. Clearly, we have $T_s = T_{FFT} + T_{GI}$.

By defining

$$g_l(t) = \sum_{k=-\infty}^{\infty} d_{k,\,l} w(t - kT_s)\exp(j2\pi f_l(-T_{GI} - kT_s)) \tag{2}$$

the baseband signal can be written as

$$x(t) = \sum_{l=1}^{N} g_l(t)\exp(j2\pi l\Delta f\, t) \tag{3}$$

where $\Delta f = 1/T_{FFT}$.

Generally, a linear filter may be used to implement PRS for the system and make intentional correlation of length L among modulated symbols on each subcarrier as shown in Figure 4.

Indeed, each subcarrier is treated by such filters individually and a carrier-by-carrier PRS scheme is organized as shown in figure 5. Then we can write

$$a_{k,\,l} = \sum_{n=0}^{L} \alpha_n^l\, d_{k-n,\,l} \tag{4}$$

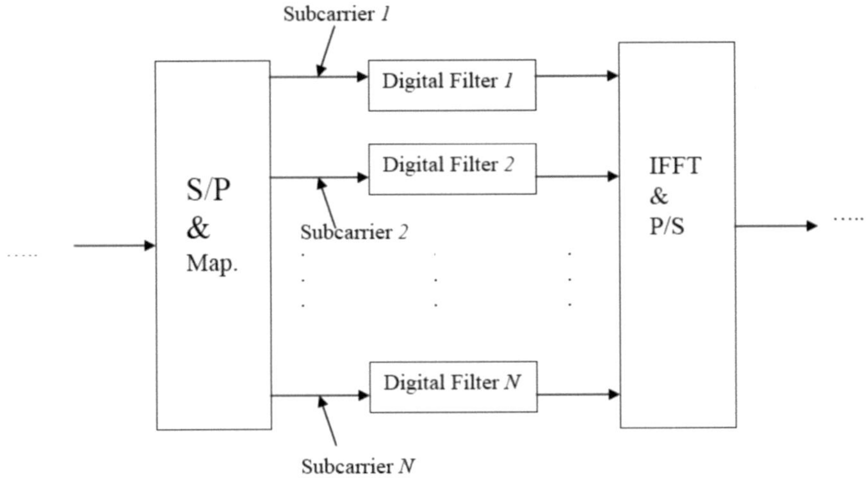

Fig. 4. Implementation of PRS in OFDM Carriers using N digital filters [Naghsh & Omidi, 2010]

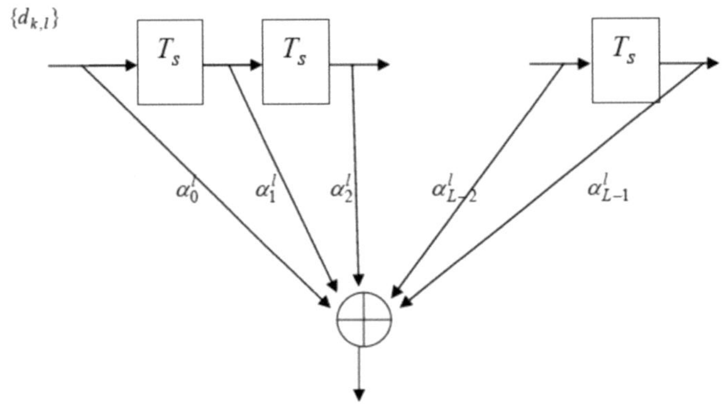

$$a_{k,\ l} = \sum_{n=0}^{L} \alpha^{l}_{n} . d_{k-n,\ l}$$

Fig. 5. PRS for *l*-th subcarrier in general case [Naghsh & Omidi, 2010]

where $\left\{\alpha^{l}_{n}\right\}$ are the linear filter coefficients for *l*-th subcarrier. Then, the baseband signal becomes

$$x(t) = \sum_{l=1}^{N} b_l(t) \exp(j2\pi l \Delta f\, t) \tag{5}$$

where $b_l(t)$ is defined as

$$b_l(t) = \sum_{k=-\infty}^{\infty} a_{k,\ l} w(t - kT_s) \exp(j2\pi f_l(-T_{GI} - kT_s)) \tag{6}$$

In [Proakis, 2007], it has been shown that the PSD of the random process $b_l(t)$ is given by

$$B_l(f)=\frac{1}{T_s}|W(f)|^2\left|\sum_{n=0}^{L}\alpha_n^l \exp(-j2\pi f nT_s)\right|^2\left[\sum_{m=-\infty}^{\infty}E\left\{d_{k,l}\,d_{k+m,l}^*\right\}\exp(-j2\pi f mT_s)\right] \tag{7}$$

where $W(f)$ is the Fourier transform of the window function, i.e., $w(t)$ and *E* denotes the expected value operator.

Under the assumption of uncorrelated data sequence we have

$$E\left\{d_{k,l}\,d_{k+m,l}^*\right\}=\delta(m)$$

and hence $B_l(f)$ can be simplified to

$$B_l(f)=\frac{1}{T_s}|W(f)|^2\left|\sum_{n=0}^{L}\alpha_n^l \exp(-j2\pi f nT_s)\right|^2 \tag{8}$$

In the following, we assume that data sequence is uncorrelated since it is the output of the source coder or scrambler. Also, in the rest of this chapter we assume that the effect of the GI can be neglected, i.e., $T_{FFT} \gg T_{GI}$. Therefore, as various subcarriers carry uncorrelated modulated symbols at *k*-th block for any *k*, the PSD of OFDM signal becomes [3]

$$X(f)=\sum_{l=1}^{N}B_l(f-l\Delta f) \tag{9}$$

It is realized from (8) and (9) that the PSD of OFDM signal depends mainly on the window function, number of subcarriers, and $\left\{\alpha_n^l\right\}$, the coefficients of linear filter used to introduce PRS on *l*-th carrier. In this chapter, we use these coefficients to shape the spectrum and reduce the OOB components.

If the common RC window is used, *w(t)* can be written as

$$w(t)=\begin{cases}\sin^2\left(\frac{\pi}{2}\left(5+\frac{t}{T_R}\right)\right) & \text{for } \frac{-T_R}{2}\le t\le\frac{T_R}{2}\\ 1 & \text{for } \frac{T_R}{2}\le t\le T_s-\frac{T_R}{2}\\ \sin^2\left(\frac{\pi}{2}\left(5+\frac{t-T_s}{T_R}\right)\right) & \text{for } T_s-\frac{T_R}{2}\le t\le T_s+\frac{T_R}{2}\end{cases} \tag{10}$$

where T_R is the transition time and $W(f)$ can be expressed as [Naghsh & Omidi, 2010]

---

[3]without loss of generality it is assumed that $E\left\{|d_{k,l}|^2\right\}=1$

$$W(f) = T_s \operatorname{sinc}(T_s f) \frac{\cos(\pi T_R f)}{1 - 4T_R^2 f^2} \exp(-j\pi T_s f) \tag{11}$$

by choosing T_R, various windows can be synthesized, e.g., choosing $T_R = 0$ results in rectangular window. The effect of the increasing of T_R is shown in Figure 6. According to this figure, higher T_R leads to lower side-lobes for $W(f)$. It should be noted that lower side-lobes level of $W(f)$ reduces OOB components of the OFDM spectrum according to (8) and (9). This also results in higher effective time duration of the window as it can be inferred from (10).

In the following, it is assumed that the rectangular window is used. Therefore, from (8), (9) and (11), $X(f)$ becomes

$$X(f) = T_s \sum_{l=1}^{N} \left[\left| \operatorname{sinc}(T_s[f - l\Delta f]) \right|^2 \left| \sum_{n=0}^{L} \alpha_n^l \exp(-j2\pi[f - l\Delta f]nT_s) \right|^2 \right] \tag{12}$$

To obtain conventional OFDM signal PSD from (12), assuming there is no PRS ($L = 0,\ \alpha_0^l = 1, \forall l$), yields

$$X(f) = T_s \sum_{l=1}^{N} \operatorname{sinc}^2(T_s[f - l\Delta f]). \tag{13}$$

In the next section, we investigate how a proper choice of $\left\{ \alpha_n^l \right\}$ can shape the spectrum and result in lower OOB in OFDM spectrum.

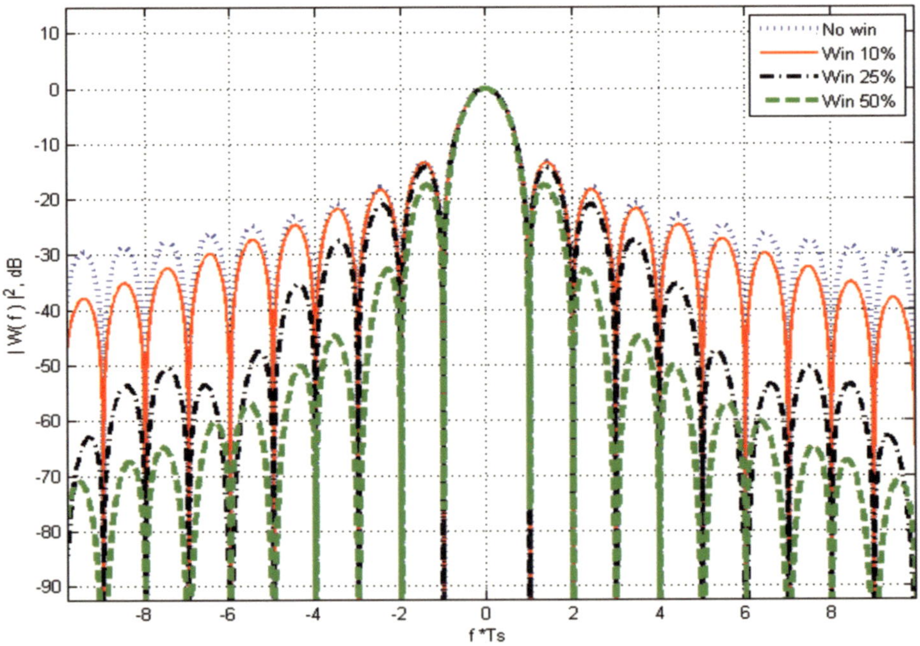

Fig. 6. The effect of the T_R on the Fourier transform of the window

5. OFDM spectral shaping using PRS

In this section the advantage of introducing proper PRS between modulated symbols on each subcarrier, or proper carrier-by-carrier partial response signaling, is investigated. Considering (8) and (9), proper choice of this controlling term

$$\left|\sum_{n=0}^{L} \alpha_n^l \exp(-j2\pi[f - l\Delta f]nT_s)\right|^2 \quad (14)$$

in (12) can be used for shaping each subcarrier's spectrum separately, and hence overall PSD shaping will be possible for OOB component reduction. As mentioned in the previous section, assume that the data sequence, $d_{k,l}$, is uncorrelated. Otherwise, each subcarrier's spectrum is expressed as (7) and our discussion still will be valid.

Now, we rewrite the PSD of OFDM signal as

$$X(f) = T_s \sum_{l=1}^{N} \left[|\text{sinc}(T_s[f - l\Delta f])|^2 \left|\sum_{n=0}^{L} \alpha_n^l \exp(-j2\pi[f - l\Delta f]nT_s)\right|^2 \right] \quad (15)$$

Figure7 shows schematic of the carrier-by-carrier PRS method and intentionally introduced correlation between modulated symbols on each subcarrier across the time.

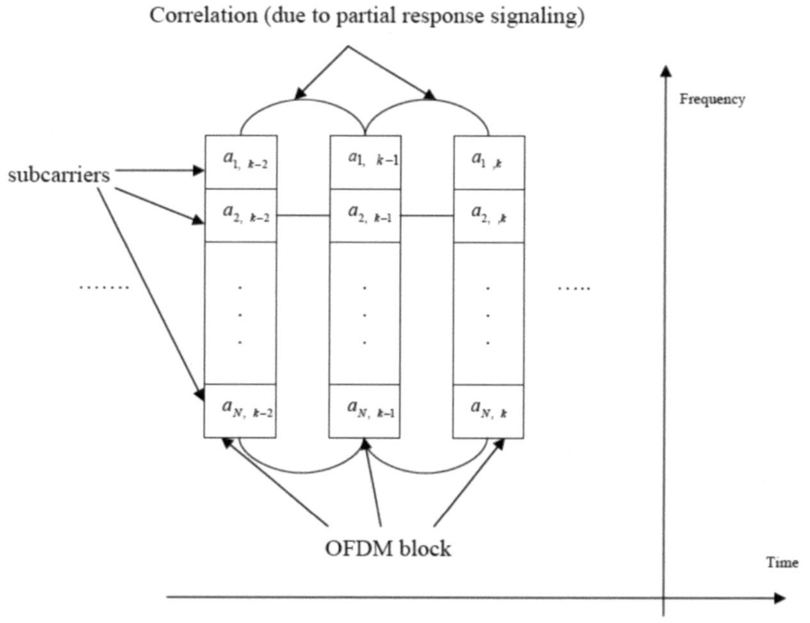

Fig. 7. Schematic of the carrier-by-carrier PRS method [Naghsh & Omidi, 2010]

Considering (15), in general, each subcarrier may be processed based on its own PRS pattern depending on α_n^l. In this case, more degrees of freedom are available and better spectral performance is achievable. But, it results in high complexity because each subcarrier must be treated individually for symbol modulation and detection. In the following, assume that all subcarriers have the same PRS pattern. It means

$$\alpha_n^l = \alpha_n \quad l = 1,2,...,N \tag{16}$$

Then, according to (15), PSD of OFDM signal becomes

$$X(f) = T_s \sum_{l=1}^{N} \left[\left|\operatorname{sinc}(T_s[f - l\Delta f])\right|^2 \left| \sum_{n=0}^{L} \alpha_n \exp(-j2\pi[f - l\Delta f]nT_s) \right|^2 \right] \tag{17}$$

In (17), the PRS pattern should be selected for shaping the PSD and specially OOB components reduction. Clearly, large L gives more degrees of freedom, but leads to more states for receiver slicer. Hence L is dictated to the transmitter based on acceptable complexity for the receiver.

In general, in the receiver, the effect of PRS on each subcarrier symbol could be removed by means of an MLSD. In other words, this controlled correlation might be assumed as known channel impulse response and thus detection could be implemented by MLSD methods such as Viterbi algorithm.

It should be noted that not every PRS scheme can improve OOB radiation and thus in this context 'proper PRS' term has been used. In fact, the controlling term in (14) is the discrete Fourier transform of the PRS sequence $\left\{\alpha_n^l\right\}$, regardless of $l\Delta f$. Therefore, it is reasonable that sequences with low-pass frequency characteristic are selected to reduce OOB components; because the PSD of conventional OFDM system is multiplied by the Fourier transform of the chosen sequence according to (17). In the next section, we introduce two common examples of these proper schemes.

Although symbols with proper PRS pattern will result in lower OOB radiation compared to the case of no PRS, but it will increase the number of equivalent signal space components, which leads to increased WER because of more neighborhoods for each component. In addition, in the proposed method, error propagation may occur because of transmitting symbols by PRS (some controlled correlation); fortunately, these effects can be reduced by pre-coding schemes [Passupathy, 1977; Kabal & Passupathy 1975]. The general form of pre-coding for M-array modulation used in PRS with integer coefficient could be written as

$$\alpha_0 d_{k,l} = c_{k,l} - \sum_{i=1}^{L} \alpha_i d_{k-i,l} \mod M \tag{18}$$

where $c_{k,l}$ is the original uncorrelated sequence and $d_{k,l}$ is the pre-coded sequence. The necessary and sufficient condition for uniquely retrieving $c_{k,l}$ from $d_{k,l}$ is that α_0 and M be relatively prime [Kobayashi, 1971]. In this condition

$$c_{k,l} = \sum_{i=0}^{L} \alpha_i d_{k-i,l} \mod M \tag{19}$$

For example in the Duobinary system, pre-coding can be implemented as

$$d_{k,\,l} = c_{k,\,l} \oplus d_{k-1,\,l} \tag{20}$$

where $\oplus$ denotes XOR. It can be inferred from (18) and (19) that pre-coding scheme for non-binary cases with " $\bmod\ M$ "calculations imposes more complexity on transceiver. In the binary case pre-coding could be easily implemented using logical devices, even for large *L*.

Based on the above discussion, the total block diagram of the transmitter and receiver can be illustrated as in figure 8 and figure 9. Note that S/P, P/S, D/A, and A/D denote serial-to-parallel, parallel-to-serial, digital-to-analog, and analog-to-digital converter, respectively.

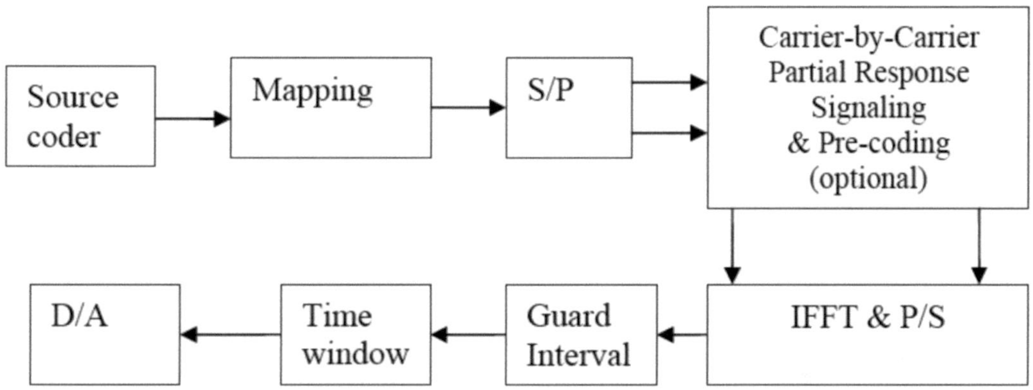

Fig. 8. Transmitter block diagram of the carrier-by-carrier PRS OFDM [Naghsh & Omidi, 2010]

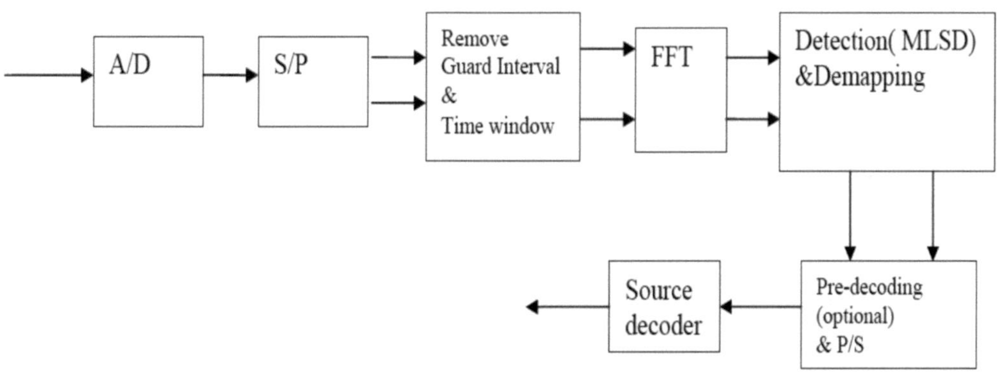

Fig. 9. Receiver block diagram of the carrier-by-carrier PRS OFDM [Naghsh & Omidi, 2010]

6. Numerical results

6.1 Basic results

In this section we assume an OFDM system with *N=64* subcarriers, similar to IEEE 802.11 standard, which are modulated using BPSK modulation and we will use 1000 consecutive blocks for simulation. Also, it is assumed that permitted normalized bandwidth is in the [0.2, 1.8] interval and 10% of the subcarriers are set to zero at the bordering of the spectrum. It should be noted that for fair comparison, PSDs are normalized such that the total transmitted power in the permitted bandwidth remains constant.

Figure 10 shows the resulting PSD of the proposed method in comparison with conventional OFDM where *L=1*. Note that all of the subcarriers have the same PRS pattern and values of α_0, α_1 have been selected based on well-known Duobinary system which has low-pass frequency characteristic [Passupathy, 1977; Kabal & Passupathy 1975; Kretzmer, 1966], $\alpha_0 = 1, \alpha_1 = 1$. In this case, the transmitted symbols are $\{-2, 0, 2\}$ and analytic PSD using (17) becomes

$$\begin{aligned} X_{\text{Duobinary}}(f) &= T_s \sum_{l=1}^{N} \left[\left|\text{sinc}(T_s[f - l\Delta f])\right|^2 \left|1 + \exp(-j2\pi T_s[f - l\Delta f])\right|^2 \right] \\ &= T_s \sum_{l=1}^{N} \left[\left|\text{sinc}(T_s[f - l\Delta f])\right|^2 (2 + 2\cos(2\pi T_s[f - l\Delta f])) \right] \end{aligned} \quad (21)$$

If no pre-coding is used in the transmitter, the receiver should detect the modulated symbols on each subcarrier by means of MLSD after performing FFT. But, by using the pre-coding of (20), receiver can be implemented easily based on the absolute value of received noisy symbol after performing FFT [Naghsh & Omidi, 2010]

$$\begin{cases} c_{k,l} = 0 & \text{if} \quad y_{k,l} = \pm 2 \\ c_{k,l} = 1 & \text{if} \quad y_{k,l} = 0 \end{cases} \Rightarrow \begin{cases} c_{k,l} = 0 & \text{if} \quad |y_{k,l}| \geq 1 \\ c_{k,l} = 1 & \text{if} \quad |y_{k,l}| \leq 1 \end{cases} \quad (22)$$

Reduced OOB components in Duobinary carrier-by-carrier PRS OFDM are observable in figure 10. Explicit ripples in the available bandwidth in figure 10 are due to introduced inter-block-interference (because of PRS) similar to frequency selectivity at multipath channel [Naghsh & Omidi, 2010].

Now, let *L=2* and choose a class-2 PRS which is another low-pass frequency characteristic sequence. In this case the coefficients are $\{\alpha_0 = 1,\ \alpha_1 = 2,\ \alpha_2 = 1\}$ and five levels will be produced as $\{-4, -2, 0, 2, 4\}$ [Passupathy, 1977; Kabal & Passupathy 1975; Kretzmer, 1966]. A typical receiver in this case will use MLSD after performing FFT. In this case, based on (17), analytic PSD expression can be written as

$$\begin{aligned} X_{\text{class}-2}(f) = T_s \sum_{l=1}^{N} \Big[& \left|\text{sinc}(T_s[f - l\Delta f])\right|^2 \\ & \times \left|1 + 2\exp(-j2\pi T_s[f - l\Delta f]) + \exp(-j4\pi T_s[f - l\Delta f])\right|^2 \Big] \end{aligned} \quad (23)$$

By using some trigonometric equality, the PSD becomes

$$X_{class-2}(f) = T_s \sum_{l=1}^{N} \Big[|\text{sinc}(T_s[f - l\Delta f])|^2 \times (6 + 8\cos(2\pi T_s[f - l\Delta f]) + 2\cos(4\pi T_s[f - l\Delta f])) \Big] \tag{24}$$

In figure 11, comparison of the PSD of a subcarrier and overall modulated signal for conventional OFDM, Duobinary carrier-by-carrier PRS OFDM, and a class-2 carrier-by-carrier PRS OFDM are illustrated. It is clear from this figure that larger *L* results in lower OOB components while it imposes more complexity. Also, as we expect, more ripples exist for larger *L* because of the existence of more (controlled) correlation in the system.

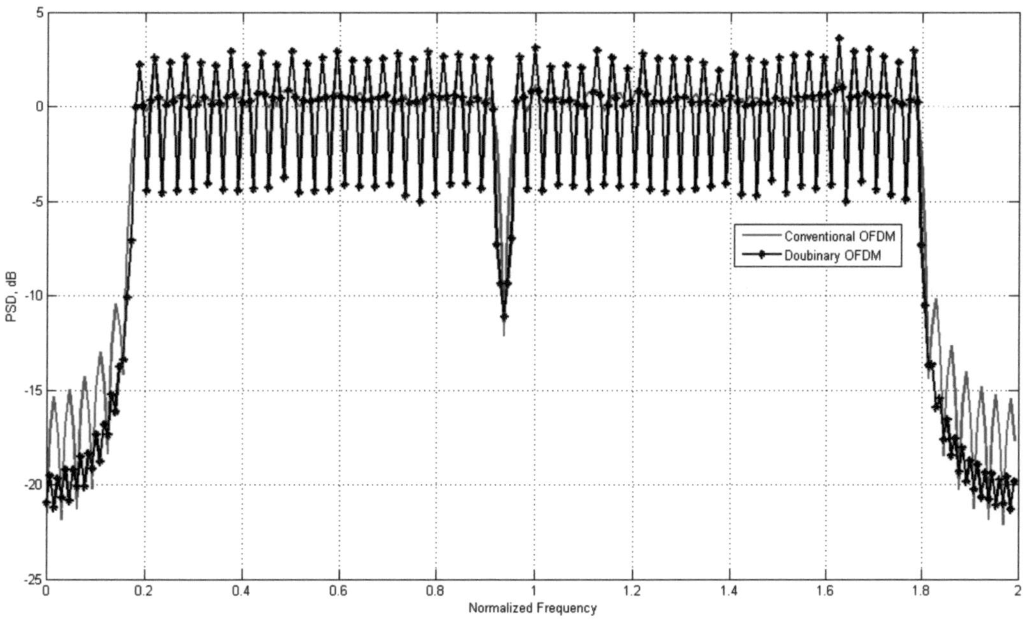

Fig. 10. PSD comparison of conventional and Duobinary carrier-by-carrier PRS OFDM for 802.11a

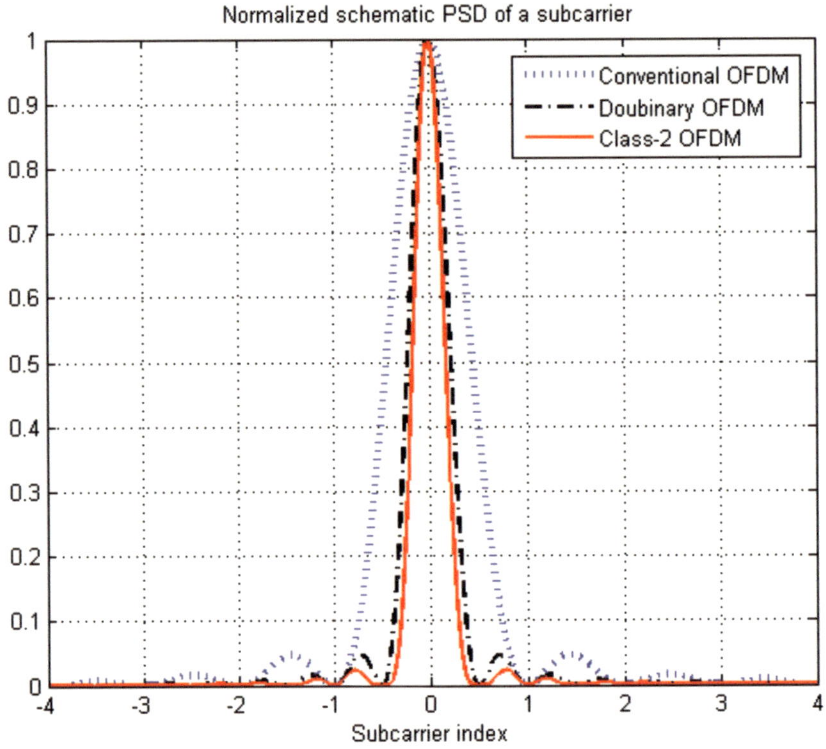

(a)

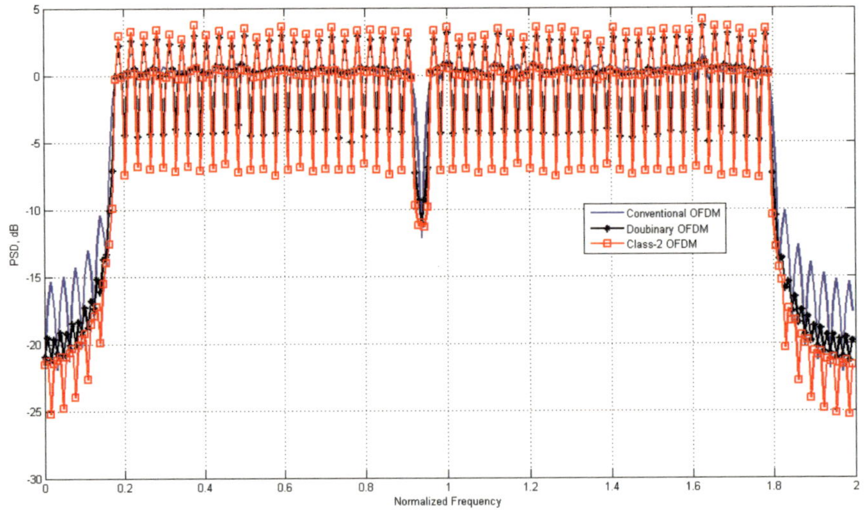

(b)

Fig. 11. a) A subcarrier, b) Overall PSD; comparison of conventional, Duobinary, and a class-2 carrier-by-carrier PRS OFDM

6.2 Comparison with other methods

As mentioned in the introduction, there are several OOB reduction methods. Some of them such as cancellation carrier insertion have high complexity and so they are not suitable for practical applications. On the other hand, some methods such as time windowing, insertion of more guard band at the border of spectrum, and frequency filtering are very common in various applications and they could be implemented with acceptable complexity. Therefore, we compare our proposed method to the latter. The results are presented in figure 12 and Table 1. In time windowing, two RC windows are used with T_R / T_s equal to *6%* and *11%* respectively. According to Figure 12 and Table 1, this method has failed to reduce the first peak of OOB components. In addition, larger T_R / T_s will result in better OOB reduction while producing more inter-block interference and leads to less spectral efficiency. In the frequency filtering approach, the OOB components are filtered by an FIR filter with length 9, about 15% of an OFDM block, in the transmitter. Similar to time windowing, this method has failed to reduce the first peak of OOB components effectively. It should be noted that although better OOB reduction is possible by frequency filtering method, it requires long FIR filter that not only imposes more complexity to the system but also extends blocks in time domain resulting in more inter-block interference. In the last method, about *20%* of subcarriers are set to zero at the border of the spectrum. Note that for all of the previous methods mentioned value was 10%. Therefore, the available bandwidth is not used efficiently[4]. Figure 12, and Table 1 show that our proposed method can reduce the main side-lobe peak more effectively in comparison with other methods while windowing and frequency filtering can reduce average OOB more effectively.

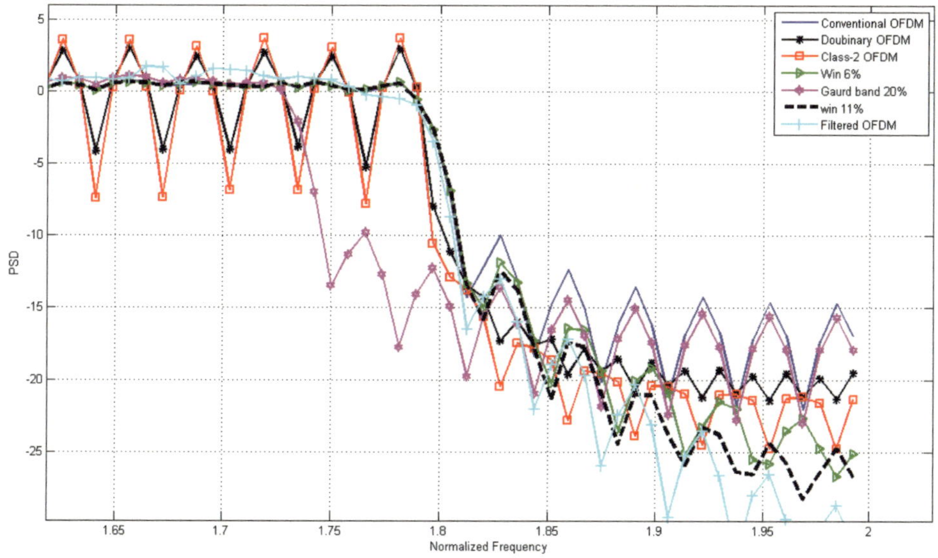

Fig. 12. PSD comparison of conventional , windowed, filtered, and guard band inserted OFDM with the proposed method

[4] In this simulation, the total transmitted power of the Guard insertion method is less than that of other methods.

Method	Peak OOB, dB	Mean OOB, dB
Conventional OFDM	*-9.98*	*-17.11*
Duobinary OFDM	*-13.75*	*-19.84*
Class-2 OFDM	*-13.78*	*-21.48*
Time window OFDM (6%)	*-11.92*	*-22.02*
Time window OFDM(11%)	*-12.46*	*-23.92*
Frequency filtered OFDM	*-13.04*	*-27.86*
Guard insertion (20%)	*-13.61*	*-18.51*

Table 1. Comparison of peak and mean values of OOB components for investigated methods

6.3 PAPR and WER analysis

Now we examine our proposed method, carrier-by-carrier PRS OFDM, in the view of PAPR as a challenge for many OFDM systems and also WER. Assume that in each OFDM block the carrying symbols, $a_{k,\,l}$, are independent and identically distributed, hence

$$E\left\{ a_{k,\,l} a^{*}_{k,\,l+n} \right\} = \delta(n) \tag{25}$$

Hence, after performing IFFT, the assumption of independency is still valid due to the orthonormality of IFFT basis [Vadde, 2001; Baxley, 2005]. Similar to conventional OFDM systems, for large *N* the Complementary Cumulative Distribution Function (CCDF) of PAPR can be written as

$$\text{prob}\{\text{PAPR} \geq z\} = 1-(1-e^{z})^{N} \tag{26}$$

Therefore, we expect that carrier-by-carrier PRS OFDM does not lead to greater PAPR compared to conventional OFDM systems.

Figure 13 shows the CCDF of PAPR for conventional OFDM, Duobinary, and a class-2 carrier-by-carrier PRS OFDM system. For the MonteCarlo simulation 10^4 OFDM blocks have been tested. This figure indicates that carrier-by-carrier PRS between consecutive OFDM blocks does not trade off higher PAPR for the better OOB reduction.

Now we investigate our proposed method in the view of WER. As mentioned in section 5, the introduced correlation between subsequent symbols on each subcarrier will cause error propagation and higher WER in the system. In [Passupathy, 1977; Kabal & Passupathy 1975], the lower band and upper band of the WER in AWGN channel for single carrier PRS systems have been calculated as follows

$$P_{e,lower} \leq P_e \leq \frac{M^{L-1} P_{e,lower}}{\frac{M}{M-1} P_{e,lower}(M^{L-1}-1)+1} \tag{27}$$

where

$$P_{e,lower} = 2\left(1-\frac{1}{M}\right)Q(\alpha_0 / \sigma) \tag{28}$$

and P_e is the WER, M is the modulation alphabet size, σ^2 is the variance of the Gaussian noise, and Q(x) is defined as

$$Q(x) = \frac{1}{\sqrt{2\pi}} \int_{x}^{\infty} \exp(-u^2 / 2) du \quad (29)$$

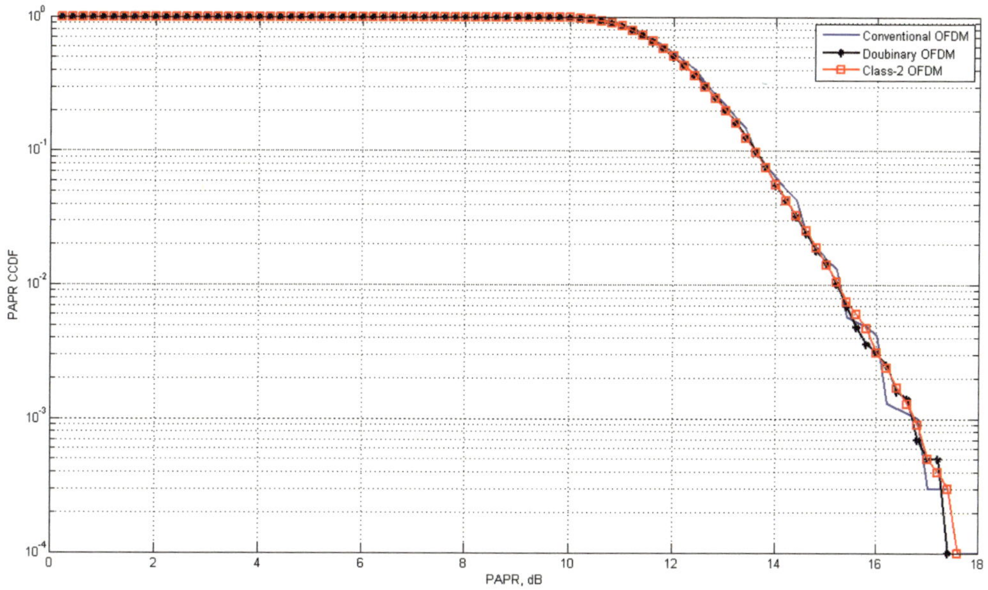

Fig. 13. The CCDF of PAPR for conventional, Duobinary, and a class-2 carrier-by-carrier PRS OFDM

Although (27) and (28) have been drawn under the implicit assumption of decision feedback detector, they can be useful for overall WER behavior of the system.

In [Kobayashi, 1971], for single carrier systems, the exact WER has been calculated for MLSD in Doubinary case for both non pre-coded data sequence and pre-coded data sequence according to (18). If a real M-array modulation is used, we will have

$$P_{e,MLSD} = 2M(M-1)Q(d) \quad (30)$$

where $P_{e,MLSD}$ is WER in non pre-coded data sequence system, $d^2 = A^2 / 2\sigma^2$, and A is the smallest distance between modulated symbols in the constellation. Similarly, the WER in the pre-coded data sequence, P_{e,p_MLSD}, becomes

$$P_{e,p\_MLSD} = 4(M-1)Q(d) \quad (31)$$

It could be realized from (30) and (31) that $P_{e,p\_MLSD} \le P_{e,MLSD}$, and the equality holds only for $M = 2$. It means that for $M > 2$, pre-coding prevents error propagation and hence it improves WER performance. Although in binary modulation, $M = 2$, pre-coding yields no WER gain, it results in very simple symbol detection according to (22).

In figure 14, the simulated WER for the same OFDM system of previous section is shown. It is reasonable that larger L would result in higher WER because of more correlation between symbols. Table 2 shows the approximated SNR loss due to PRS in comparison with conventional OFDM system for the two investigated cases.

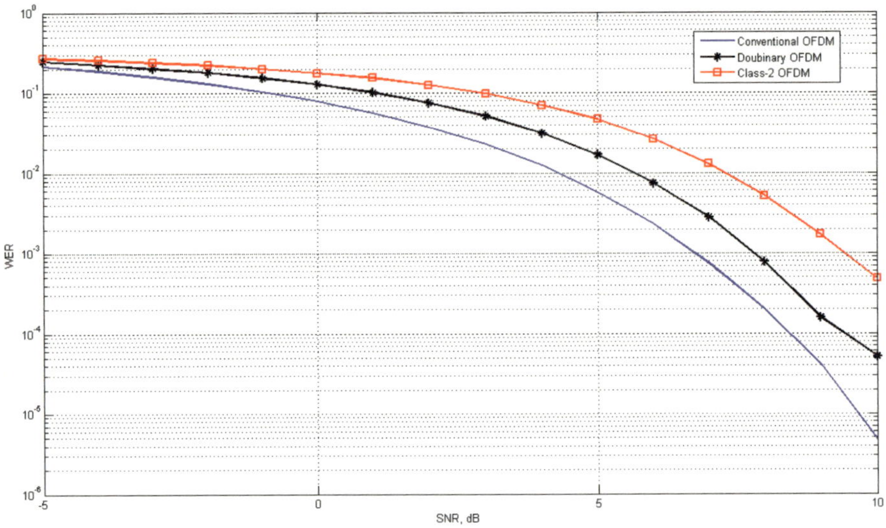

Fig. 14. WER comparison for conventional, Duobinary, and a class-2 carrier-by-carrier PRS OFDM

Method	Approximated SNR loss @ WER= 10^{-3}
Duobinary OFDM	0.9dB
Class-2 OFDM	1.7dB

Table 2. Approximated SNR loss @ WER=10^{-3} for two investigated systems compared to conventional OFDM

6.4 Application to CR PHY

In this subsection we examine our proposed method for a hypothetical CR scenario. Figure 15 shows the PSD of conventional OFDM, Duobinary, and a class-2 carrier-by-carrier PRS OFDM for the hypothetical cognitive radio physical layer with three available frequency notches.

The OOB reduction of carrier-by-carrier PRS OFDM and less interference by SUs in CR application can be realized clearly from this. As mentioned above, larger L can make this situation more attractive. It is worth noting that carrier-by-carrier PRS OFDM method can be generally implemented with acceptable complexity by means of a simple digital filter in the transmitter and an MLSD in the receiver. Note that in Duobinary case with pre-coding and BPSK modulation, receiver can be implemented by a simple slicer rather than the MLSD. Furthermore, this method will not limit the system. In other words, in the proposed method, many other techniques of OOB radiation reduction can be used such as windowing, cancellation carrier insertion, frequency filtering, and so on in addition to the PAPR reduction techniques. Also, GI can be added to signal as before to improve inter-block-interference robustness and obtain better synchronization performance. Consequently, the carrier-by-carrier PRS OFDM method is a more appropriate candidate for PHY of CR networks than conventional OFDM.

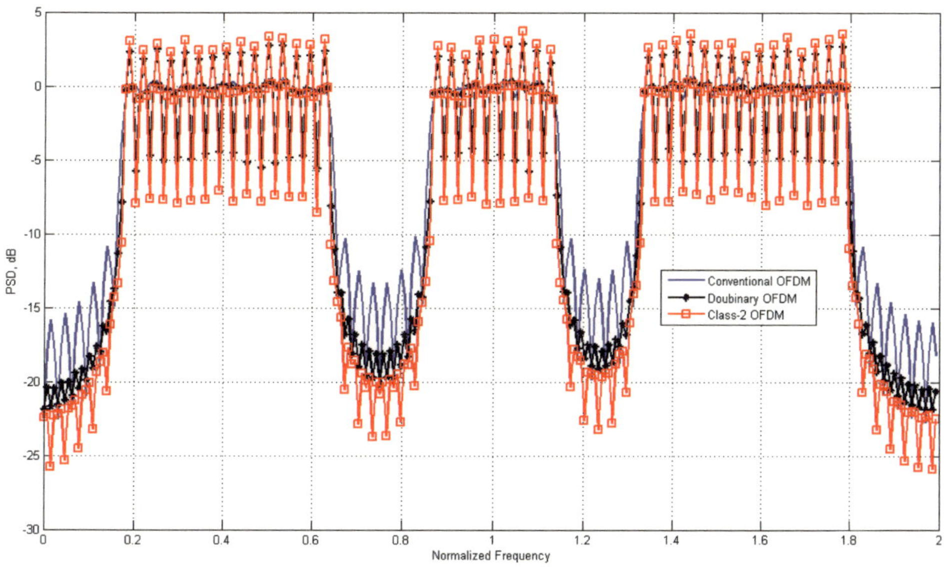

Fig. 15. PSD comparison of conventional, Duobinary, and a class-2 carrier-by-carrier PRS OFDM for a hypothetical CR system

7. Conclusion

The reduction of OOB components of an OFDM spectrum is a challenge both in conventional applications and in CR networks. In this chapter using carrier-by-carrier PRS between consecutive OFDM blocks to reduce OOB radiation was introduced. This method neither increases effective time duration of signal nor decreases the bandwidth efficiency. Also, it is compatible with many other techniques and can be implemented in the existing systems by means of digital filtering in the transmitter and an MLSD, or in special case by a simple slicer, in the receiver. However, error propagation, and thus higher WER, may occur because of introduced controlled correlation which can be reduced by means of pre-coding. In addition, in the proposed system many other techniques of OOB radiation and PAPR level reduction can be used. Furthermore, GI can be added to signal as before. Simulation results show that this method can reduce OOB radiation effectively by an acceptable added complexity while it increases WER and PAPR remains unchanged. Also, it was showed that introducing correlation between more modulated symbols leads to better OOB radiation reduction, more complexity, higher WER, and the same PAPR level. Investigation of the combination of the proposed method with existing methods such as frequency filtering, cancellation carrier insertion, and etc. can be the topic of the future works. Also, derivation of the optimum correlation pattern, not necessarily PRS patterns, via an optimization problem considering OOB components and WER is an attractive subject.

8. Acknowledgment

The authors would like to thank Ms. Z. Naghsh for editing help and advice.

9. References

Amini, R. Kempter, R-R. Chen, L. Lin & B. Farhang-Boroujeny (2005), "Filter Bank Multitone: A Candidate for Physical Layer of Cognitive Radio," Presented in the SDR Forum Technical Conference, Hyatt Regency, Orange County, California, 14-17

Baxley, R. J. (2005). Analyzing Selected Mapping for Peak-to-Average Power Reduction in OFDM. *M. Sc. Thesis,* Georgia Institute of Technology, (2005)

Brandes, S.; Cosovic, I. & Schnell, M. (2006). Reduction of Out-of-Band Radiation in OFDM Systems by Insertion of Cancellation Carriers. *IEEE communication LETTER,* vol.10, No.6,(2006), pp. 420–422

Cosovic, I. & Mazzoni, T. (2006). Suppression of sidelobes in OFDM systems by multiple-choice sequences. *European Trans. on Telecommunication,* vol.17, No.6, (2006), pp. 623–630

Cosovic, I.; Brandes, S. & Schnell, M. (2006). Subcarrier Weighting: A Method for Sidelobe Suppression in OFDM Systems. *IEEE communication LETTER,* vol.10, No.6, (2006), pp. 444–446

ETSI DVB-RCT (EN 301 958). (2001). Interaction Channel for Digital Terrestrial Television (RCT) Incorporating Multiple Access OFDM. *S. Antipolis, France,* (2001)

Farhang-Boroujeny, B. & Kempter, R. (2008). Multicarrier Communication Techniques for Spectrum Sensing and Communication in Cognitive Radios. *IEEE Communications Magazine,* (April 2008), pp .80-85.

Ghassemi, A.; Lampe, L.; Attar, A. & Gulliver, T.A. (2010). Joint Sidelobe and Peak Power Reduction in OFDM-Based Cognitive Radio. IEEE 72 nd Conf. on vehicular technology, (VTC Fall-2010)

Haykin, S. (2005). Cognitive radio: brain-empowered wireless communications. IEEE Journal on selected areas topics in communications, Vol. 23, Issue 2, pp.201-220.

IEEE 802.11.a-1999. (1999). IEEE standard for wireless LANs part 11: Wireless LAN Medium Access Control (MAC) and Physical Layer (PHY) specifications: High-Speed Physical Layer in the 5 GHz Band. (R2003).

IEEE 802.16-2004. (2004). IEEE Standard for Local and Metropolitan Area Networks Part 16: Air Interface for Fixed Broadband Wireless Access Systems. IEEE standard, (October 2004).

Kabal, P. & Passupathy, S. (1975). Partial-response signaling. *IEEE Trans. on Communication,* vol.23,(1975), pp. 921-934

Kim, K. H. & Km, H. M. (2005). A Suppression Scheme of the ICI caused by Phase Noise based on Partial Response Signaling in OFDM systems. *Fifth international conference on information, communication, and signal processing,* pp. 253-257, Bangkok, 2005

Kobayashi, H. (1971). Correlative level coding and maximum-likelihood decoding. *IEEE Trans. on Information theory,* vol.IT-17, (1971), pp. 586-594

Kretzmer, E. R. (1966). Generalization of a technique for binary data communication. *IEEE Trans. on communication,* vol.14, (1966), pp. 67-68

Lander, A. (1966). Correlative level coding for binary data transmission. IEEE spectrum, Vol. 3, pp. 104-115.

Mahmoud, H. A. & Arsalan, H. (2008). Sidelobe Suppression in OFDM-Based Spectrum Sharing Systems Using Adaptive Symbol Transition. *IEEE communication LETTER,* vol.12, No.2, (2008), pp. 133-135

Mahmoud, H.A. & Arslan, H. (2008). Spectrum Shaping of OFDM-based Cognitive Radio Signals. IEEE symposium on radio and wireless, (RWS 2008)

Mitola, J. & Maguire Jr. (1999). Cognitive radio: making software radios more personal. *IEEE Personal Communication,* vol.6, No.4, (Aug.1999), pp. 13–18.

Naghsh, M.M. & Omidi, M.J. (2010). Reduction of out of band radiation using carrier-by-carrier partial response signaling in orthogonal frequency division multiplexing. *IET communications,* Vol.4, (2010), Iss.12, pp. 1433-1442

Noreen, S. & Azeemi, N.Z. (2010). A Technique for Out-of-Band Radiation Reduction in OFDM-Based Cognitive Radio. IEEE 17th International Conf. on telecommunications, (ICT 2010)

Passupathy, S. (1977). A bandwidth-efficient signaling scheme. *IEEE communication society magazine,* (July 1977), pp. 4-10

Phillip, E. (2001). Adaptive Techniques for Multiuser OFDM. *Ph.D Thesis,* James Cook University, (December 2001)

Proakis, J. G. (2007). *Digital communication,* Fifth edition, McGraw-Hill, 2007

Renhui Xu, Hai Wang *y,* Chen, M. (2009). On the Out-of-Band Radiation of DFT-based OFDM using Pulse Shaping. *IEEE International Conf. on wireless communication and signal processing,* (WCSP 2009)

Safavi, S.M. (2010). Sidelobe Suppression in OFDM-based Cognitive Radio Systems. *10th International Conference on Information Science, Signal Processing and their Applications,* (ISSPA 2010)

Syed-yusof, S. K.; Fisal, N. & Muladi. (2006). Integer Coefficients Partial Response Signaling in OFDM System. *International RF and microwave conference,* pp. 326-328, Putra Jaya, 2006

Vadde, V. & Gray, S. (2001), Partial Response Signaling for Enhanced Spectral Efficiency and RF Performance in OFDM systems. *IEEE global telecommunication conference,* vol.5, pp. 3120-3124, USA, 2001.

Vadde, V. (2001). PAPR reduction by envelope stabilization using partial response signaling in OFDM systems. *IEEE radio and wireless conference,* pp. 197-201, USA, 2001

Van Nee, R. & Prasad, R. (2000). *OFDM for Wireless Multimedia Communications,* Artech House

Weiss T.A. & F.K. Jondral (2004). Spectrum pooling: an innovative strategy for the enhancement of spectrum efficiency, *IEEE Communications Magazine,* Vol. 42, No. 3, pp. S8 - S14.

Weiss, T.; Hillenbrand, J.; Krohn, A. & Jondral, F. K. (2004). Mutual interference in OFDM-based spectrum pooling systems, *in Proc. IEEE Veh.Technol. Conf.,* vol.4, pp. 1873–1877, Italy, May 2004.

Yu, L.; Rao, B.D.; Milstein, L.B. & Proakis, J.G. (2010). Reducing out-of-band radiation of OFDM-based cognitive radios . Eleventh *International workshop on signal processing advances in wireless communication,* (SPAWC 2010)

Yuan, Z. & Wyglinski, A.M. (2009). Cognitive Radio-Based OFDM Sidelobe Suppression Employing Modulated Filter Banks and Cancellation Carriers. IEEE Conf. on military communications, (MILCOM 2009)

Yuan, Z. & Wyglinski, A.M. (2010). On Sidelobe Suppression for Multicarrier-Based Transmission in Dynamic Spectrum Access Networks. *IEEE Transaction on vehicular technology*, Vol.59, NO.4, (May 2010)

Zhou, X.; Li, G.Y. & Sun, G. (2011). Low-Complexity Spectrum Shaping for OFDM-based Cognitive Radios. *IEEE Conf. on communications and networking,* (WCNC 2011)

8

Reconfigurable Multirate Systems in Cognitive Radios

Amir Eghbali and Håkan Johansson
Division of Electronics Systems, Department of Electrical Engineering, Linköping University Sweden

1. Introduction

One perspective in communication systems is to increase the spectrum utilization using cognitive radios. A cognitive radio is a network of intelligent co-existing radios which senses the environment to find the available frequency slots, white spaces, or the spectrum holes as noted in Akyildiz et al. (2008); Haykin (2005). Then, it modifies its transmission characteristics to use that particular frequency slot. Figure 1 illustrates the overlay spectrum sharing outlined in Cabric et al. (2006), or the opportunistic spectrum access discussed in Zhao & Sadler (2007); Zhao & Swami (2007), or the dynamic spectrum access which is considered in Sherman et al. (2008). Here, secondary users occupy the frequency slots which are not used[1] by the primary users. One of the main tasks in a cognitive radio is thus the spectrum mobility as in Akyildiz et al. (2006; 2008), or the dynamic frequency allocation as in Haykin (2005), or the dynamic spectrum allocation as in Leaves et al. (2004); Zhao & Swami (2007). This chapter uses the term dynamic frequency-band allocation (DFBA). Being dynamic means that the transmission parameters, e.g., bandwidth, center frequency, transmission power, and communication standard etc., may vary with time according to Akyildiz et al. (2006). One should at least be able to change the center frequency and bandwidth although other parameters may also change. This is also referred to as the reconfigurability according to Akyildiz et al. (2008); Haykin (2005); Jondral (2005); Leaves et al. (2004); Ramacher (2007); Sherman et al. (2008).

Another perspective in communication systems calls for satellites to play a complementary role in supporting various wideband services as proposed by Arbesser-Rastburg et al. (2002); Evans et al. (2005); Farserotu & Prasad (2000); Lucente et al. (2008); Nguyen et al. (2002); Re & Pierucci (2002); Wittig (2000). For this purpose, the European space agency has proposed three major network structures for broadband satellite-based communication systems as in Arbesser-Rastburg et al. (2002). This requires an efficient use of the limited available frequency spectrum by satellite on-board signal processing as discussed in Abdulazim & Göckler (2005a;b; 2006); Abdulazim et al. (2007); Arbesser-Rastburg et al. (2002); Eghbali et al. (2007a;b; 2009a); Eghbali, Johansson, Löwenborg & Göckler (2011); Evans et al. (2005); Farserotu & Prasad (2000); Göckler & Abdulazim (2005; 2007);

[1] Under certain conditions, the secondary users need not wait for a vacant channel. This allows a simultaneous transmission over the same time or frequency as noted in Devroye et al. (2006).

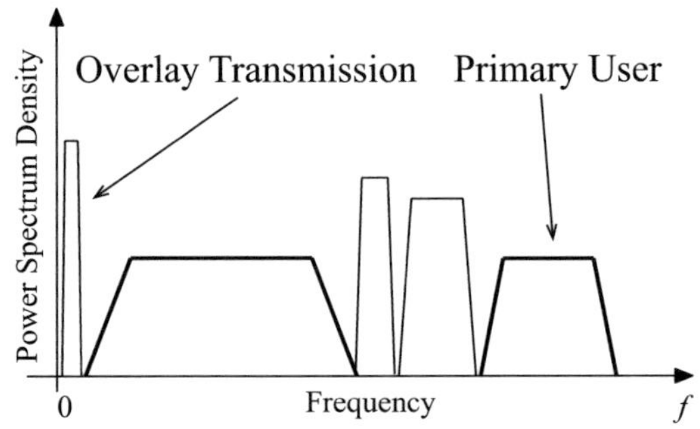

Fig. 1. Overlay approach for spectrum sharing.

Johansson & Löwenborg (2005; 2007); Lippolis et al. (2004); Lucente et al. (2008); Nguyen et al. (2002); Re & Pierucci (2002); Rosenbaum et al. (2006); Wittig (2000).

Like satellite-based communication systems which require both on-ground DFBA and on-board dynamic frequency-band reallocation (DFBR), the ad hoc- or infrastructure-based cognitive radios can also utilize DFBA and DFBR. In the ad hoc-based networks, individual users can utilize DFBA while DFBR can be performed by the base stations of infrastructure-based networks. The DFBA can also be deployed by the individual users of infrastructure-based networks. Both DFBA and DFBR need interpolation/decimation with variable parameters. For large sets of variable conversion factors, the implementation complexity increases. Complexity reduction can be achieved using reconfigurable structures which perform various tasks by simple modifications and without hardware changes. Also, the filter coefficients do not change thereby enabling us to solve the filter design problem only once and offline. Specifically, one must be able to reprogram the same hardware.

This chapter discusses the structure, reconfiguration, and the parameter selection when adopting the DFBA and DFBR for cognitive radios. Two approaches, i.e., Approach I and II, are discussed. They are appropriate based on the availability of (i) a composite signal comprising several user signals, or (ii) the individual user signals. Combinations of Approaches I and II provide increased freedom to allocate and reallocate the user signals.

2. Basics of multirate signal processing

This section treats some basics of sampling rate conversion (SRC), filter banks (FBs), perfect reconstruction (PR), and transmultiplexers (TMUXs).

2.1 Sampling rate conversion: conventional

Different parts of a multirate system operate at different sampling frequencies thereby necessitating interpolation (decimation) to increase (decrease) the sampling frequency of digital signals as outlined in Mitra (2006); Vaidyanathan (1993). Interpolation and decimation comprise lowpass filters as well as upsamplers and downsamplers whose block diagrams are shown in Fig. 2. In Fig. 2(a),

$$y(n) = x(nM) \iff Y(z) = \frac{1}{M}\sum_{k=0}^{M-1} X(z^{\frac{1}{M}} W_M^k), \;\; W_M = e^{-j\frac{2\pi}{M}}. \tag{1}$$

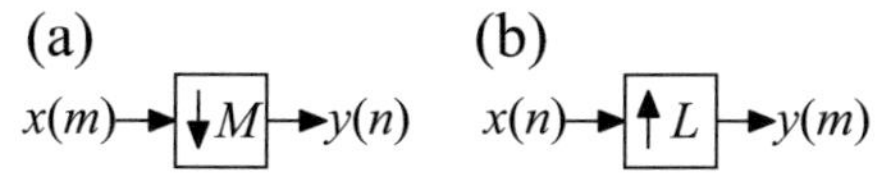

Fig. 2. (a) M-fold downsampler. (b) L-fold upsampler.

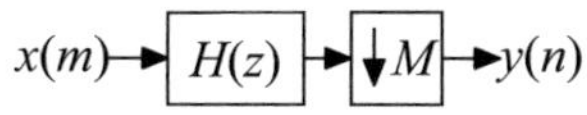

Fig. 3. Decimation by M.

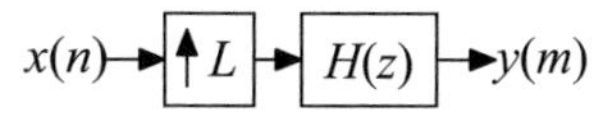

Fig. 4. Interpolation by L.

Note that $X(z^{\frac{1}{M}})$ is not periodic by 2π but adding the shifted versions gives a signal $Y(z)$ with a period of 2π such that the Fourier transform can be defined. In Fig. 2(b),

$$y(n) = \begin{cases} x(\frac{n}{L}) & \text{if } n = 0, \pm L, \pm 2L, \ldots \\ 0 & \text{otherwise} \end{cases} \iff Y(z) = X(z^L). \tag{2}$$

The upsampler and downsampler are linear time-varying systems. Unless $x(n)$ is lowpass[2] and bandlimited, downsampling results in aliasing and decimation thus requires an additional filter as in Fig. 3. This anti-aliasing filter $H(z)$ limits the bandwidth of $x(n)$. In Fig. 3,

$$y(n) = \sum_{k=-\infty}^{+\infty} x(k)h(nM - k). \tag{3}$$

As upsampling causes imaging, interpolation requires a filter as in Fig. 4. This lowpass anti-imaging filter $H(z)$ removes the images and, as in Vaidyanathan (1993), we have

$$y(n) = \sum_{k=-\infty}^{+\infty} x(k)h(n - kL). \tag{4}$$

For SRC[3] by a rational ratio $\frac{M}{L}$, interpolation by L must be followed by decimation by M. Consequently, the cascade of the anti-imaging and anti-aliasing filters results in one filter, say $G(z)$, where the output, according to Vaidyanathan (1993), is

$$y(n) = \sum_{k=-\infty}^{+\infty} x(k)g(nM - kL). \tag{5}$$

Generally, $G(z)$ is a lowpass filter with a stopband edge at, as in Mitra (2006); Vaidyanathan (1993),

$$\omega_s T = \min(\frac{\pi}{M}, \frac{\pi}{L}) = \frac{\pi}{\max(M, L)}. \tag{6}$$

In practice, there is a roll-off factor $0 \le \rho \le 1$ so that $\omega_s T = \frac{\pi(1+\rho)}{\max(M,L)}$. If M and L are mutually coprime numbers, a decimator can be obtained by transposing the interpolator. The noble identities, defined as in Fig. 5, help move the filtering operations inside a multirate structure.

---

[2] This is not necessary to avoid aliasing. For example, if $X(e^{j\omega T})$ is nonzero only at $\omega T \in [\omega_1 T, \omega_1 T + \frac{2\pi}{M}]$ for some $\omega_1 T$, there is no aliasing.

[3] If $L > M$ ($L < M$), we have interpolation (decimation) by a rational ratio $\frac{L}{M} > 1$ ($\frac{M}{L} > 1$). This chapter frequently refers to SRC by a rational ratio $R_p > 1$.

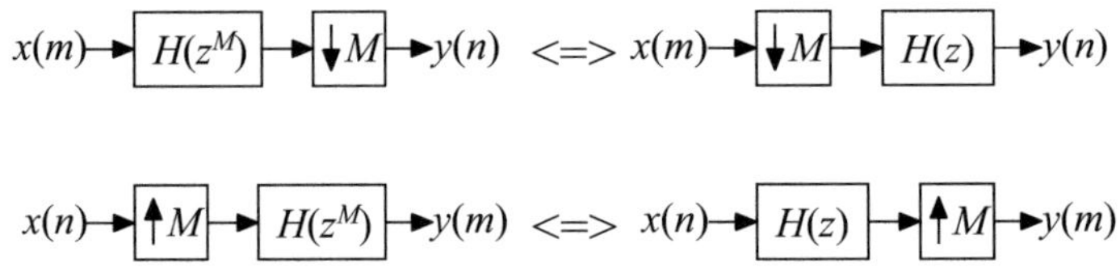

Fig. 5. Noble identities which allow us to move the arithmetic operations to the lower sampling frequency.

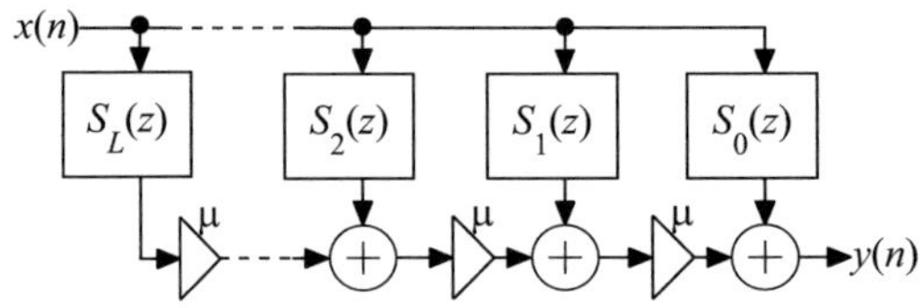

Fig. 6. Farrow structure with fixed subfilters $S_k(z)$ and variable fractional delay (FD) of μ.

2.2 Sampling rate conversion: Farrow structure

In a conventional SRC and if the SRC ratio changes, we need new filters thereby reducing the flexibility. The Farrow structure, introduced in Farrow (1988) and shown in Fig. 6, obtains flexibility in an elegant way. The Farrow structure is traditionally composed of linear-phase finite-length impulse response (FIR) subfilters $S_k(z)$, $k = 0, 1, \ldots, L$, with either a symmetric (for k even) or antisymmetric (for k odd) impulse response[4].

When $S_k(z)$ are linear-phase FIR filters, the Farrow structure is sometimes referred to as the modified Farrow structure, e.g., Vesma & Saramäki (1996), but we simply refer to it as the Farrow structure. The Farrow structure is efficient for interpolation whereas, for decimation, it is better to use the transposed Farrow structure, as discussed in Babic et al. (2002); Hentschel & Fettweis (2000).

The transfer function of the Farrow structure is

$$H(z,\mu) = \sum_{k=0}^{L} S_k(z)\mu^k = \sum_{k=0}^{L}\sum_{n=0}^{N_k} s_k(n)z^{-n}\mu^k = \sum_{n=0}^{N}\sum_{k=0}^{L} s_k(n)\mu^k z^{-n} = \sum_{n=0}^{N} h(n,\mu)z^{-n}. \quad (7)$$

Here, $|\mu| \leq 0.5$ and N is the order of the overall impulse response $h(n,\mu)$. Further, μ is the FD value, i.e., the time difference between each input sample and its corresponding output sample. In the rest of the chapter, we will use $h(n)$ and $H(z)$ instead of $h(n,\mu)$ and $H(z,\mu)$, respectively. If μ is constant for all input samples, the Farrow structure delays a bandlimited signal by a fixed μ. In general, SRC amounts to delaying every input sample with a different μ. Thus, by controlling μ for every input sample, the Farrow structure can perform SRC.

Generally, $S_k(z)$ are designed so that $H(z)$ approximates an allpass transfer function with FD of μ over the frequency range of interest according to Babic et al. (2002); Johansson & Hermanowicz (2006); Johansson & Löwenborg (2003); Pun et al. (2003); Tseng (2002); Vesma & Saramäki (1996; 1997; 2000). The desired causal magnitude and unwrapped

[4] With infinite-length impulse response (IIR) filters, care must be taken to avoid transients as μ changes for every sample.

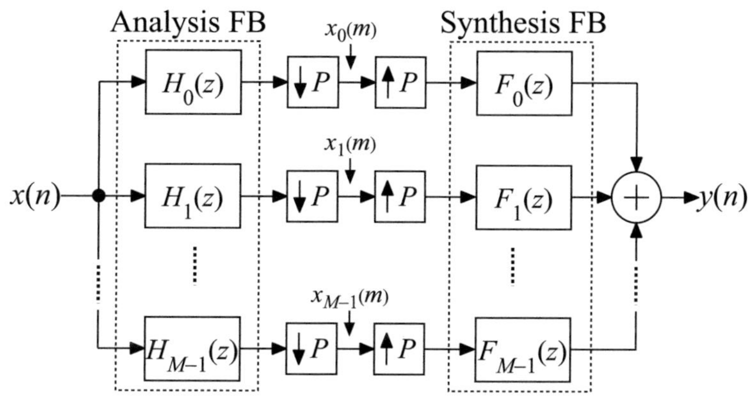

Fig. 7. General M-channel FB.

phase responses are

$$H_{\text{des}}(e^{j\omega T}) = e^{-j(\Delta+\mu)\omega T}, \tag{8}$$

$$\Phi_{\text{des}}(\omega T) = -(\Delta+\mu)\omega T, \tag{9}$$

where

$$\Delta = \frac{\max_k(N_k)}{2}. \tag{10}$$

The main advantage of the Farrow structure is its ability to perform rational SRC using only one set of $S_k(z)$ and by simple adjustments of μ.

2.3 General M-channel FBs

An M-Channel filter bank (FB), shown in Fig. 7, splits $X(z)$ into the M subbands $X_m(z)$, $m = 0, 1, \ldots, M-1$, using the analysis filter bank (AFB) filters $H_m(z)$. To reconstruct $X(z)$, we need the synthesis filter bank (SFB) filters $F_m(z)$. Furthermore, upsamplers and downsamplers by P are also required as in Fig. 7. The output of a general M-channel FB is

$$Y(z) = \frac{1}{P}\sum_{n=0}^{P-1} X(zW_P^n)\sum_{m=0}^{M-1} H_m(zW_P^n)F_m(z). \tag{11}$$

Ideally, $Y(z)$ is a scaled (by α) and delayed (by β) version of $X(z)$, i.e., $y(n) = \alpha x(n-\beta)$. Such a system is referred to as PR. In near perfect reconstruction (NPR) FBs, α is frequency dependent and the distortion transfer function is

$$V_0(z) = \frac{1}{P}\sum_{m=0}^{M-1} H_m(z)F_m(z), \tag{12}$$

whereas the aliasing transfer functions are

$$V_l(z) = \frac{1}{P}\sum_{m=0}^{M-1} H_m(zW_P^l)F_m(z),\ l = 1, 2, \ldots, P-1. \tag{13}$$

These FBs are generally linear periodic time-varying systems with a period M. Without aliasing, we have a linear time-invariant (LTI) system as defined in Vaidyanathan (1993). In a PR FB,

$$V_0(e^{j\omega T}) = c, \ c > 0 \tag{14}$$
$$V_l(e^{j\omega T}) = 0, \ l = 1, 2, \ldots, P-1. \tag{15}$$

If $P = M$, the FB is maximally decimated but $P < M$ leads to oversampled FBs as in Vaidyanathan (1993). If $V_0(z)$ is allpass (has linear-phase), we have zero amplitude (phase) distortion.

2.4 Modulated FBs

To obtain the AFB and SFB filters, one can modulate a single Nth-order linear-phase FIR prototype filter $G(z) = \sum_{n=0}^{N} g(n)z^{-n}$. With cosine modulation, which is outlined in Chen & Chiueh (2008); Ihalainen et al. (2007); Saramäki & Bregović (2001a), we have

$$h_m(n) = 2g(n)\cos[(m+0.5)\frac{\pi}{M}(N-n+\frac{M+1}{2})], \tag{16}$$
$$f_m(n) = 2g(n)\cos[(m+0.5)\frac{\pi}{M}(n+\frac{M+1}{2})] = h_m(N-n). \tag{17}$$

In a PR cosine modulated filter bank (CMFB), $N = 2KM - 1$ where K is an integer overlapping factor as defined in Viholainen et al. (2006). For complex modulated FBs,

$$h_m(n) = g(n)W_M^{-mn}, \tag{18}$$
$$f_m(n) = h_m(n). \tag{19}$$

In the maximally decimated case, we can use modified discrete Fourier transform (MDFT) FBs, as in Bregović & Saramäki (2005); Karp & Fliege (1999). An M-channel MDFT FB can be realized as either of the following, according to Fliege (1995),

- Two SRC stages with ratios $\frac{M}{2}$ and 2 while adding some phase offset between these stages.
- Two separate FBs where the phase offset is applied outside the AFB and SFB.

If an MDFT FB is PR, N is an integer as $KM + s$ where $0 \leq s < M$. With the AFB and SFB filters, having uniform or nonuniform passbands, we have uniform or nonuniform FBs as in Vaidyanathan (1993). Nonuniform FBs can also be obtained by modulation, as discussed in Princen (1994), where

$$h_m(n) = a_m g_m(n) e^{-\frac{j\pi\alpha_m}{M_m}(n-\frac{L_m-1}{2})} + a_m^* g_m^*(n) e^{\frac{j\pi\alpha_m}{M_m}(n-\frac{L_m-1}{2})}, \tag{20}$$
$$f_m(n) = b_m g_m(n) e^{-\frac{j\pi\alpha_m}{M_m}(n-\frac{L_m-1}{2})} + a_m^* g_m^*(n) e^{\frac{j\pi\alpha_m}{M_m}(n-\frac{L_m-1}{2})}. \tag{21}$$

Here, $\alpha_m = (K_m + 0.5)$ and $g_m(n)$ is the (possibly complex) prototype filter of length L_m with M_m being the decimation factor in each branch. Each branch has a center frequency as $\pm\frac{\pi\alpha_m}{M_m}$ with K_m being an integer where a_m and b_m define the modulation phase. The nonuniform FBs achieve a more general time and frequency tiling. The sine modulated filter bank (SMFB) is obtained similar to (16) and (17). The exponentially modulated filter bank (EMFB), with

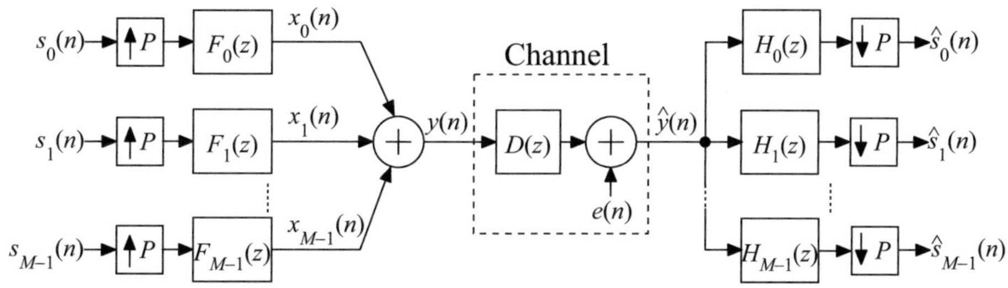

Fig. 8. General M-channel TMUX.

complex filters, is a combination of SMFB and CMFB, as outlined in Chen & Chiueh (2008); Ihalainen et al. (2007).

2.4.1 Filter design for modulated FBs

To design the prototype filter $G(z)$, we can use any standard filter design technique, e.g., Bregović & Saramäki (2005); Fliege (1995); Heller et al. (1999); Martin-Martin et al. (2008; 2005); Mirabbasi & Martin (2003); Mitra (2006); Saramäki & Bregović (2001a); Saramäki & Bregovic (2001b); Vaidyanathan (1993); Viholainen et al. (1999). The MDFT FB has a lowpass $G(z)$ with a stopband edge at $\omega_s T = \frac{2\pi}{M}$ according to Fliege (1995). The CMFB has a lowpass $G(z)$ with a stopband edge at $\omega_s T = \frac{\pi(1+\rho)}{2M}$ and a 3-dB cutoff frequency at $\omega T = \frac{\pi}{2M}$ as discussed in Diniz et al. (2004); Martin-Martin et al. (2005). If $0 < \rho \leq 1$, only the adjacent branches overlap. With $1 < \rho \leq 2$ (or $\rho > 2$), two (or at least three) adjacent branches overlap. In both FBs, $G(z)$ satisfies the power complementary property.

2.5 General M-channel TMUXs

A transmultiplexer (TMUX) converts the time multiplexed components of a signal into a frequency multiplexed version and back so that several users transmit and receive over a common channel, as noted in Vaidyanathan & Vrcelj (2004). A TMUX is also referred to as a FB transceiver, e.g., Beaulieu & Champagne (2009); Bianchi & Argenti (2007); Borna & Davidson (2007); Chiang et al. (2007); Lin & Phoong (2001).

Assume that we want to transmit a series of symbol streams $s_k(n)$, $k = 0, 1, \ldots, M-1$, through a channel. As in Fig. 8, we can pass $s_k(n)$ through the transmitter filters $F_k(z)$. Then, (4) gives

$$x_k(n) = \sum_{m=-\infty}^{\infty} s_k(m) f_k(n - mP). \tag{22}$$

Here, the channel is described by a possibly complex LTI filter $D(z) = \sum_{n=0}^{L_D} d(n) z^{-n}$ followed by an additive noise source $e(n)$. The receiver filters $H_k(z)$ separate the signals and only a downsampling by P is needed to retrieve the original symbol streams. Ignoring the channel and for $i = 0, 1, \ldots, M-1$, we have

$$\hat{S}_i(z) = \sum_{k=0}^{M-1} S_k(z) T_{ki}(z^P), \quad T_{ki}(z^P) = \frac{1}{P} \sum_{l=0}^{P-1} F_k(zW_P^l) H_i(zW_P^l). \tag{23}$$

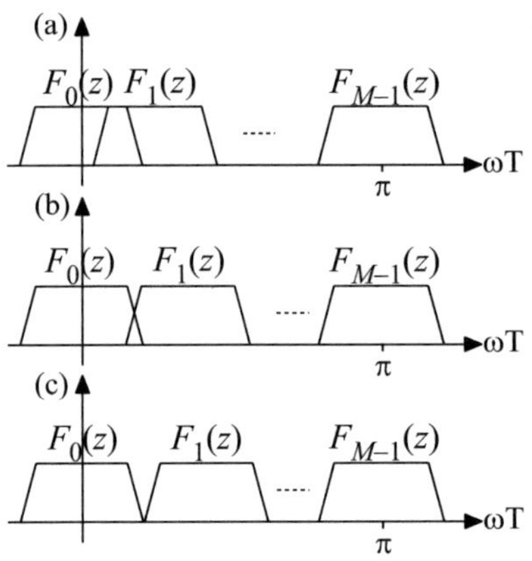

Fig. 9. M-channel TMUX filters. (a) Overlapping. (b) Marginally overlapping. (c) Non-overlapping.

Typical characteristics of $F_k(z)$ and $H_k(z)$ are shown in Fig. 9. Similar to FBs, TMUXs can be redundant ($P > M$) or critically sampled ($P = M$). To avoid inter-symbol interference (ISI), a level of redundancy may be needed such that $P - M \geq L_D$, according to Bianchi et al. (2005). The output of the TMUX in (23) is

$$\hat{S}_i(z) = T_{ii}(z)S_i(z) + \sum_{k=0,k\neq i}^{P-1} T_{ki}(z)S_k(z) \tag{24}$$

where $T_{ii}(z)$ and $T_{ki}(z)$ represent the ISI and the inter-carrier interference (ICI), respectively, as in Furtado et al. (2005). The ISI (ICI) is sometimes also referred to as interband (cross-band) ISI, e.g., Chiang et al. (2007).

If an LTI filter is placed between an upsampler and a downsampler of ratio P, the overall system is equivalent to the decimated (by P) version of its impulse response, as mentioned in Vaidyanathan (1993). In this case, designing $F_k(z)$ and $H_k(z)$ so that the decimated (by P) version of $F_k(z)H_m(z)$ becomes a pure delay if $k = m$ and zero otherwise, the TMUX becomes PR. In terms of (24), this means

$$T_{ii}(z) = \frac{1}{P}\sum_{l=0}^{P-1} F_i(z^{\frac{1}{P}}W_P^l)H_i(z^{\frac{1}{P}}W_P^l) = \alpha z^{-\beta}, \tag{25}$$

$$T_{ki}(z) = \frac{1}{P}\sum_{l=0}^{P-1} F_k(z^{\frac{1}{P}}W_P^l)H_i(z^{\frac{1}{P}}W_P^l) = 0. \tag{26}$$

In a PR system, $\hat{s}_k(n) = \alpha s_k(n-\beta)$. The PR properties can be satisfied for both critically sampled and redundant TMUXs. For the critically sampled case, there may not always exist FIR or stable IIR solutions. Therefore, some redundancy makes the solutions feasible, as in de Barcellos et al. (2006); Kovačević & Vetterli (1993); Li et al. (1997); Xie, Chan & Yuk (2005); Xie, Chen & Sho (2005), and it also simplifies the PR conditions. Duality of TMUXs and FBs allows one to obtain a TMUX from its corresponding FB, as noted in Vaidyanathan (1993).

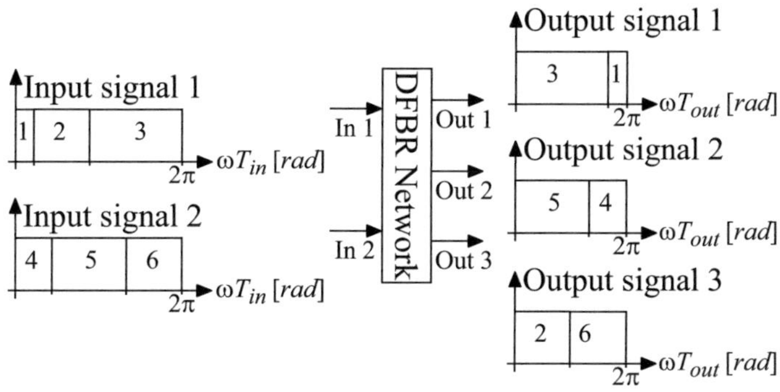

Fig. 10. Approach I: DFBR networks process composite signals to reallocate users from one composite input signal to another composite output signal.

This duality applies to both critically sampled and redundant systems. If a FB is free from aliasing, the corresponding TMUX is free from ICI, according to Fliege (1995).

3. Approach I: use of DFBR networks

For DFBR, we assume that signals from several users, e.g., mobile handsets in a cellular network or computers in a wireless local area network (WLAN), have been added into a composite signal at a main station, e.g., a base station in a cellular network or an access point in a WLAN. This main station then finds available frequency slots and reallocates each user to one of them. Such a main station is similar to a bentpipe satellite payload, as outlined in Nguyen et al. (2002), with its idea of operation shown in Fig. 10. The composite signals are processed by the DFBR network and the users are reallocated to new frequency slots. These slots could be different antenna beams of a satellite payload or different cells in a cellular network. Multiple antennas of a satellite payload perform signal filtering in spatial rather than frequency domain. This is similar to the techniques utilizing multiple antennas for cognitive radios which are discussed in Cabric & Brodersen (2005). The DFBR networks could also be useful for the centralized cooperative cognitive radios, as in Ganesan & Li (2005), and they can also be considered as secondary base stations in licensed band cognitive radios. In licensed band networks, the DFBR can coexist with the primary networks so as to opportunistically operate in an overlay transmission.

The DFBR network can be a mutli-input multi-output system as it can have a number of composite input and output signals. The dynamic nature of the DFBR networks allows the users to occupy any suitable[5] frequency slot in a time-varying manner. Each user can be sent in contiguous or separate frequency bands leading to contiguous or fragmented DFBR which is outlined in Leaves et al. (2004). The separate frequency bands can be considered as a multi-spectrum transmission. Specifically, as white spaces are mostly fragmented, according to Yuan, Bahl, Chandra, Chou, Ferrell, Moscibroda, Narlanka & Wu (2007), the user signals can be transmitted in several non-contiguous frequency bands.

[5] The frequency slot depends on spatial and temporal parameters, e.g., the number of available slots, user movement, and primary user activity, etc. Akyildiz et al. (2006) but the DFBR network is independent of these parameters.

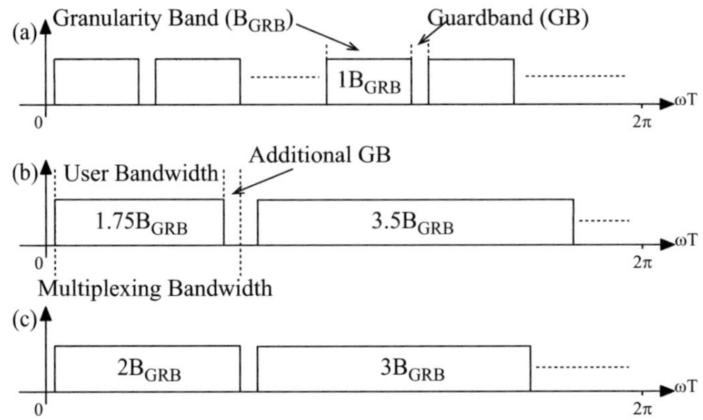

Fig. 11. User bandwidth versus multiplexing bandwidth. The multiplexing bandwidth is an integer multiple of the GRB and it contains a user bandwidth as well as some additional GB.

3.1 Structure of the DFBR network

This chapter uses the term DFBR which is essentially the same as the flexible frequency-band reallocation (FFBR) in, e.g., Abdulazim & Göckler (2006); Eghbali et al. (2009a); Johansson & Löwenborg (2007); Rosenbaum et al. (2006). For the illustrations, we will use the FFBR network, in Johansson & Löwenborg (2007), but one can use any other FFBR network as well.

3.2 User bandwidth versus multiplexing bandwidth

The DFBR networks divide the user signals into a number of granularity bands (GRBs) on which the frequency shifts are performed. As the DFBR networks utilize FBs, the multiplexing bandwidth must be an integer multiple of the granularity band (GRB). The DFBR networks perform frequency shifts on users whose bandwidths are, in general, rational multiples of the GRB. An important issue is to ensure that the users do not share a GRB. This can be achieved by allowing some additional guardband (GB). However, the additional GB affects the spectrum efficiency resulting in a trade-off. As in Fig. 11, a multiplexing bandwidth contains a user bandwidth and some additional GB.

3.3 Reconfigurability

A cognitive radio should adjust its operating parameters without hardware modifications as discussed in Jondral (2005). It is built on the platform for a software defined radio where the processing is mainly in the digital domain, according to Zhao & Sadler (2007). There are several reconfigurable parameters such as operating frequency, modulation method, transmission power, and communication standard etc. For adaptable operating frequency, or flexible frequency carrier tuning, as in Leaves et al. (2004), a cognitive radio should change its operating frequency without restricting the system throughput and hardware.

The DFBR networks can perform any frequency shift of any user having any bandwidth, using a channel switch. This switch seamlessly directs different FB channels to their desired outputs without any arithmetic complexity. In addition, the system parameters are determined and fixed only once, offline. The reconfigurable operation is then performed by reconfiguring

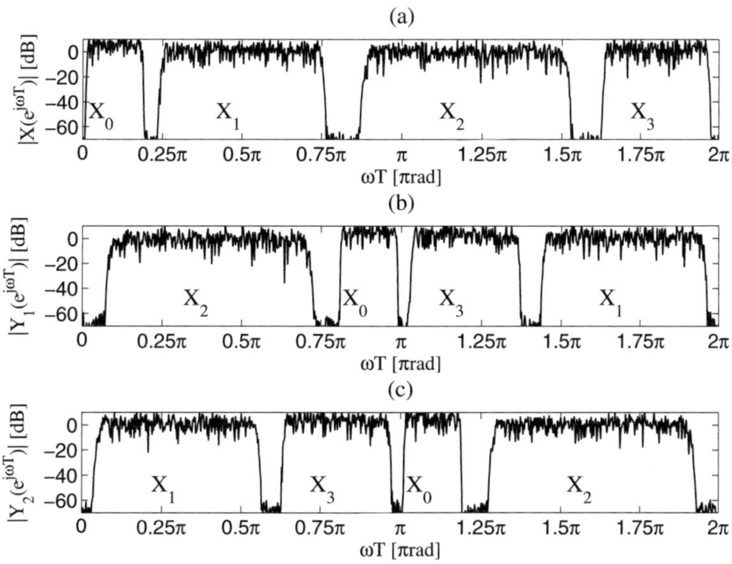

Fig. 12. Input and the reallocated outputs using the channel switch configurations in Figs. 14(a) and 14(b).

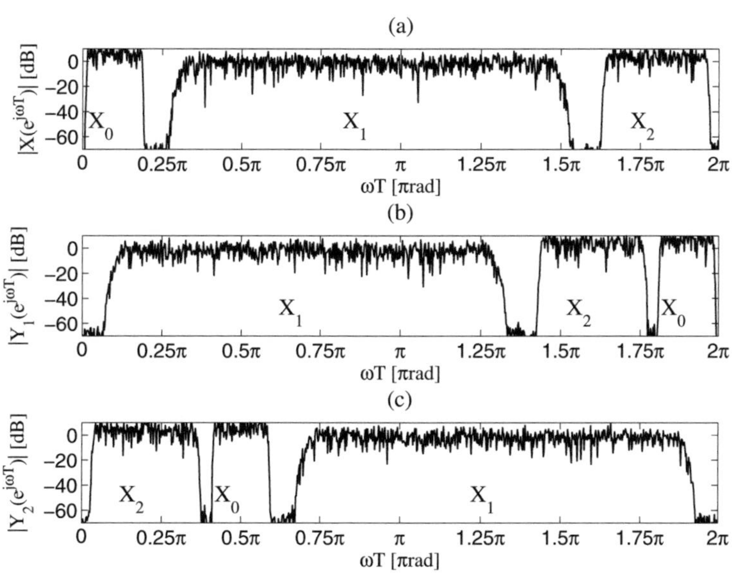

Fig. 13. Input and the reallocated outputs using the channel switch configurations in Figs. 14(c) and 14(d).

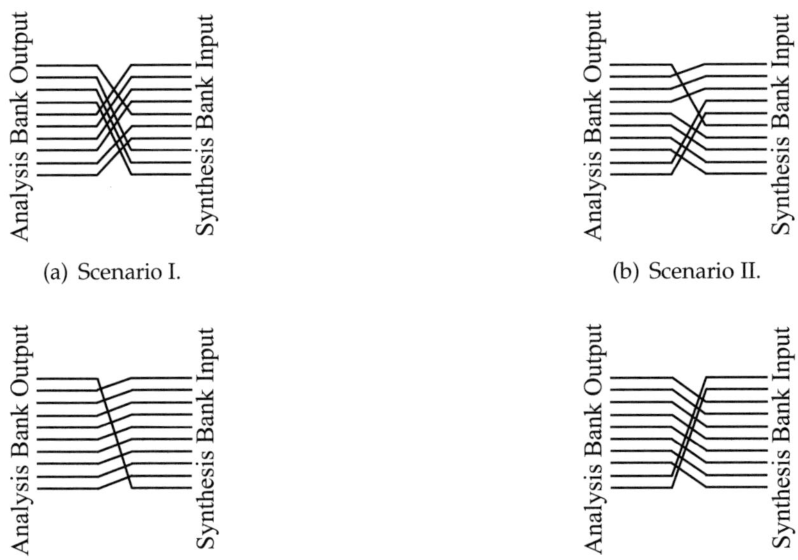

(a) Scenario I. (b) Scenario II.

(c) Scenario III. (d) Scenario IV.

Fig. 14

the channel switch, online. Here, the user bandwidths are predetermined but they can be arbitrary. The DFBR network makes a hand off by changing the operating frequency.

Figures 12 and 13 show two cases where, respectively, four and three users have occupied the whole frequency band. To generate these user signals, the multimode TMUX of Eghbali et al. (2008b) has been used. In Fig. 12(a), the user signals $\{X_0, X_1, X_2, X_3\}$ occupy, respectively, user bandwidths of $\{1, 2.9, 3.6, 1.9\}$ GRBs. Each GRB has a spectral width of $\frac{2\pi}{Q} - \frac{2\epsilon\pi}{Q}$ with $Q = 10$ and $\epsilon = 0.125$. In Fig. 13(a), the user signals $\{X_0, X_1, X_2\}$ occupy $\{1, 6.9, 1.9\}$ GRBs, respectively. To ensure that the users do not share a GRB, one can add some additional GB. This difference in the amount of the GB, between different users, can be recognized from Figs. 12 and 13.

These examples assume the DFBR network to operate on the same antenna beam. By having several DFBR networks, the users can be reallocated between different antenna beams according to Johansson & Löwenborg (2007). This requires a duplication of DFBR networks and a channel switch which directs the user signals between different DFBR networks. Each branch of the channel switches, in Figs. 14(a)–14(d), represents the operation of two FB channels as each GRB contains two FB channels. Specifically, the values of N, M, and L, in Johansson & Löwenborg (2007), are 20, 10, and 2, respectively, for the examples above.

3.4 Modifications

The use of DFBR networks in cognitive radios needs some modifications, mainly in the system parameters. For different system parameters, the implementation complexity may be different but once the parameters are chosen, the implementation complexity remains constant and the system can be easily reconfigured on the same hardware platform. For the DFBR networks, the width of a GRB must be proportional to that of the spectrum holes. Thus, one requires to

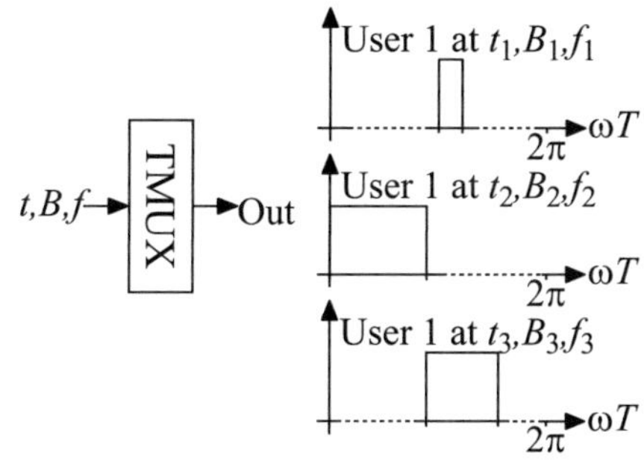

Fig. 15. Approach II: TMUXs to perform DFBA. At any time t_k, each user decides its bandwidth B_k and operating frequency f_k.

choose a value for the $B_{GRB} = \frac{2\pi(1-\epsilon)}{Q}$, in Fig. 11, so as to represent any spectrum hole as a rational multiple of B_{GRB}.

4. Approach II: use of TMUXs

Using TMUXs, each user terminal can adjust its operating frequency and bandwidth. The basic idea is depicted in Fig. 15 where different bandwidths and center frequencies can be generated using multirate signal processing techniques. These TMUXs can also be regarded as the time-spectrum blocks, discussed in Yuan, Bahl, Chandra, Moscibroda, Narlanka & Wu (2007), which can transmit any amount of data at any time interval and on any portion of the frequency spectrum. This applies if the licensed users choose frequency division multiple access and/or time division multiple access as their spectrum access mode. Then, the spectrum holes are identified in the time/frequency plane, as outlined in Jondral (2007). As shown in Fig. 16, the interpolation part represents the transmitter with a variable filter placing the desired user signal at the desired center frequency. The receiver, i.e., the decimation part, retrieves the input signal.

Similar to straightforward DFBR solutions, one can use conventional nonuniform TMUXs to place users, with different bandwidths, at different center frequencies. This becomes inefficient when simultaneously considering the increased number of communication scenarios and the desire to support dynamic communications. In this context, TMUX structures of the general form shown in Fig. 17 can be used. In the SFB, the system C_p performs interpolation by a rational ratio R_p whereas the system $\hat{C}_p$ in the AFB performs decimation by a rational ratio R_p.

4.1 Structure of the TMUX

Any of the TMUXs, in the references Eghbali et al. (2007a; 2008a;b; 2009b; 2010); Eghbali, Johansson & Löwenborg (2011a;b), can be used here. The TMUX, in Eghbali et al. (2010); Eghbali, Johansson & Löwenborg (2011b), has a rather different structure. Instead of variable lowpass filters and frequency shifters as, in Eghbali et al. (2007a; 2008a;b; 2009b); Eghbali, Johansson & Löwenborg (2011a), it performs bandpass rational SRC using flexible commutators and fixed bandpass filters. However, one can generally describe it in terms of Fig. 17.

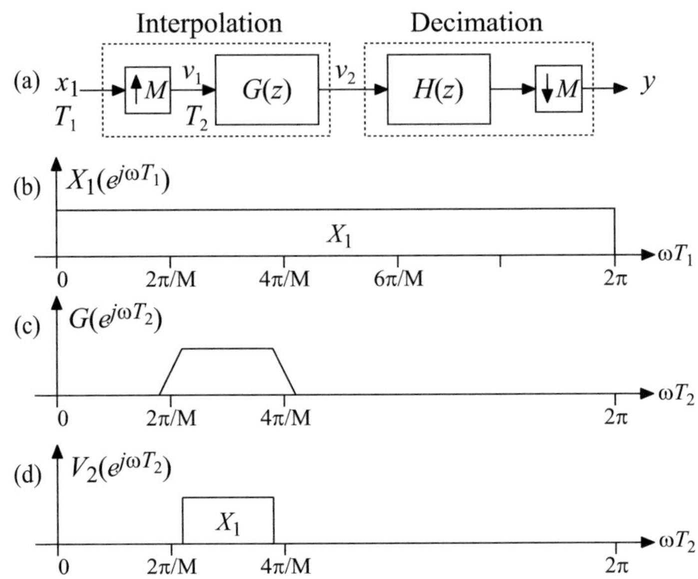

Fig. 16. Principle of TMUXs using multirate building blocks. The interpolation (decimation) part represents the transmitter (receiver). Variable filters place the desired user signal at the desired center frequency.

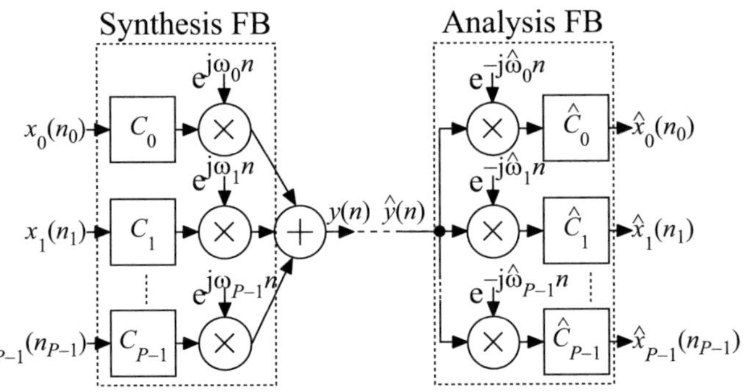

Fig. 17. General structure of a multimode TMUX where systems C_p and $\hat{C}_p$ perform rational SRC.

4.2 Reconfigurability

A cognitive radio must adjust its operating frequency and bandwidth without hardware modifications. The DFBR networks partially provide this capability but they have no control over the user bandwidth. In contrast, the TMUX-based approaches add reconfigurability to the user bandwidth as well. Furthermore, they bring flexible receiver signal filtering, outlined in Leaves et al. (2004), by changing the transmitter and receiver filters. As can be seen from Figs. 12 and 13, the TMUX allows different numbers of users, e.g., four and three, with different user bandwidths to occupy the whole frequency band. These TMUXs provide this full reconfigurability without any hardware changes.

4.3 Modifications

Similar to the DFBR networks, we should have certain system parameters to eliminate the need for hardware changes while having simple reconfigurability. Regarding DFBA, there are different ways to perform SRC which could be useful in different scenarios. The TMUX, in Eghbali et al. (2007a; 2008b), generates a GRB through integer interpolation by, e.g., W, thereby resulting in $B_{GRB} = \frac{2\pi(1+\rho)}{W}$ where ρ is the roll-off. Then, rational R_p multiples of B_{GRB} can be created using the Farrow structure.

The TMUX, in Eghbali et al. (2008a; 2009b); Eghbali, Johansson & Löwenborg (2011a), assumes no GRBs and it allows the users to occupy any portion of the spectrum. It utilizes the Farrow structure to perform general rational SRC by, e.g., $R_p = \frac{A_p}{B_p}$. Here, one can also assume a GRB of size $B_{GRB} = \frac{2\pi(1+\rho)}{A_p}$. Then, users can have bandwidths which are integer B_p multiples of B_{GRB}.

Although references Eghbali et al. (2010); Eghbali, Johansson & Löwenborg (2011b) propose a slightly different TMUX, one can also assume $B_{GRB} = \frac{2\pi(1+\rho)}{M}$. Then, users have bandwidths which are integer M_p multiples of B_{GRB}. This applies to the case with MDFT FBs but for a CMFB, similar formulae can be derived.

5. Choice of frequency shifters

To perform a hand off without information loss, the DFBR network requires the users not to share a GRB. Consequently, a lossless reallocation requires to (i) generate appropriate frequency division multiplexed (FDM) input patterns, and (ii) determine proper parameters for the DFBR networks. To generate the input patterns, the reconfigurability of the TMUXs in Fig. 17 can be used. After generating the user signals with desired bandwidths, the frequency shifters ω_p, $p = 0, 1 \ldots, P-1$, can be computed to allow some additional GB. Here, an example using the TMUX in Eghbali et al. (2008b) is provided. Assuming some bandwidths which are rational, e.g., R_p, multiples of B_{GRB}, the subcarrier ω_p for user p is

$$\omega_p = \begin{cases} \frac{F_0}{2} & \text{if } p = 0 \\ \sum_{k=0}^{p-1} F_k + \frac{F_p}{2} & \text{if } p \neq 0. \end{cases} \tag{27}$$

where $F_p = \lceil R_p \rceil \frac{2\pi}{Q}$, $p = 0, 1, \ldots, k$, with $\lceil . \rceil$ being the ceiling operation. Here, F_p is the multiplexing bandwidth and the ceiling operation ensures that the users do not share a GRB. This formulation applies to the case where DFBA and DFBR are simultaneously used. Otherwise, similar formulae can be used but one may anyhow require some additional GB due to the design margins.

In Figs. 12 and 13, the users occupy $R_p = \{2.9, 3.6, 1.9, 6.9\}$ GRBs. This necessitates an additional GB which is $E_p = \{0.1, 0.4, 0.1, 0.1\}$ multiples of $\frac{2\pi}{Q}$. Therefore, the spectrum efficiency decreases. For a set of values R_p, $p = 0, 1, \ldots, P-1$, about

$$\eta_{dec} = \frac{\frac{2\pi}{Q} \sum_{p=0}^{P-1} (\lceil R_p \rceil - R_p)}{2\pi} = \frac{\sum_{p=0}^{P-1} \lceil R_p \rceil - R_p}{Q} \tag{28}$$

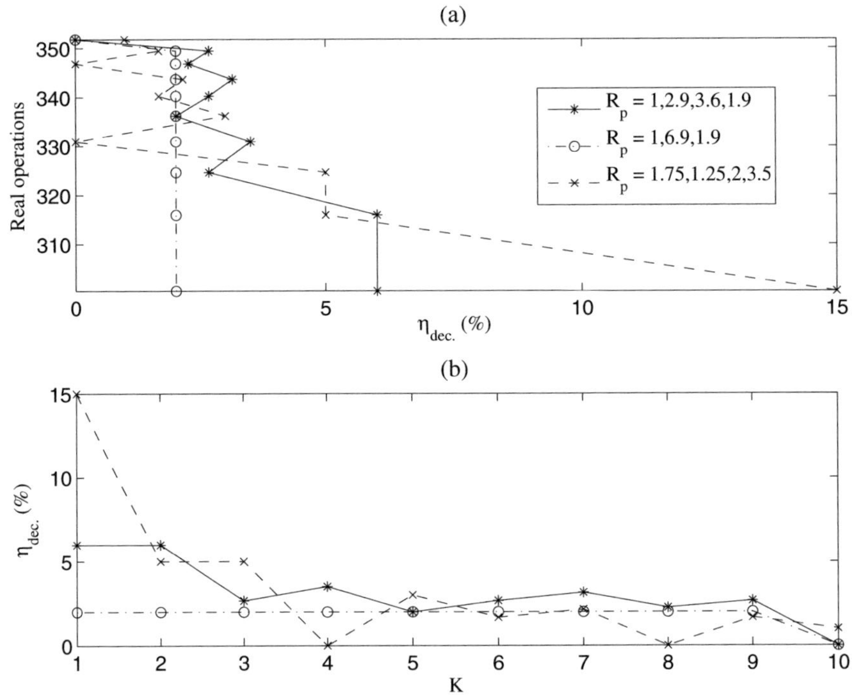

Fig. 18. Trade-off between spectrum efficiency and arithmetic complexity. (a) Decrease in spectrum efficiency versus per-sample arithmetic complexity. (b) Trend of spectrum efficiency versus different K in (29).

percent of the spectrum in $[0, 2\pi]$ is not used. In the examples of Figs. 12 and 13, about 6% and 2% of the total spectrum is not used due to the additional GB. To decrease η_{dec}, one can increase Q by, e.g., K times, so that (28) becomes

$$\eta_{dec} = \frac{\sum_{p=0}^{P-1} \lceil KR_p \rceil - KR_p}{KQ}. \tag{29}$$

However, increasing Q would increase the order of the prototype filter of the corresponding FB. For each K, the prototype filter of the DFBR network would have a transition band of $\frac{2\pi\epsilon}{KQ}$ according to Eghbali et al. (2009a); Johansson & Löwenborg (2007). As the order of a linear-phase FIR filter is inversely proportional to the width of its transition band, there is a trade-off between the spectrum efficiency and the arithmetic complexity.

With a K-fold increase in Q, the length of the prototype filter and the number of FB channels increase proportional to K. Figure 18 shows the trend in spectrum efficiency with respect to the per-sample arithmetic complexity of the DFBR network in Johansson & Löwenborg (2007). Here, the examples of Figs. 12 and 13 as well as that of Eghbali et al. (2008b) with $R_p = \{1.75, 1.25, 2, 3.5\}$ are considered. As can be seen, a larger K increases the per-sample arithmetic complexity but it decreases η_{dec}. The values of R_p mainly determine the maximum and minimum amounts of η_{dec}. Hence, for every set of R_p, one can determine a K such that η_{dec} and the per-sample arithmetic complexity are within the acceptable ranges.

6. Conclusion

This chapter discussed two approaches for the baseband processing in cognitive radios based on DFBR and DFBA. They can support different bandwidths and center frequencies for a large set of users while being easily reconfigurable.

In DFBR networks, composite FDM signals are processed and the users are reallocated to new center frequencies. They are applicable to cognitive radios with multiple antennas, centralized cooperative cognitive radios, and secondary base stations in licensed band cognitive radios. In DFBA networks, each user controls its operating frequency and bandwidth. These networks can be regarded as the time-spectrum blocks.

The reconfigurability of DFBA and DFBR is performed either by a channel switch, in DFBR, or by variable multipliers/commutators, in DFBA. The examples in Figs. 12 and 13 show the increased flexibility to allocate and reallocate any user to any center frequency by simultaneous utilization of DFBA and DFBR. In this case, the individual users can occupy any available frequency slot and be reallocated by the base station.

Basically, utilizing any of Approaches I and II only requires modifications imposed by the special choice of the system parameters. After choosing these parameters once, we must design the filters to satisfy any desired level of error. Then, the same hardware can be reconfigured in a simple manner.

7. References

Abdulazim, M. N. & Göckler, H. G. (2005a). Design options of the versatile two-channel SBC-FDFMUX filter bank, *Proc. Eur. Conf. Circuit Theory Design*, Cork, Ireland.

Abdulazim, M. N. & Göckler, H. G. (2005b). Efficient digital on-board de- and remultiplexing of FDM signals allowing for flexible bandwidth allocation, *Proc. Int. Commun. Satellite Syst. Conf.*, Rome, Italy.

Abdulazim, M. N. & Göckler, H. G. (2006). Flexible bandwidth reallocation MDFT SBC-FDFMUX filter bank for future bent-pipe FDM satellite systems, *Proc. Int. Workshop Signal Processing Space Commun.*, Noordwijk, Netherlands.

Abdulazim, M. N., Kurbiel, T. & Göckler, H. G. (2007). Modified DFT SBC-FDFMUX filter bank systems for flexible frequency reallocation, *Proc. Eur. Signal Processing Conf.*, Poznan, Poland, pp. 60–64.

Akyildiz, I. F., Lee, W. Y., Vuran, M. C. & Mohanty, S. (2006). Next generation/dynamic spectrum access/cognitive radio wireless networks: A survey, *Comput. Networks J. (Elsevier)* 50: 2127–2159.

Akyildiz, I. F., Lee, W. Y., Vuran, M. C. & Mohanty, S. (2008). A survey on spectrum management in cognitive radio networks, *IEEE Commun. Mag.* 46(4): 40–48.

Arbesser-Rastburg, B., Bellini, R., Coromina, F., Gaudenzi, R. D., del Rio, O., Hollreiser, M., Rinaldo, R., Rinous, P. & Roederer, A. (2002). R&D directions for next generation broadband multimedia systems: An ESA perspective, *Proc. AIAA Int. Commun. Satellite Syst. Conf. Exhibit*, Montreal, Canada.

Babic, D., Vesma, J., Saramäki, T. & Renfors, M. (2002). Implementation of the transposed Farrow structure, *Proc. IEEE Int. Symp. Circuits Syst.*, Vol. 4, pp. 5–8.

Beaulieu, F. D. & Champagne, B. (2009). Design of prototype filters for perfect reconstruction DFT filter bank transceivers, *Signal Processing* 89(1): 87–98.

Bianchi, T. & Argenti, F. (2007). SVD-Based techniques for zero-padded block transmission over fading channels, *IEEE Trans. Signal Processing* 55(2): 594–604.

Bianchi, T., Argenti, F. & Re, E. D. (2005). Performance of filterbank and wavelet transceivers in the presence of carrier frequency offset, *IEEE Trans. Commun.* 53(8): 1323–1332.

Borna, B. & Davidson, T. N. (2007). Biwindowed discrete multitone transceiver design, *IEEE Trans. Signal Processing* 55(8): 4217–4226.

Bregović, R. & Saramäki, T. (2005). A systematic technique for designing linear-phase FIR prototype filters for perfect-reconstruction cosine-modulated and modified DFT filter banks, *IEEE Trans. Signal Processing* 53(8): 3193–3201.

Cabric, D. & Brodersen, R. W. (2005). Physical layer design issues unique to cognitive radio systems, *Proc. IEEE Personal Indoor Mobile Radio Commun.*

Cabric, D., ÓDonnell, I. D., Chen, M. S. W. & Brodersen, R. W. (2006). Spectrum sharing radios, *IEEE Circuits Syst. Mag.* 6(2): 30–45.

Chen, K.-H. & Chiueh, T.-D. (2008). A cognitive radio system using discrete wavelet multitone modulation, *IEEE Trans. Circuits Syst. I* 55(10): 3246–3258.

Chiang, H.-T., Phoong, S.-M. & Lin, Y.-P. (2007). Design of nonuniform filter bank transceivers for frequency selective channels, *EURASIP J. Adv. Signal Processing* 2007, Article ID 61396.

de Barcellos, L. C. R., Diniz, P. S. R. & Netto, S. L. (2006). A generalized oversampled structure for cosine-modulated transmultiplexers and filter banks, *Circuits Syst. Signal Processing* 25(2): 131–151.

Devroye, N., Mitran, P. & Tarokh, V. (2006). Achievable rates in cognitive radio channels, *IEEE Trans. Inform. Theory* 52(5): 1813–1827.

Diniz, P. S. R., de Barcellos, L. C. R. & Netto, S. L. (2004). Design of high-resolution cosine-modulated transmultiplexers with sharp transition band, *IEEE Trans. Signal Processing* 52(5): 1278–1288.

Eghbali, A., Johansson, H. & Löwenborg, P. (2007a). An arbitrary bandwidth transmultiplexer and its application to flexible frequency-band reallocation networks, *Proc. Eur. Conf. Circuit Theory Design*, Seville, Spain, pp. 248–251.

Eghbali, A., Johansson, H. & Löwenborg, P. (2007b). Flexible frequency-band reallocation MIMO networks for real signals, *Proc. Int. Symp. Image Signal Processing Anal.*, Istanbul, Turkey, pp. 75–80.

Eghbali, A., Johansson, H. & Löwenborg, P. (2008a). A Farrow-structure-based multi-mode transmultiplexer, *Proc. IEEE Int. Symp. Circuits Syst.*, Seattle, Washington, USA, pp. 3114–3117.

Eghbali, A., Johansson, H. & Löwenborg, P. (2008b). A multimode transmultiplexer structure, *IEEE Trans. Circuits Syst. II* 55(3): 279–283.

Eghbali, A., Johansson, H. & Löwenborg, P. (2009a). Flexible frequency-band reallocation: complex versus real, *Circuits Syst. Signal Processing* 28(3): 409–431.

Eghbali, A., Johansson, H. & Löwenborg, P. (2009b). On the filter design for a class of multimode transmultiplexers, *Proc. IEEE Int. Symp. Circuits Syst.*, Taipei, Taiwan, pp. 89–92.

Eghbali, A., Johansson, H. & Löwenborg, P. (2010). Reconfigurable nonuniform transmultiplexers based on uniform filter banks, *Proc. IEEE Int. Symp. Circuits Syst.*, Paris, France, pp. 2123–2126.

Eghbali, A., Johansson, H. & Löwenborg, P. (2011a). A class of multimode transmultiplexers based on the Farrow structure, *Circuits Syst. Signal Processing* . accepted.

Eghbali, A., Johansson, H. & Löwenborg, P. (2011b). Reconfigurable nonuniform transmultiplexers using uniform modulated filter banks, *IEEE Trans. Circuits Syst. I* 58(3): 539–547.

Eghbali, A., Johansson, H., Löwenborg, P. & Göckler, H. G. (2011). Flexible dynamic frequency-band reallocation and allocation: from satellite-based communication systems to cognitive radios, *J. Signal Processing Syst.* 62(2): 187–203.

Evans, B., Werner, M., Lutz, E., Bousquet, M., Corazza, G. E., Maral, G. & Rumeau, R. (2005). Integration of satellite and terrestrial systems in future multimedia communications, *IEEE Wireless Commun. Mag.* 12(5): 72–80.

Farrow, C. W. (1988). A continuously variable digital delay element, *Proc. IEEE Int. Symp. Circuits Syst.*, Vol. 3, Espoo, Finland, pp. 2641–2645.

Farserotu, J. & Prasad, R. (2000). A survey of future broadband multimedia satellite systems, issues and trends, *IEEE Commun. Mag.* 38(6): 128–133.

Fliege, N. J. (1995). *Multirate Digital Signal Processing*, John Wiley & Sons.

Furtado, M. B. J., Diniz, P. S. R., Netto, S. L. & Saramäki, T. (2005). On the design of high-complexity cosine-modulated transmultiplexers based on the frequency-response masking approach, *IEEE Trans. Circuits Syst. I* 52(11): 2413–2426.

Ganesan, G. & Li, Y. G. (2005). Cooperative spectrum sensing in cognitive radio networks, *Proc. IEEE Int. Dynamic Spectrum Access Networks*, pp. 137–143.

Göckler, H. G. & Abdulazim, M. N. (2005). Joint oversampling FDM demultiplexing and perfectly reconstructing SBC filter bank for two channels, *Proc. Eur. Signal Processing Conf.*, Antalya, Turkey.

Göckler, H. G. & Abdulazim, M. N. (2007). Tree-structured MIMO FIR filter banks for flexible frequency reallocation, *Proc. Int. Symp. Image Signal Processing Anal.*, Istanbul, Turkey.

Haykin, S. (2005). Cognitive radio: Brain-empowered wireless communications, *IEEE J. Select. Areas Commun.* 23(2): 201–220.

Heller, P. N. ., Karp, T. & Nguyen, T. Q. (1999). A general formulation of modulated filter banks, *IEEE Trans. Signal Processing* 47(4): 986–1002.

Hentschel, T. & Fettweis, G. (2000). Sample rate conversion for software radio, *IEEE Commun. Mag.* 38(8): 142–150.

Ihalainen, T., Stitz, T. H., Rinne, M. & Renfors, M. (2007). Channel equalization in filter bank based multicarrier modulation for wireless communications, *EURASIP J. Appl. Signal Processing* 2007, Article ID 49389.

Johansson, H. & Hermanowicz, E. (2006). Adjustable fractional-delay filters utilizing the Farrow structure and multirate techniques, *Proc. Int. Workshop Spectral Methods Multirate Signal Processing*, Florence, Italy.

Johansson, H. & Löwenborg, P. (2003). On the design of adjustable fractional delay FIR filters, *IEEE Trans. Circuits Syst. II* 50(4): 164–169.

Johansson, H. & Löwenborg, P. (2005). Flexible frequency band reallocation network based on variable oversampled complex-modulated filter banks, *Proc. IEEE Int. Conf. Acoust. Speech Signal Processing*, Philadelphia, USA.

Johansson, H. & Löwenborg, P. (2007). Flexible frequency-band reallocation networks using variable oversampled complex-modulated filter banks, *EURASIP J. Adv. Signal Processing* 2007, Article ID 63714.

Jondral, F. K. (2005). Software defined radio: Basics and evolution to cognitive radio, *EURASIP J. Wireless Commun. Networking* .

Jondral, F. K. (2007). Cognitive radio: A communications engineering view, *IEEE Wireless Commun. Mag.* 14(4): 28–33.

Karp, T. & Fliege, N. J. (1999). Modified DFT filter banks with perfect reconstruction, *IEEE Trans. Circuits Syst. II* 46(11): 1404–1414.

Kovačević, J. & Vetterli, M. (1993). Perfect reconstruction filter banks with rational sampling factors, *IEEE Trans. Signal Processing* 41(6): 2047–2066.

Leaves, P., Moessner, K., Tafazolli, R., Grandblaise, D., Bourse, D., Tonjes, R. & Breveglieri, M. (2004). Dynamic spectrum allocation in composite reconfigurable wireless networks, *IEEE Commun. Mag.* 42(5): 72–81.

Li, J., Nguyen, T. Q. & Tantaratana, S. (1997). A simple design method for near-perfect-reconstruction nonuniform filter banks, *IEEE Trans. Signal Processing* 45(8): 2105–2109.

Lin, Y.-P. & Phoong, S.-M. (2001). ISI-free FIR filterbank transceivers for frequency-selective channels, *IEEE Trans. Signal Processing* 49(11): 2648–2658.

Lippolis, G., Simone, L., Comparing, M. C., Gelfusa, D., Piloni, V. & Novello, R. (2004). Overview on band-pass sampling approaches for on-board processing, *Proc. IEEE Int. Symp. Signal Processing Inf. Theory*, pp. 543–548.

Lucente, M., Re, E., Rossi, T., Sanctis, M. D., Stallo, C., Cianca, E. & Ruggieri, M. (2008). Future perspectives for the new European data relay system, *Proc. IEEE Aerospace Conf.*, pp. 1–7.

Martin-Martin, P., Bregovic, R., Martin-Marcos, A., Cruz-Roldan, F. & Saramäki, T. (2008). A generalized window approach for designing transmultiplexers, *IEEE Trans. Circuits Syst. I* 55(9): 2696–2706.

Martin-Martin, P., Cruz-Roldan, F. & Saramäki, T. (2005). Optimized transmultiplexers for multirate systems, *Proc. IEEE Int. Symp. Circuits Syst.*, Vol. 2, Kobe, Japan, pp. 1106–1109.

Mirabbasi, S. & Martin, K. (2003). Overlapped complex-modulated transmultiplexer filters with simplified design and superior stopbands, *IEEE Trans. Circuits Syst. II* 50(5): 1170–1183.

Mitra, S. K. (2006). *Digital Signal Processing: A Computer Based Approach*, McGraw-Hill.

Nguyen, T., Hant, J., Taggart, D., Tsang, C.-S., Johnson, D. M. & Chuang, J.-C. (2002). Design concept and methodology for the future advanced wideband satellite system, *Proc. IEEE Military Commun. Conf.*, Vol. 1, USA, pp. 189–194.

Princen, J. (1994). The design of nonuniform modulated filter banks, *Proc. IEEE Int. Symp. Time-Frequency Time-Scale Anal.*, pp. 112–115.

Pun, C. K. S., Wu, Y. C., Chan, S. C. & Ho, K. L. (2003). On the design and efficient implementation of the Farrow structure, *IEEE Signal Processing Lett.* 10(7): 189–192.

Ramacher, U. (2007). Software-defined radio prospects for multistandard mobile phones, *IEEE Computer* 40(10): 62–69.

Re, E. D. & Pierucci, L. (2002). Next-generation mobile satellite networks, *IEEE Commun. Mag.* 40(9): 150–159.

Rosenbaum, L., Johansson, H. & Löwenborg, P. (2006). Oversampled complex-modulated causal IIR filter banks for flexible frequency-band reallocation networks, *Proc. Eur. Signal Processing Conf.*, Florence, Italy.

Saramäki, T. & Bregović, R. (2001a). An efficient approach for designing nearly perfect-reconstruction cosine-modulated and modified DFT filter banks, *Proc. IEEE Int. Conf. Acoust. Speech Signal Processing*.

Saramäki, T. & Bregovic, R. (2001b). An efficient approach for designing nearly perfect-reconstruction cosine-modulated and modified DFT filter banks, *Proc. IEEE Int. Conf. Acoust. Speech Signal Processing*, Vol. 6, pp. 3617–3620.

Sherman, M., Mody, A. N., Martinez, R., Rodriguez, C. & Reddy, R. (2008). IEEE standards supporting cognitive radio and networks, dynamic spectrum access, and coexistence, *IEEE Commun. Mag.* 46(7): 72–79.

Tseng, C. C. (2002). Design of variable fractional delay FIR filter using differentiator bank, *Proc. IEEE Int. Symp. Circuits Syst.*, Vol. 4, pp. 421–424.

Vaidyanathan, P. P. (1993). *Multirate Systems and Filter Banks*, Prentice-Hall, Englewood Cliffs, NJ.

Vaidyanathan, P. P. & Vrcelj, B. (2004). Transmultiplexers as precoders in modern digital communications: A tutorial review, *Proc. IEEE Int. Symp. Circuits Syst.*, Vol. 5, pp. 405–412.

Vesma, J. & Saramäki, T. (1996). Interpolation filters with arbitrary frequency response for all-digital receivers, *Proc. IEEE Int. Symp. Circuits Syst.*

Vesma, J. & Saramäki, T. (1997). Optimization and efficient implementation of FIR filters with adjustable fractional delay, *Proc. IEEE Int. Symp. Circuits Syst.*, Vol. IV, Hong Kong, pp. 2256–2259.

Vesma, J. & Saramäki, T. (2000). Design and properties of polynomial-based fractional delay filters, *Proc. IEEE Int. Symp. Circuits Syst.*, Geneva, Switzerland.

Viholainen, A., Alhava, J. & Renfors, M. (2006). Efficient implementation of complex modulated filter banks using cosine and sine modulated filter banks, *EURASIP J. Appl. Signal Processing* 2006, Article ID 58564.

Viholainen, A., Saramäki, T. & Renfors, M. (1999). Nearly perfect-reconstruction cosine-modulated filter banks for VDSL modems, *Proc. IEEE Int. Conf. Electron. Circuits Syst.*, pp. 373–376.

Wittig, M. (2000). Satellite on-board processing for multimedia applications, *IEEE Commun. Mag.* 38(6): 134–140.

Xie, X. M., Chan, S. C. & Yuk, T. I. (2005). Design of perfect-reconstruction nonuniform recombination filter banks with flexible rational sampling factors, *IEEE Trans. Circuits Syst. I* 52(9): 1965–1981.

Xie, X. M., Chen, X. Y. & Sho, G. M. (2005). A simple design method of linear-phase nonuniform filter banks with integer decimation factors, *Proc. IEEE Int. Midwest Symp. Circuits Syst.*, pp. 7–10.

Yuan, Y., Bahl, P., Chandra, R., Chou, P. A., Ferrell, J. I., Moscibroda, T., Narlanka, S. & Wu, Y. (2007). KNOWS: Cognitive radio networks over white spaces, *Proc. IEEE Int. Dynamic Spectrum Access Networks*, pp. 416–427.

Yuan, Y., Bahl, P., Chandra, R., Moscibroda, T., Narlanka, S. & Wu, Y. (2007). Allocating dynamic time-spectrum blocks in cognitive radio networks, *Proc. ACM Int. Symp. Mobile Ad Hoc Networking Comput.*

Zhao, Q. & Sadler, B. M. (2007). A survey of dynamic spectrum access, *IEEE Signal Processing Mag.* 24(3): 79–89.

Zhao, Q. & Swami, A. (2007). A survey of dynamic spectrum access: Signal processing and networking perspectives, *Proc. IEEE Int. Conf. Acoust. Speech Signal Processing*, Vol. 4, pp. 1349–1352.

9

Opportunistic Spectrum Access in Cognitive Radio Network

Waqas Ahmed[1], Mike Faulkner[1] and Jason Gao[2]
[1]*Centre for Telecommunication and Microelectronics, Victoria University*
[2]*School of Computer & Information Engineering, Shanghai University of Electrical Power*
[1]*Australia*
[2]*China*

1. Introduction

Cognitive radio (Mitola III, 2000) also known as opportunistic spectrum access (OSA) has emerged as a promising solution to increase the spectrum efficiency Haykin (2005). In OSA, the SU finds spectrum holes (white space) by sensing the radio frequency spectrum. The presence of spectrum holes in the PU channels are highlighted in Fig. 1. These spectrum holes are used by the SU for its transmission. This scheme is often referred to as opportunistic spectrum access (OSA). No concurrent transmission of the PU and the SU is allowed. The SU must vacate the channel as soon as the PU reappears, which leads to the forced termination of the SU connection. Since the SU has no control over the resource availability, the transmission of the SU is blocked when the channel is occupied by the PU. The forced termination and blocking of a SU connection is shown in Fig. 2. The forced termination probability and blocking probability are the key parameters which determine the throughput of the SU, and thus its viable existence. The forced termination depends on the traffic behaviour of the PUs and the SUs (e.g. arrival rates, service time etc.). In the case of multiple SU groups with different traffic statistics, the forced termination and blocking probabilities lead to unfairness among the SU groups. The QoS provisioning task becomes difficult.

2. Related work and contributions

In the existing literature, several authors Weiss & Jondral (2004)-Ahmed et al. (2010) have studied the forced termination and blocking probabilities for one and two groups of SUs. In these papers, spectrum pooling Weiss & Jondral (2004) is used as a base system model. Spectrum pooling refers to an OSA paradigm which enables the PU network to rent out its idle spectrum bands to the SU group. It is assumed that the SU group will be able to perform wideband sensing and during their transmissions will introduce spectral nulls in the frequency bands where they find the PU active. It is suggested that in order to accommodate simultaneous SU connections, a wideband PU channel can be divided into multiple narrowband subchannels for SU access. The Continuous Time Markov Chain (CTMC) Mehdi (1991) is extensively used in modelling spectrum sharing scenarios based on interweave access. To simplify mathematical analysis it is commonly assumed that the traffic behaviour of the PU and the SU groups obey a Markov (memoryless) property i.e. the arrival rates follow a Poisson distribution and the service rates follow an exponential distribution. In

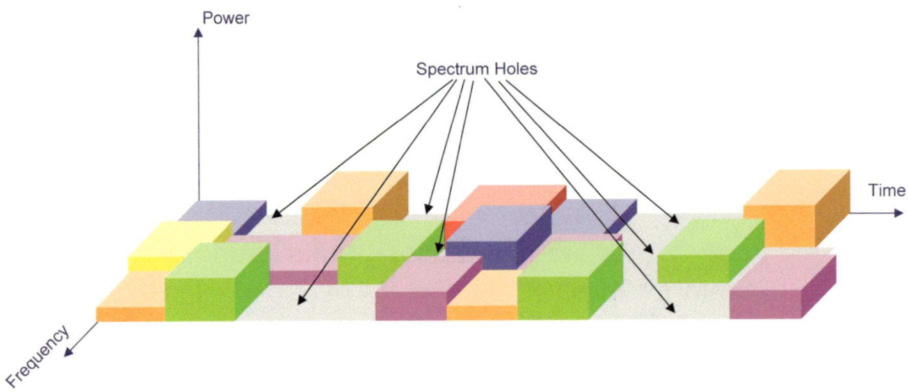

Fig. 1. Conceptual view of spectrum holes for interweave approach

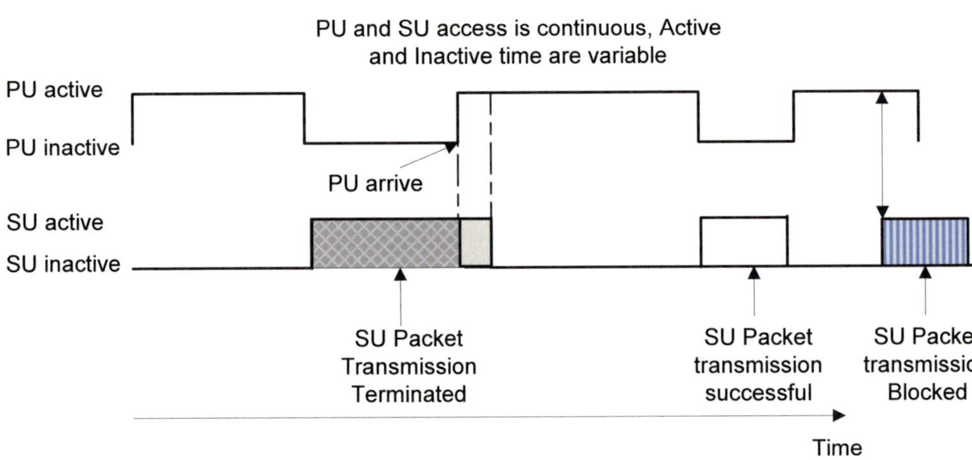

Fig. 2. Illustration of forced termination and blocking

addition, the SU connections are coordinated through a central entity (centralised secondary network) which ensures no collision between the SU connections i.e, no concurrent SU connections on the same frequency band or subchannel. The SU connections have negligible access, switching delays and perfect sensing information is available to them at all times. In a CTMC model, the number of connections from the user groups are represented by states (written as $n - tuple$), such as (x_1,x_2), where x_1 and x_2 may represent the number of PU and SU connections. The transition of one state to another is based on the Markov propertyMehdi (1991) which assumes memoryless arrival and departure. Under stationarity assumptions, the rate of transition from and to a state is equal. This fundamental fact is used to calculate the state probabilities under the constraint that the sum of all state probabilities is equal to 1. These state probabilities are further used to calculate the parameters of interest. The papers dealing with QoS analysis can be divided into two scenarios: without forced termination; with forced termination, described in the following.

2.1 Without forced termination

The main aim in this type of system model is to understand the blocking probability tradeoff between two users with different bandwidth requirements. In Raspopovic et al. (2005) while investigating the blocking probability tradeoff between equal priority users, the authors concluded that the blocking probability of the wideband user is lower bounded by the blocking probability of the narrowband user. The best probability tradeoff can only be achieved by changing the arrival rates of both the wideband and the narrowband user Raspopovic & Thompson (2007). This fact is used in Xing et al. (2007) where it is shown that the optimal arrival rates which achieve airtime fairness can be derived using the Homo-Egualis model. Furthermore, the blocking probability in the network can be reduced when the narrowband users pack the channels (Xing et al., 2007, Fig. 6) by employing spectrum handoff[1]. The authors in Chou et al. (2007) developed an upper bound on the throughput that can be achieved using spectrum agility with a listen before talk rule[2]. It was found that a fixed channel assignment strategy yields better results than spectrum agility under heavy load conditions. Although, the above mentioned schemes provide a fair understanding of the blocking probability and airtime behaviour in a CRN, these analyses are only applicable to users in open unlicensed networks (e.g. ISM bands in 900MHz, 2.4 GHz and 5GHz), when the SU group utilizes the unlicensed channels for coordination or backup (in case the PU channels are fully occupied Xing et al. (2007)).

2.2 With forced termination

Initial investigation carried out in Capar et al. (2002), focused on the advantages of primary assisted SU spectrum sharing; termed controlled spectrum sharing. In controlled spectrum sharing, the PU network assigns channels to its users (PUs) so as to avoid termination of SU connections. It is concluded that this approach increases bandwidth utilization without causing any significant increase in the blocking probability of the SU connections. No analytical expressions were derived for the forced termination probability and the throughput. In another controlled spectrum sharing scenario Tang et al. (2006); Tang & Chew (2010), a number of channels are allocated for SU connections by the PU network. Under saturated condition an incoming PU connection can also occupy the secondary allocated channels. Although this scheme decreases the forced termination probability, it comes with a cost of an increased SU blocking probability. To counter the effect of the forced termination the authors in Huang et al. (2008) have proposed and analyzed random access schemes employing different sensing and backoff mechanisms. However, the analysis is limited to a single channel and saturated SU traffic conditions.

To avoid this foreseeable termination, spectrum handoff techniques have also been investigated from the forced termination and blocking probability perspective. Generally, spectrum handoff techniques can be categorised as either reactive or proactive. In the reactive approach, the SU moves to another vacant channel only when the current channel is reoccupied by a PU, whereas in the proactive approach the SU avoids collision and switches

[1] Spectrum handoff allows the SU connection to move to another vacant frequency channel during its transmission.

[2] Spectrum agility refers to the ability of a user to access multiple channels. Please note that spectrum agility is a stepping stone for spectrum handoff.

to another vacant channel before a PU reappears. The channel switching is performed based on the observed statistics of the PU channels[3]. To further ensure the continuity of SU transmission, inband channel reservation (number of channels are reserved exclusively for handoff calls within the primary channels) and out-of-band channel reservation have also been suggested Zhu et al. (2007), Al-Mahdi et al. (2009).

2.2.1 Reactive spectrum handoff

In Zhu et al. (2007), authors investigated the impact of spectrum handoff and channel reservation on the forced termination and blocking probabilities. However, the paper presented an incomplete CTMC analysis. In addition the expressions are derived based on a false assumption of completion probability from each Markov state. The corrections on this work is provided in Ahmed et al. (2009); Zhang et al. (2008); Bauset et al. (2009). The work in Ahmed et al. (2009) forms part of this chapter and discussed in chapter 3. Although, the authors in Zhang et al. (2008) included the queuing of SU connections, the analysis is limited to only handoff scenario. The investigations in Zhang et al. (2008); Bauset et al. (2009) do not address the incomplete Markov analysis and throughput. The approach presented in Ahmed et al. (2009) can be scaled to address the inaccurate probability analysis in Al-Mahdi et al. (2009). It has been shown in Tzeng (2009) that an adaptive channel reservation scheme gives better performance than the fixed reservation policy. This adaption is based on traffic statistics of the PU. A relatively similar conclusion has been drawn in Bauset et al. (2010). Extending the work in Zhu et al. (2007)-Tzeng (2009), the authors in Kannappa & Saquib (2010) showed that the forced termination probability can be reduced by assigning multiple subchannels to each SU connection (reducing their transmission time). The forced termination probability expression derived in this paper is the termination probability seen by an incoming SU (an incoming SU call may be blocked) rather than the termination probability on the channel. The throughput in Bauset et al. (2009)-Kannappa & Saquib (2010) is defined as the average completion rate of SU connections which does not include the average duration of completed SU connections. In order to address this problem, the duration of completed SU connections is calculated in Heo (2008), however, the analytical results derived are not very accurate. A more compact and accurate expression of throughput is derived in Ahmed et al. (2009). The authors in Xue et al. (2009) consider a PU network which has three different user classes. The users are admitted into the network based on a Guarded Threshold channel reservation Policy Ramjee et al. (1996). One of the classes is treated as a SU group, while the remaining two have PU status. The results indicate that such a scheme is only useful when the collision probability between the SU class and the two PU classes is high.

Based on a simplified system model, the link maintenance probability and the forced termination probability of delay sensitive SU connections are discussed in Zhang (2009)-Willkomm et al. (2005). Including imperfect sensing, the probability derivations have been carried out in Tang & Mark (2009), where authors derive the collision probability rather than the forced termination probability[4]. In Wang & Anderson (2008); Willkomm et al. (2005)

[3] This chapter only considers reactive spectrum handoff techniques. Some interesting proactive spectrum handoff strategies are provided in Yuan (2010)-Yoon& Ekici (2010).

[4] Without spectrum handoff the collision probability and forced termination probability is the same. However in the case of spectrum handoff, the SU connection can collide with the PU multiple times without being terminated. In essence collision probability calculation is performed on the basis of a single PU arrival, whereas the forced termination calculation are performed on the basis of multiple PU arrivals in the SU connection's life time Ahmed et al. (2009)Zhang et al. (2008)

it has been found that reactive spectrum handoff based on dynamic channel selection is a better strategy than the static strategy, however, it results in an increased sensing overhead. To decrease this overhead, a handoff reduction policy is proposed in Khalil et al. (2010) using connection success rate as a key metric Lin et al. (2009). Based on our findings Ahmed et al. (2009), the authors in Chung et al. (2010) proposed a channel allocation technique which yields almost similar performance to a spectrum handoff system.

From the perspective that CRN will be able to support multiple SU services, call admission control is investigated in Wang et al. (2009)-Tumuluru et al. (2011). The authors in Wang et al. (2009) derived optimal access probabilities to achieve proportional fairness among two SU groups. These results are derived based on a loose definition of throughput without considering the forced termination. A channel packing scheme which provides 10%- 15% gain over random channel access has been presented in Luo & Roy (2009). The optimal access rates are derived only for a single wideband channel and three different type of users: one wideband PU: one wideband SU group; and one narrowband SU group. Although overall forced termination probability is also shown, it is strictly numerical and does not provide any insight into the individual forced termination probabilities of the two SU groups. A fair opportunistic spectrum access scheme with emphasis on two SU groups has been presented in Ma et al. (2008). Although the effect of collision probability is included, the results are only valid when the SU groups have the same connection length. The prioritisation among SU traffic from a physical layer and network integration perspective have been studied in Wiggins et al. (2008); Gosh et al. (2009). Moreover, as identified in Tumuluru et al. (2011), these works do not include the effect of spectrum handoff. The probability aspects of prioritisation between two SU groups have been recently analyzed in Tumuluru et al. (2011). Subchannel reservation policies were investigated to achieve the QoS of the prioritised SU group. The probability derivations were very similar to those previously presented by us in Ahmed et al. (2010); Ahmed et al. (2009) and now described in this Chapter. Specifically, in this chapter

- A complete and exact CTMC analyses is presented (compared to Zhu et al. (2007); Zhang et al. (2008); Bauset et al. (2009)). The QoS parameter are obtained for spectrally agile single SU group operating in multichannel PU network. These derivations also include SU spectrum handoff and channel reservation scenarios.
- Spectrum sharing between two SU groups is investigated in terms of probability tradeoff gains and airtime fairness. It is shown that compared to non-termination scenarios Raspopovic et al. (2005)-Chou et al. (2007); Wang et al. (2009) the access rates have very little impact on airtime fairness. A channel partitioning approach is developed to achieve airtime fairness.

The remainder of the chapter is organised as follows. Section 3 presents the system model and key assumptions. Section 4 extends the discussion to include multiple PU channels and two SU groups. A single SU group is treated as a special case. The airtime fairness among the two SU groups is analyzed in Section 5. Section 6 concludes the chapter.

3. System model

In this chapter, a widely acceptable spectrum pooling model (Weiss & Jondral, 2004, Fig. (3)) is adopted, in which a vacant wideband PU channel is divided into multiple narrowband subchannels. These narrowband subchannels are used by SU groups for their opportunistic transmissions. Fig. 3 shows the system model, in which there are K available PU channels

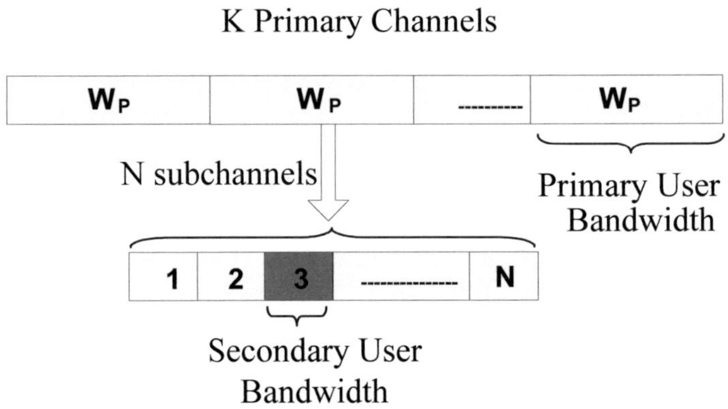

Fig. 3. Channel arrangements of PU and SU channels.

Symbol	Parameter
k	Number of PU connections
i	Number of SU connections of S^1 and S^A
j	Number of SU connections of S^B
l	Number of terminated SU connections S^1 and S^A
m	Number of terminated SU connections S^B

Table 1. List of Symbols for CTMC

$k \in \{1, 2.., K\}$. Each of the PU channels has a fixed bandwidth W_p, which is subdivided into N subchannels of bandwidth, $W_s = W_p/N$. The service duration $\frac{1}{\mu_p}$ of the PU connections are assumed to be exponentially distributed. The arrival rate λ_p of new connections from the PU group follow an independent Poisson process. It is assumed that the PU network is an M/M/m/m loss network, where channel occupancy only depends on the mean service rate of PU group Mehdi (1991). The exponentially distributed mean service duration of S^1, S^A and S^B are denoted by $\frac{1}{\mu_{se}^1}$, $\frac{1}{\mu_{se}^A}$ and $\frac{1}{\mu_{se}^B}$ respectively. Similarly, the Poisson arrival rates are λ_s^1, λ_s^A, and λ_s^B.

4. Multiple PU channels ($K \geq 1$, $N \geq 1$) and two SU groups

In this section, we first investigate the impact of spectrum sharing between two SU groups in a CRN with spectrum agility. We refer to this as the "basic system". In a basic system, the SU connection can access one of the available PU channels, however, no spectrum handoff is allowed (Fig. 4). Each of the two SU groups has different traffic statistics. Second, we also investigate the impact of horizontal handoff and channel reservation. The single SU group is treated as a special case, by letting $\theta_{se} = \frac{\lambda_s}{\mu_{se}}$ tend to 0 for one of the SU groups. Such a spectrum sharing scenario has partially been considered in Wang et al. (2009), where the effect of forced termination was neglected. This section includes the forced termination aspect, and therefore gives a more realistic result. For mathematical convenience it is assumed that the SU header length is 0.

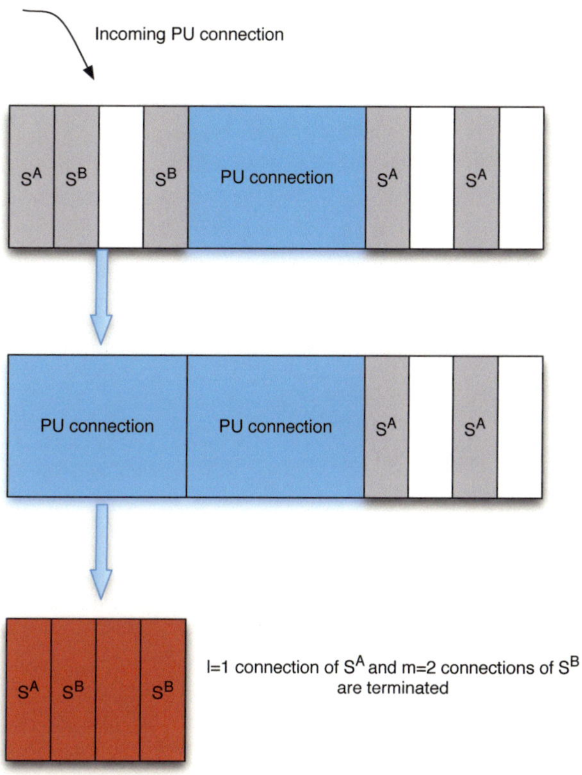

Fig. 4. Illustration of the Basic system.

4.1 Basic system

Fig. 5 shows the state transition diagram of a basic system. The states in the model are given by the number of active connections in the system, i.e., (i, j, k), whereas i, j are the number of active SU connections of S^A and S^B overlaying k active PU connections. The transition rate from state to state is given by the labels on the arrows. An example of a forced termination is the corner state $(i-l, j-m, k+1)$ from state (i, j, k) (shown as a dashed arrow in Fig. 5), where l and m represent the number of terminated active connections from S^A and S^B respectively. Following (Zhu et al., 2007, Eq. (1)), the termination of $q = l+m$ out of $x = i+j$ SU connections follows a hypergeometric distribution, and can be written as

$$P^{(i+j,k)}_{((i+j)-(l+m),k+1)} = \frac{\binom{N}{l+m}\binom{(K-k-1)N}{(i+j)-(l+m)}}{\binom{(K-k)N}{i+j}}. \tag{1}$$

Given i and j, the probability of exactly l and m can be written as

$$P^{(i,j)}_{(i-l,j-m)} = \frac{\binom{i}{l}\binom{j}{m}}{\binom{i+j}{l+m}}. \tag{2}$$

t

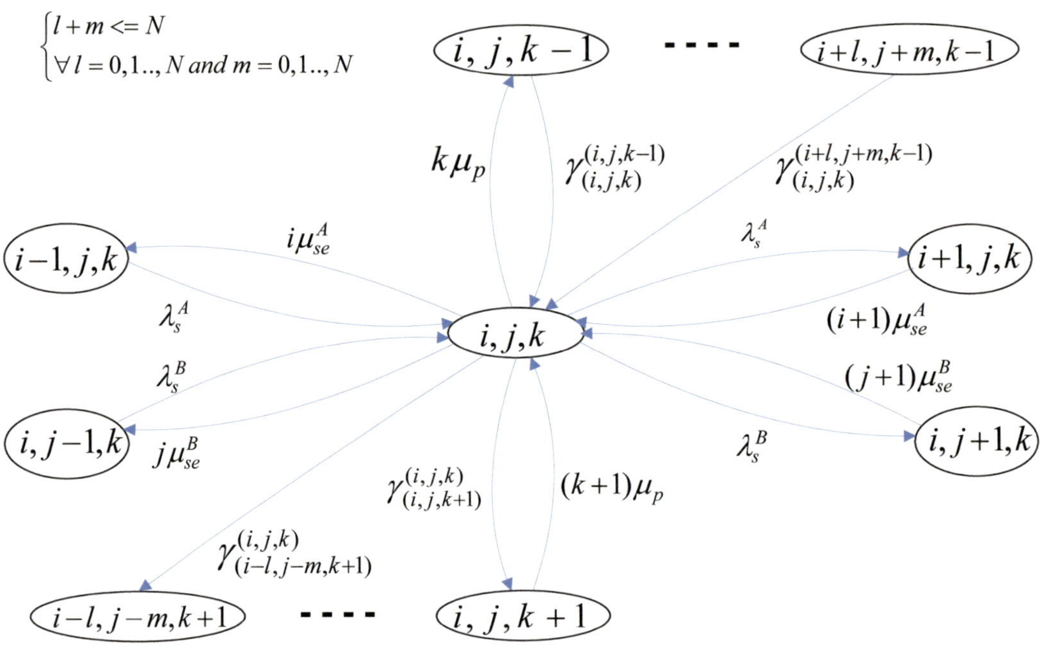

Fig. 5. CTMC for the basic system.

Combining (1) and (2) gives the probability of having exactly l and m SU connections terminated from the state (i, j, k), and can be calculated from

$$P_{(i-l,j-m,k+1)}^{(i,j,k)} = P_{(i-l,j-m)}^{(i,j)} P_{((i+j)-(l+m),k+1)}^{(i+j,k)} \tag{3}$$

The termination of l and m SU connections occurs at the arrival of each PU connection. Therefore, given the arrival rate of a PU connection λ_P, the state transition *rate* $\gamma_{(i-l,j-m,k+1)}^{(i,j,k)}$ can be written as

$$\gamma_{(i-l,j-m,k+1)}^{(i,j,k)} = P_{(i-l,j-m,k+1)}^{(i,j,k)} \lambda_P \tag{4}$$

By substituting (1) - (3) in (4), the state transition rate from (i, j, k) to $(i-l, j-m, k+1)$ is given as

$$\gamma_{(i-l,j-m,k+1)}^{(i,j,k)} = \frac{\dfrac{\binom{i}{l}\binom{j}{m}}{\binom{i+j}{l+m}} \binom{N}{l+m} \binom{(K-k-1)N}{i+j-l-m}}{\binom{(K-k)N}{i+j}} \lambda_P, \tag{5}$$

for $0 \le l+m \le N$. Setting j and m to zero, (5) is the same as (Zhu et al., 2007, Eq. (1)). From Fig. 5, a set of balance equations of the CTMC model for all $0 \le i, j \le KN$ and $0 \le k \le K$ can be written as

$$\lambda_s^A P_\phi(i-1,j,k) + (i+1)\mu_{se}^A P_\phi(i+1,j,k) + \lambda_s^B P_\phi(i,j-1,k) + (j+1)\mu_{se}^B P_\phi(i,j+1,k)$$
$$+(k+1)\mu_P P_\phi(i,j,k+1) + \sum_{l=0}^{N}\sum_{m=0}^{N} \gamma_{(i,j,k)}^{(i+l,j+m,k-1)} P_\phi(i+l,j+m,k-1)\delta(l+m \le N)$$
$$= \left(\lambda_s^A + i\mu_{se}^A + \lambda_s^B + j\mu_{se}^B + k\mu_P + \sum_{l=0}^{N}\sum_{m=0}^{N} \gamma_{(i-l,j-m,k+1)}^{(i,j,k)} \delta(l+m \le N)\right) P_\phi(i,j,k), \quad (6)$$

where, $P_\phi(i,j,k) = P(i,j,k)\phi(i,j,k)$, $\phi(i,j,k)$ is one for all valid states and zero for all non-valid states. Mathematically,

$$\phi(i,j,k) = \begin{cases} 1 & i+j \le (K-k)N, \\ 0 & \text{otherwise,} \end{cases} \quad (7)$$

and $P(i,j,k)$ denotes the probability that the system is in the state (i,j,k). The state probability must satisfy the following constraint

$$\sum_{i=0}^{NK}\sum_{j=0}^{NK}\sum_{k=0}^{K} P_\phi(i,j,k) = 1. \quad (8)$$

The set of equations expressed by (6) can be written as a multiplication of state transition matrix Q and state probability vector P

$$QP = 0. \quad (9)$$

Replacing the last row of Q, with the constraint in (8), the state probabilities can be solved by

$$P = Q^{-1}\epsilon, \quad (10)$$

where, $\epsilon^T = [0\,,\,0\,,\,..\,,\,1]$ and $(\cdot)^T$ is the transpose operator. For the basic system, the blocking of a new SU connection occurs when all PU channels are fully occupied. The state (i,j,k) is a blocking state if $i+j+Nk = NK$. The probabilities of all blocking states are summed to calculate the blocking probability which is given by

$$P_{Be}^{AB} = \sum_{i=0}^{NK}\sum_{j=0}^{NK}\sum_{k=0}^{K} \delta(i+j+Nk-NK)P_\phi(i,j,k). \quad (11)$$

From its definition and using the fact that the total termination rate (sum of forced and unforced termination rates) equals the connection rate in Fig. 5, the forced termination probability can be written as

$$\text{Forced Termination Probability} = \frac{\text{Total SU forced termination rate}}{\text{Total SU connection rate}} \quad (12)$$

For the given state (i,j,k), $\gamma_{(i-l,j-m,k+1)}^{(i,j,k)} P_\phi(i,j,k)$ is the termination rate of l and m SU connections. From the state (i,j,k), the total termination rate $F(i,j,k)$ can be written as

$$F(i,j,k) = \sum_{l=0}^{NK}\sum_{m=0}^{NK} (l+m)\gamma_{(i-l,j-m,k+1)}^{(i,j,k)} P_\phi(i,j,k)\delta_f, \quad (13)$$

where δ_f is given as

$$\delta_f = \begin{cases} 1 & 0 \leq l+m \leq N \\ 0 & \text{otherwise,} \end{cases} \tag{14}$$

The total connection rate is given by $(1 - P_{Be}^{AB})(\lambda_s^A + \lambda_s^B)$. Using (12) and (13) the combined forced termination probability P_{Fe}^{AB} of a SU connection can be written as

$$P_{Fe}^{AB} = \frac{\sum_{i=0}^{NK}\sum_{j=0}^{NK}\sum_{k=0}^{K}\sum_{l=0}^{N}\sum_{m=0}^{N}(l+m)\gamma_{(i-l,j-m,k+1)}^{(i,j,k)}P_{\phi}(i,j,k)\delta_f}{(1-P_{Be}^{AB})(\lambda_s^A+\lambda_s^B)}. \tag{15}$$

Similarly, the individual forced termination probabilities for user groups S^A and S^B can be respectively calculated as

$$P_{Fe}^{A} = \frac{\sum_{i=0}^{NK}\sum_{j=0}^{NK}\sum_{k=0}^{K}\sum_{l=0}^{N}\sum_{m=0}^{N}l\gamma_{(i-l,j-m,k+1)}^{(i,j,k)}P_{\phi}(i,j,k)\delta_f}{(1-P_{Be}^{AB})\lambda_s^A}, \tag{16}$$

and

$$P_{Fe}^{B} = \frac{\sum_{i=0}^{NK}\sum_{j=0}^{NK}\sum_{k=0}^{K}\sum_{l=0}^{N}\sum_{m=0}^{N}m\gamma_{(i-l,j-m,k+1)}^{(i,j,k)}P_{\phi}(i,j,k)\delta_f}{(1-P_{Be}^{AB})\lambda_s^B}. \tag{17}$$

From (15),(16) and (17), it can be shown that

$$P_{Fe}^{AB} = \frac{\lambda_s^A}{\lambda_s^A+\lambda_s^B}P_{Fe}^{A} + \frac{\lambda_s^B}{\lambda_s^A+\lambda_s^B}P_{Fe}^{B}. \tag{18}$$

Note that P_{Fe}^{AB} is not the average of P_{Fe}^{A} and P_{Fe}^{B} except for the special case where $\lambda_s^A = \lambda_s^B$.

4.1.1 Special case-single SU group

In this section, the forced termination and blocking probabilities for a single SU group are treated as a special case of two SU groups. Let P_{Fe}^1 and P_{Be}^1 be the forced termination and blocking probabilities of a single SU group respectively. Setting $j = 0$, $m = 0$, $\lambda_s^B = 0$, $\mu_{se}^B = 0$ in (5) and (6), we get the same equations as in (Zhu et al., 2007, Eqs. (1-2)) i.e., $\gamma_{(i-l,0,k+1)}^{(i,0,k)} = \gamma_{(i-l,k+1)}^{(i,k)}$ and $P_{\phi}(i,0,k) = P_{\phi}(i,k)$. Under this condition, the expression in (11) for the blocking probability is identical to that of the single SU group.

Similarly, it can be shown that the forced termination probability in (15) can be simplified to

$$P_{Fe}^{1} = \frac{\sum_{i=0}^{NK}\sum_{k=0}^{K}\sum_{l=1}^{N}l\gamma_{(i-l,k+1)}^{(i,k)}P_{\phi}(i,k)}{(1-P_{Be}^{1})\lambda_s^1}. \tag{19}$$

The term $\sum_{l=1}^{NK} l\gamma_{(i-l,k+1)}^{(i,k)}P_{\phi}(i,k)$ in (19) is the termination rate of the SU connections from the state (i,k). By summing over all valid states $\phi(i,0,k)$, we get the total number of terminated connections per unit time. Note that the forced termination probability P_{Fe}^1 from (Zhu et al., 2007, Eq. (4)) differs from the above, because it describes a rate rather than a probability.

4.2 Basic system with spectrum handoff and channel reservation

In this section, we extend the results derived in the previous section to system with spectrum handoff and channel reservation. As described previously, spectrum handoff allows an active SU connection to move to another vacant subchannel, rather being terminated by an incoming PU connection. This phenomena is illustrated in Fig. 6. To further ensure the continuity of existing SU connections, a small number of subchannels (marked R) are reserved exclusively for spectrum handoff purpose. These subchannels improve the forced termination probability at the expense of increased blocking probability.

In order to quantify the performance using spectrum handoff and channel reservation, the following measures are defined:

$$G_F(r) = \frac{P_{Fe}^{AB}(\text{Basic System}) - P_{Fe}^{AB}(r)}{P_{Fe}^{AB}(\text{Basic System})}, r = 0 \ldots N, \tag{20}$$

and

$$G_B(r) = \frac{P_{Be}^{AB}(r) - P_{Be}^{AB}(\text{Basic System})}{P_{Be}^{AB}(\text{Basic System})}, r = 0 \ldots N, \tag{21}$$

where, $G_F(r)$ expresses the improvement (fractional reduction) in forced termination probability of the basic system when r reserved channels are used, and $G_B(r)$ describes the degradation (fractional increase) in blocking probability with r reserved channels. Since N is the maximum number of terminations per PU connection, we only consider $r \leq N$. If $r > N$ the complexity of the CTMC increases significantly as it has to cater for multiple PU arrivals rather than a single arrival. We define a measure, *tradeoff gain* $L(r)$, which relates improvement in forced termination probability to the degradation in terms of the blocking probability as

$$L(r) = \frac{G_F(r)}{G_B(r)}. \tag{22}$$

The condition $L(r) >> 1$ indicates that the SU network has a greater improvement in forced termination probability compared to its increase in blocking probability.

Fig. 7 shows the state transitions of the CTMC with r reserved channels $0 \leq r < N$. The state diagram for $r = N$ requires slight modification. In Fig. 7(a), spectrum handoff ensures no termination when a new PU connection is made. Fig. 7(b)-(e) depicts the condition when the total number of vacant sub-channels is less than N. On the arrival of a new PU connection, we have $i + j + (k+1)N > KN$, therefore forced termination occurs. Fig. 7(b) allows new SU connections because the number of free subchannels is greater than r. Contrary to Fig. 7(b), Figs. 7(c)-(e) will allow no new SU connections since the number of free channels are less than or equal to r. Note that there is a number of states which have a single arrow to/from state (i, j, k) . These states include the cases where either forced termination occurs, or a new SU connection is blocked because there is no vacant subchannel except the reserved subchannels. The condition Figs. 7(d)-(e) also signifies that the SUs are utilising the reserved channels. The state (i, j, k) in Fig. 7(e) can also result from other states $(i + l, j - m, k - 1)$. In these transitions, forced termination occurs.

With spectrum handoff and channel reservation, the forced termination will only occur when $i + j + Nk > (K - 1)N$. The arrival of a new PU connection will cause the existing SU connections to pack themselves in $(K - (k + 1))N$ subchannels, i.e., the transition of

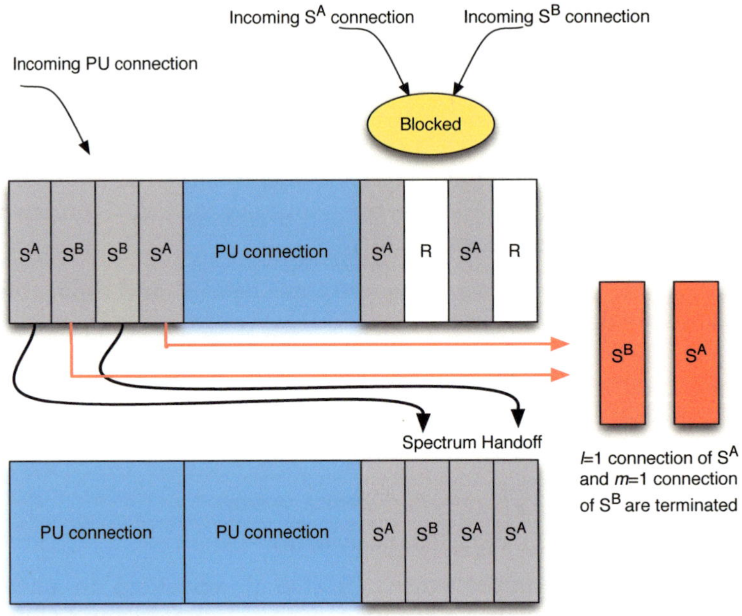

Fig. 6. Illustration of the system with spectrum handoff and channel reservation. The number of reserved channels $r = 2$.

current state (i, j, k) to state $(i - l, j - m, k + 1)$ with the *packing condition* $(i - l) + (j - m) = (K - k - 1)N$. The rate of this transition is expressed as

$$\gamma^{(i,j,k)}_{(i-l,j-m,k+1)} = \frac{\binom{i}{l}\binom{j}{m}}{\binom{i+j}{l+m}}\lambda_p. \tag{23}$$

The condition of a valid state $\phi_r(i, j, k)$ for all $i, j = \{0, \ldots, KN - r\}$ and $k = \{0, \ldots K\}$ is given as

$$\phi_r(i,j,k) = \begin{cases} 1 \; i + j \leq (K-k)N \quad \text{and i + j} \leq \text{KN} - \text{r} \\ 0 \qquad\qquad\qquad\qquad \text{otherwise.} \end{cases} \tag{24}$$

Similar to the basic system in the previous section, the set of balance equations for the CTMC are written. Together with the constraint (8), the state probabilities $P_{\phi_r}(i, j, k) = P(i, j, k)\phi_r(i, j, k)$ can be solved using (9) and (10). With spectrum handoff and channel reservation, the blocking of a new SU connection occurs when the number of vacant subchannels is less than or equal to r. Mathematically,

$$P^{AB}_{Be}(r) = \sum_{i=0}^{NK-r} \sum_{j=0}^{NK-r} \sum_{k=0}^{K} \delta(i + j + Nk \geq KN - r) P_{\phi_r}(i, j, k). \tag{25}$$

Using (12), the combined forced termination probability with r reserved subchannels can be calculated from,

$$P^{AB}_{Fe}(r) = \frac{1}{(1 - P^{AB}_{Be}(r))} \sum_{i=0}^{NK-r} \sum_{j=0}^{NK-r} \sum_{k=0}^{K} \sum_{l=0}^{N} \sum_{m=0}^{N} \frac{(l+m)}{(\lambda^A_s + \lambda^B_s)} \gamma^{(i,j,k)}_{(i-l,j-m,k+1)} P_{\phi_r}(i, j, k)\delta_{fr}. \tag{26}$$

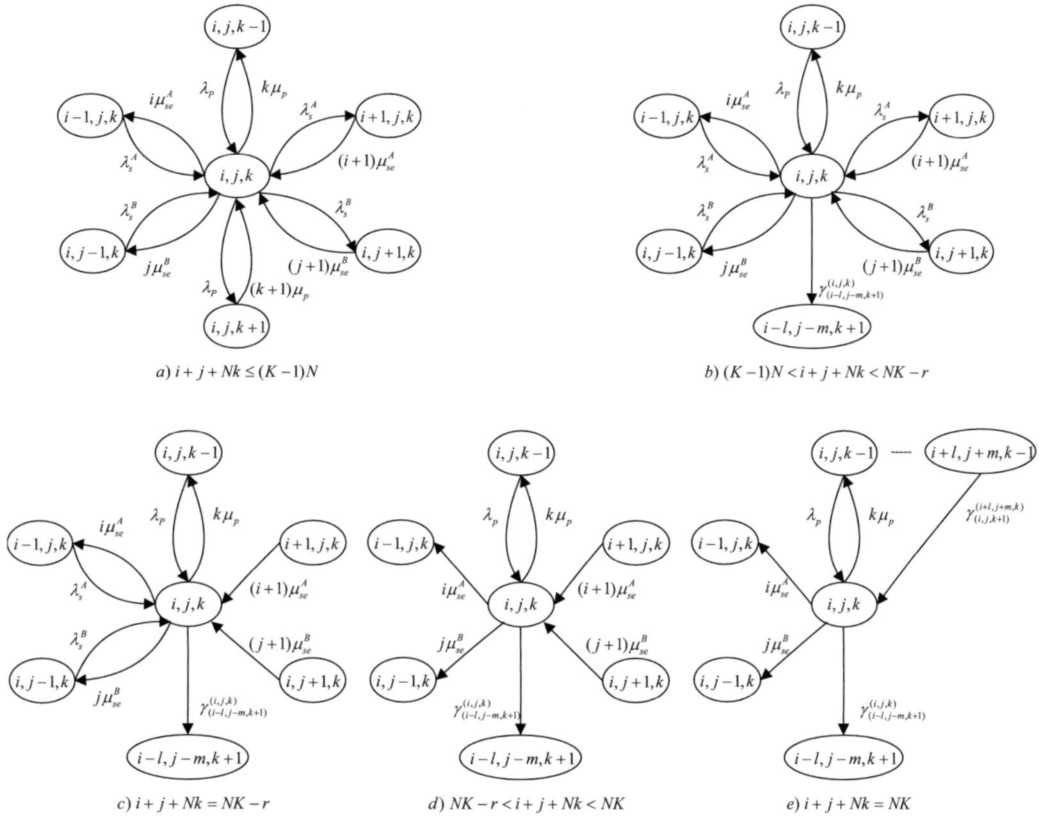

Fig. 7. A CTMC for the system with spectrum handoff and channel reservation $r < N$. The states $(i-l, j-m, k+1)$ satisfy packing condition i.e., $(i-l)+(j-m) = (K-k-1)N$ for $0 < l+m \le N$.

The individual forced termination probabilities for user groups S^A and S^B are

$$P_{Fe}^{A}(r) = \frac{1}{(1-P_{Be}^{AB}(r))} \sum_{i=0}^{NK-r} \sum_{j=0}^{NK-r} \sum_{k=0}^{K} \sum_{l=0}^{N} \sum_{m=0}^{N} \frac{l}{\lambda_s^A} \gamma_{(i-l,j-m,k+1)}^{(i,j,k)} P_{\phi_r}(i,j,k) \delta_{fr}, \tag{27}$$

and

$$P_{Fe}^{B}(r) = \frac{1}{(1-P_{Be}^{AB}(r))} \sum_{i=0}^{NK-r} \sum_{j=0}^{NK-r} \sum_{k=0}^{K} \sum_{l=0}^{N} \sum_{m=0}^{N} \frac{m}{\lambda_s^B} \gamma_{(i-l,j-m,k+1)}^{(i,j,k)} P_{\phi_r}(i,j,k) \delta_{fr}, \tag{28}$$

where, δ_{fr} is defined as

$$\delta_{fr} = \delta((i-l)+(j-m)-(K-k-1)N)\ \delta(i+j+Nk > (K-1)N)\ \delta(0 < l+m \le N). \tag{29}$$

Note that δ_{fr} consists of three conditions, i.e., the packing, termination and a valid number of dropped SU connections.

4.2.1 Special case-single SU group

In the case of a single SU group, we set the respective parameters for S^B in Figs. 7(a)-(e) to 0. Note that the resulting state transition diagrams are different from (Zhu et al., 2007, Fig. 4). The latter did not include the proper state transitions for the reserved channels, which gave over optimistic result. In Figs. 7(a)-(e) all five state transition scenarios are included which gives a complete CTMC analysis.

The state probabilities $P_{\phi_r}(i,k)$ are calculated by following the basic system's approach(). For a single SU group, the blocking probability with r reserved subchannels can be written as

$$P_{Be}^{1}(r) = \sum_{i=0}^{NK-r} \sum_{k=0}^{K} \delta(i + Nk \geq KN - r) P_{\phi_r}(i,k). \tag{30}$$

The above expression states that the blocking of a new SU connection can only occur when the number of vacant subchannels are less than or equal to r. For a single SU group, the forced termination probability is given by

$$P_{Fe}^{1}(r) = \frac{\sum_{i=0}^{NK-r} \sum_{k=0}^{K} \sum_{l=1}^{N} \gamma_{(i-l,k+1)}^{(i,k)} P_{\phi_r}(i,k) \delta_f^1(r)}{(1 - P_B(r)^1)\lambda_s^1}, \tag{31}$$

where, $\delta_f^1(r) = \delta(i + Nk > (K-1)N)\ \delta(i - (K-k-1)N - l)$. Note that $\delta_f^1(r)$ refers to two conditions. Firstly, the forced termination occurs only when $(i + Nk) > (K-1)N$. Under this condition $\gamma_{(i-l,k+1)}^{(i,k)} = \lambda_P$. Secondly, $(i-l)$ SU connections are packed into $(K-k-1)N$ available subchannels and the remaining l connections are terminated.

4.3 Network throughput

The network throughput is defined as the products of the connection completion rate, the average service duration per connection and the data rate. For unit data rate, the theoretical throughput can be expressed as

$$\rho_{se}^{x_0} = (1 - P_{Be}^{x_0})(1 - P_{Fe}^{x_0})^2 \theta_{se}^{x_0} \tag{32}$$

where $\theta_{se}^{x_0} = (\lambda_s^{x_0} / \mu_{se}^{x_0})$ is called traffic intensity Mehdi (1991) and $x_0 \in \{1, A, B\}$ for single or two user groups (S^A and S^B), respectively.

For a fixed $\theta_{se}^{x_0}$, the throughput ρ_{se}^{x} reaches a maximum when the SU service rate (μ_{se}^{x}) approaches infinity, i.e., $\rho_{se}^{x_0*} = \lim_{\mu_{se}^{x_0} \to \infty} \rho_{se}^{x_0}$. A simple proof is given as follows. When $\mu_{se}^{x_0} \to \infty$, we have $P_{Fe}^{x_0} \to 0$. The blocking probability in (32) consists of two parts i.e., $P_{Be}^{x_0} = P_{Be}^{x_0}(pri) + P_{Be}^{x_0}(sec)$, where $P_B(pri)$ is the blocking probability due to the situation in which PUs occupy all the subchannels, and $P_{Be}^{x0}(sec)$ is the blocking probability which can be computed from the probability for a given number of vacant subchannels. It is known that $P_{Be}^{x_0}(sec)$ is constant for a fixed $\theta_{se}^{x_0}$ Mehdi (1991). Therefore, the maximum achievable throughput ρ_{se}^{x} can be written as

$$\rho_{se}^{x_0*} = \lim_{\mu_{se}^{x_0} \to \infty} (1 - P_{Be}^{x_0})(1 - P_{Fe}^{x_0})\theta_{se}^{x_0} = (1 - P_{Be}^{x}(\infty))\theta_{se}^{x_0}, \tag{33}$$

where, $P_{Be}^{x0}(\infty)$ is the blocking probability when $P_{Fe}^{x_0} = 0$.

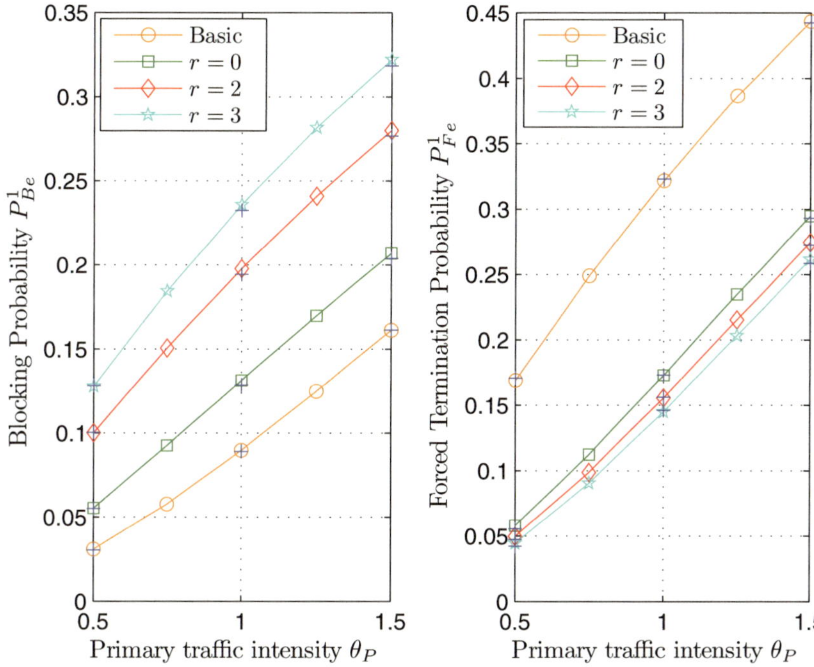

Fig. 8. Single SU group, **Left**: Blocking probability, and **Right**: Forced termination probability versus PU traffic intensity given $\theta_{se}^1 = 8$ and $\hat{\mu}_{se}^1 = 1$. Simulation results are shown with "+".

4.4 Numerical results

In this section we give some numerical examples of $P_{Fe}^{x_0}$, $P_{Be}^{x_0}$, and the throughput $\rho_{se}^{x_0}$ for SUs. In all of the following simulations, we set $K = 3, N = 6$ and $P_{Fe}^{x_0}$, $P_{Be}^{x_0}$, $\rho_{se}^{x_0}$ are plotted against PU (SU) traffic intensity $\theta_p(\theta_{se}^{x_0})$. In addition, service rates of the SU groups are normalised with respect to the PUs rate i.e., $\hat{\mu}_{se}^{x_0} = (\mu_{se}^{x_0}/\mu_p)$, where $x_0 \in \{1, A, B\}$. In addition, $r = 0$ indicates the spectrum handoff only condition without channel reservation.

4.4.1 Single SU group

In Fig. 8, P_{Fe}^1 and P_{Be}^1 are plotted against $\theta_p \in [0.5, 1.5]$, for given $\theta_{se}^1 = 8$ and $\hat{\mu}_{se}^1 = 1$. It can be calculated that the range of θ_p corresponds to the PU channel occupancy from 16.43% to 43.28%. Compared to systems with spectrum handoff and channel reservation, the basic system has the lowest blocking and the highest forced termination probability. Spectrum handoff results in a significant drop in P_{Fe}^1 for a moderate increase in P_{Be}^1. The introduction of reserved channels are not particularly effective in this instance.

Fig. 9 shows the impact of θ_{se}^1 on P_{Fe}^1 and P_{Be}^1, for a given $\theta_p = 1$ and $\hat{\mu}_{se}^1 = 1$. The behaviour of P_{Be}^1 is similar to that in Fig. 8. For the basic system, P_{Fe}^1 decreases slightly as θ_{se}^1 increases. This is counter intuitive. At high θ_{se}^1, due to fixed N, the forced termination rate $\sum_{i=0}^{NK}\sum_{k=0}^{K} F(i,0,k)$ will saturate faster than the connection rate $(1 - P_{Be}^1)\lambda_s^1$, implying an increased SU occupancy.

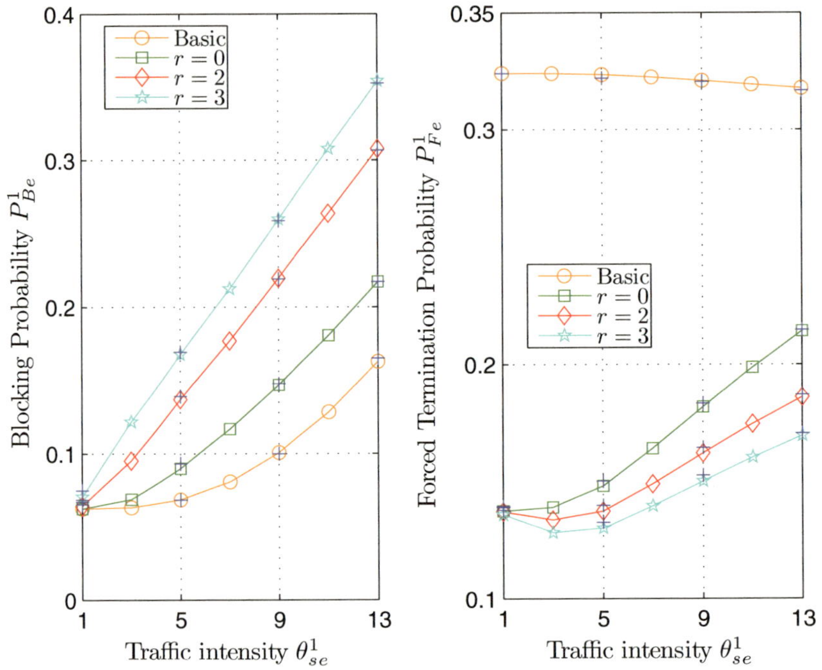

Fig. 9. Single SU group, **Left**: Blocking probability, and **Right**: Forced termination probability versus SU traffic intensity given $\theta_p = 1$ and $\hat{\mu}_{se}^1 = 1$. Simulation results are shown with "+".

Note that P_F is the probability of termination per occupied channel. With spectrum handoff and channel reservation P_{Fe}^1 initially decreases slightly, before increasing with θ_{se}^1. This can be explained as follows. At very low θ_{se}^1 values, the existing SU connections readily find vacant subchannels. As θ_{se}^1 increases the number of occupied channels increases, but there is still enough vacant subchannels to handle the arrival of a new PU connection and the probability of forced termination (for an existing connection) decreases. Eventually, at high θ_{se}^1 there are fewer vacant subchannels to accomodate the displaced subchannels and forced termination probability increases.

The throughput ρ_{se}^1 and the probability tradeoff gain $L(r)$ are shown in Fig. 10 for a given $\theta_p = 1$ and $\hat{\mu}_{se}^1 = 1$. At low values of θ_{se}^1, the network throughput with spectrum handoff and channel reservation is higher than the basic system. However, the same cannot be said for high values of θ_{se}^1. The curves of tradeoff gain $L(r)$ show that more reserved channel give less gain in probability tradeoff. Also, for the same r, the effectiveness of the tradeoff reduces as θ_{se}^1 increases.

4.4.2 Two SU groups

In Fig. 11, we compare the throughput and probability tradeoff gain $L(r)$ of two user groups with a single SU group having the same traffic intensity i.e., $\theta_{se}^1 = \theta_{se}^A + \theta_{se}^B$. In this example the abscissa is the SU service rate $\mu_{se}^B(\mu_{se}^1)$ and we assume that $\theta_p = 1$, $\theta_{se}^A = \theta_{se}^B = 6$, $\theta_{se}^1 = 12$, $r = 0$.

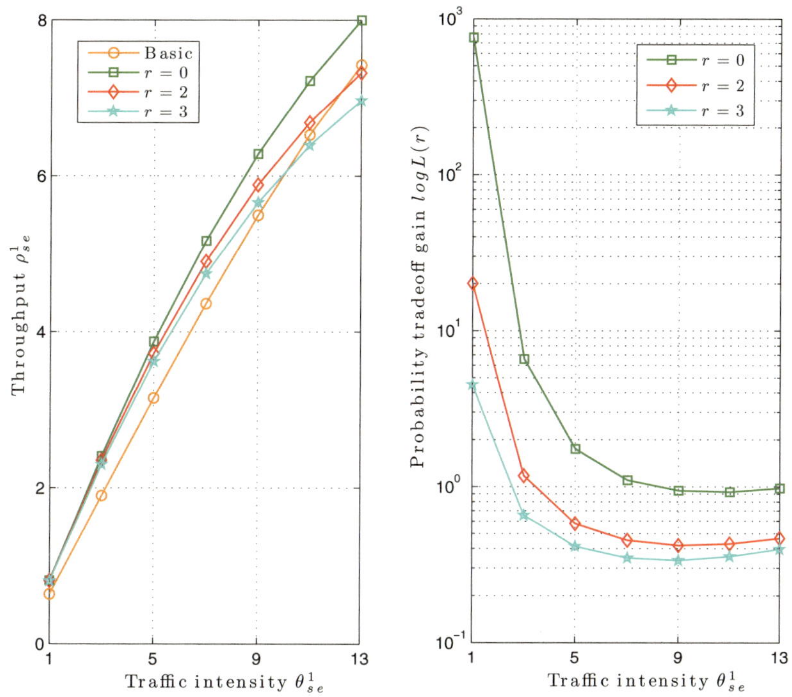

Fig. 10. Single SU group, **Left**: Aggregate throughput, and **Right**: Probability tradeoff gain versus SU traffic intensity given $\theta_P = 1$ and $\hat{\mu}_{se}^1 = 1$.

The throughput $\rho_{se}^{AB} = \rho_{se}^A + \rho_{se}^B$ and ρ_{se}^1 are monotonically increasing functions of both $\hat{\mu}_{se}^A$, $\hat{\mu}_{se}^B$ and μ_{se}^1 respectively. When $\hat{\mu}_{se}^1 = \hat{\mu}_{se}^A = \hat{\mu}_{se}^B$ the curves ρ_{se}^{AB} intersect with single user throughput ρ_{se}^1. This is expected, since at the intersection points SUs from both S^A and S^B are arriving at the same rate which is half of that of single user group S^1. The maximum achievable throughputs for both single and two user groups are identical i.e., $\rho_{max} = \rho_{se}^{AB*} = \rho_{max}^{1*}$. The aggregate throughput ρ_{AB} and ρ^1 approach maximum, when both single and two groups operate at much higher service rates than the PU arrival rate i.e., $\{\mu_{se}^1, \mu_{se}^A, \mu_{se}^B\} >> \lambda_p$.

For probability tradeoff gain $L(0)$, all the curves exhibit a U-shaped behaviour. This can be explained as follows. For relatively low service rates (long service duration), blocking probability increases at a higher rate than the rate of reduction in forced termination probability, whereas the opposite happens at high service rates (short service duration). This example demonstrates that the SU service duration relative to the PU counterpart has significant impact on the tradeoff, and the tradeoff is much more effective when the SU service rates μ_{se}^x is much larger or smaller than the PUs.

5. Airtime fairness

This section investigates the throughput fairness among two SU groups. For brevity, we limit our discussion to the spectrum handoff case only i.e., $r = 0$.

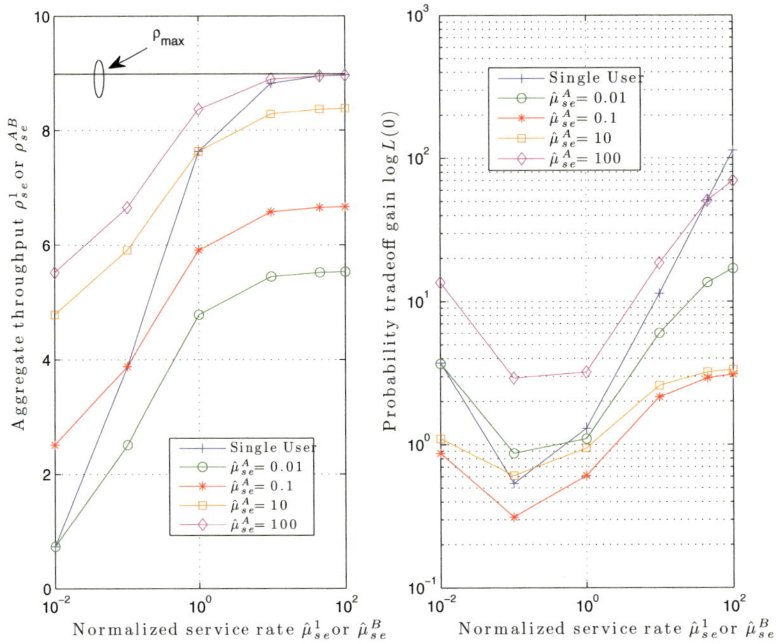

Fig. 11. Comparison of single and two SU groups, **Left**: Aggregate Throughput, and **Right**: tradeoff gain versus normalized SU service rate given $\theta_P = 1, \theta^1_{se} = 12, \theta^A_{se} = 6, \theta^B_{se} = 6$ and $r = 0$.

Fairness:*We define a fairness scheme in which the fractional throughput loss of both SU groups (with respect to their offered traffic θ^z_{se}) is equal, where $z \in \{A, B\}$*

$$\begin{aligned} G^A_\rho &= \frac{\theta^A_{se} - \rho^A_{se}}{\theta^A_{se}} \\ G^B_\rho &= \frac{\theta^B_{se} - \rho^B_{se}}{\theta^B_{se}} \end{aligned} \tag{34}$$

Based on the above equations, the fair metric is

$$F = \frac{G^A_\rho}{G^B_\rho} \tag{35}$$

The scheme is optimally fair when $F^* = 1$. In Fig. 12(a), the no constraint sharing gives an ideal fairness $F^* = 1$, only for $\mu^A_{se} = \mu^B_{se}$. It applies to any values of λ^A_s and λ^B_s(Fig. 12 has $\lambda^A_s = 12, \lambda^B_s = 20$). When $\mu^A_{se} = \mu^B_{se}$, the two SU groups act as a single SU group due to the superposition of Poisson arrivals. Mathematically, $P^A_{Fe} = P^B_{Fe}$, and P^{AB}_{Be} is constant for both S^A and S^B. As the difference in the service rates between the SU groups increases, the fairness F deviates from its ideal value.

In Fig. 12(b), we investigate the fairness F and throughput ρ for a given $\mu^A_{se} = 0.8$ and $\mu^B_{se} = 2$. The fairness curves at low to medium values of arrival rate $\lambda^A_s(\lambda^B_s)$ show that

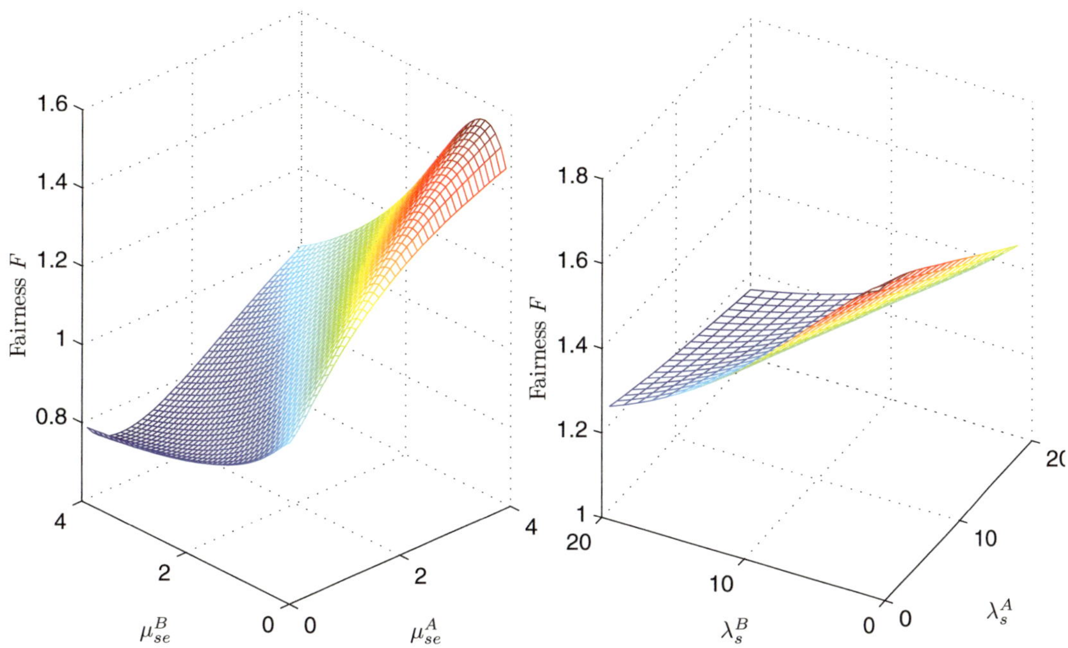

Fig. 12. **Left**: Fairness $\digamma$ versus SU service rate $\mu_{se}^{A}(\mu_{se}^{B})$ given $\lambda_s^A = 12$, $\lambda_s^B = 20$, $\mu_p = 1$ and $\lambda_p = 1$. **Right**: Fairness $\digamma$, versus SU arrival rate $\lambda_s^A(\lambda_s^B)$ given $\mu_{se}^{A} = 0.8$, $\mu_{se}^{B} = 2$, $\mu_p = 1$ and $\lambda_p = 1$

spectrum sharing results in an unfair distribution of channel resources among the two SU groups. The percentage loss in throughput of S^A(large connection length) is significantly higher. The arrival rate has less effect on fairness.

5.1 Fairness through channel partitioning

In the previous section, the results show that dissimilar SU groups can lead to unfairness. In the following we investigate the potential of a channel partitioning policy to address unfairness. In essence, we restrict the maximum number of allowable active connections to N^A and N^B for S^A and S^B respectively, where $0 \leq N^A \leq NK$ and $0 \leq N^B \leq NK$. The variation of N^A and N^B gives rise to the following 3 conditions. The first condition $N^A = NK - N^B$ represents a channel partitioning scenario, the second condition $N^A + N^B > NK$ is a weaker channel partitioning scenario, whereas the condition $N^A = N^B = NK$ is the no constraint scenario described in the previous subsection. Here, we only consider the channel partitioning scenario because it does not require the calculation of state probabilities in which the transmission of the SU groups overlap. A channel partitioning scenario is shown in Fig. 13, in which an incoming S^B connection is allowed to access the channel, while an incoming S^A connection is blocked even though there is a vacant subchannel. In the channel partitioning scenario, the objective is to find a feasible number of channels $\{N^A, N^B\}$ such that the access scheme is optimally fair i.e., $\digamma^* \approx 1$. Due to integer values of $\{N^A, N^B\}$ the optimally fair

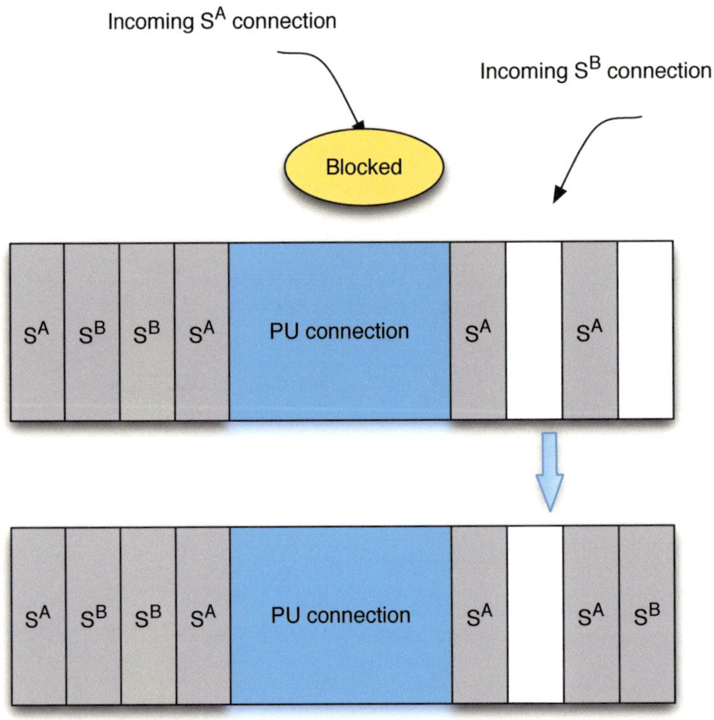

Fig. 13. Illustration of Channel partitioning scenario, $K = 3$, $N = 4$, $N^A = 4$, $N^B = 8$.

value of 1 may not always be possible. From (35) it is also obvious that the calculation of $\{N^A, N^B\}$ requires the knowledge of termination and blocking probabilities for both SU groups. In the following, we model this channel partitioning scenario as a simple extension of the CTMC in Section 3.3.3.

The maximum number of allowable connections of S^A and S^B are N^A and N^B i.e., $i = \{0, \ldots, N^A\}, j = \{0, \ldots, N^B\}$ and the state probabilities can be solved under the fundamental probability constraint (similar to the section 3.3). The blocking probabilities and forced termination probabilities are given as follows.

$$P_{Be}^{A} = \sum_{i=0}^{N^A-1} \sum_{j=0}^{N^B} \sum_{k=0}^{K} \delta(i + j + Nk - NK) P_{\phi_0}(i, j, k) + \sum_{j=0}^{N^B} \sum_{k=0}^{K} P_{\phi_0}(N^A, j, k), \tag{36}$$

$$P_{Be}^{B} = \sum_{i=0}^{N^A} \sum_{j=0}^{N^B-1} \sum_{k=0}^{K} \delta(i + j + Nk - NK) P_{\phi_0}(i, j, k) + \sum_{j=0}^{N^A} \sum_{k=0}^{K} P_{\phi_0}(i, N^B, k). \tag{37}$$

The combined blocking probability of SU groups P_B^{AB} seen by PU network is given as

$$P_{Be}^{AB} = \frac{\lambda_s^A}{\lambda_s^A + \lambda_s^B} P_{Be}^{A} + \frac{\lambda_s^B}{\lambda_s^A + \lambda_s^B} P_{Be}^{B}, \tag{38}$$

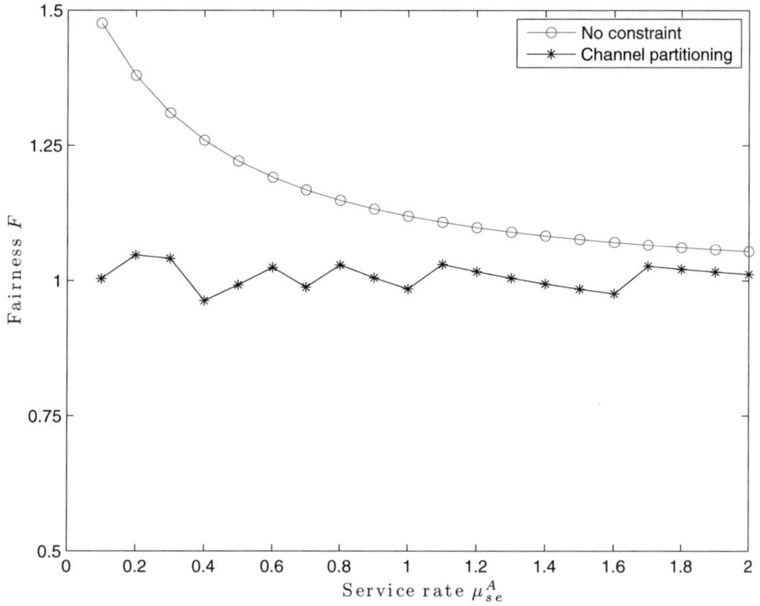

Fig. 14. Fairness F versus SU service rate μ_{se}^A given $\theta_{se}^A = 4$, $\theta_{se}^B = 7$, $\mu_{se}^B = \mu_{se}^A + 0.3$, $\mu_p = 1$ and $\lambda_p = 1$.

$$P_{Fe}^A = \frac{1}{(1 - P_{Be}^A)} \sum_{i=0}^{N^A} \sum_{j=0}^{N^B} \sum_{k=0}^{K} \sum_{l=0}^{min\{N^A,N\}} \sum_{m=0}^{min\{N^B,N\}} \frac{l}{\lambda_s^A} \gamma_{(i-l,j-m,k+1)}^{(i,j,k)} P_{\phi 0}(i,j,k) \delta_{f0}, \tag{39}$$

and

$$P_{Fe}^B = \frac{1}{(1 - P_{Be}^B)} \sum_{i=0}^{N^A} \sum_{j=0}^{N^B} \sum_{k=0}^{K} \sum_{l=0}^{min\{N^A,N\}} \sum_{m=0}^{min\{N^B,N\}} \frac{m}{\lambda_s^B} \gamma_{(i-l,j-m,k+1)}^{(i,j,k)} P_{\phi 0}(i,j,k) \delta_{f0}. \tag{40}$$

5.2 Scenario 1 $\theta_{se}^A \neq \theta_{se}^B$, $\mu_{se}^A \neq \mu_{se}^B$ and $\lambda_s^A \neq \lambda_s^B$

In Fig. 14 and Fig. 15, the fairness F and throughput ρ are plotted against μ_{se}^A respectively, for given $\mu_{se}^B = \mu_{se}^A + 0.3$, $\theta_{se}^A = 4$ and $\theta_{se}^B = 7$. In this condition, $P_{Fe}^A \neq P_{Fe}^B$, and $P_{Be}^A \neq P_{Be}^B$. The fairness curves at low to medium values of service rate μ_s^A show that the no constraint scenario results in an unfair distribution of channels among two SU groups. The percentage loss in throughput of S^A is significantly higher (due to large P_{Fe}^A). At large values of service rate μ_{se}^A the fairness improves and throughput saturates due to the lower termination probability of both SU groups. On the other hand, the channel partitioning achieves fairness at a cost of the decrease in aggregate and individual throughput i.e., by assigning more subchannels to S^A than S^B. At higher values of $\mu_{se}^A(\mu_{se}^B)$ there is almost no difference between the fairness and throughput in both strategies. The fluctuation in fairness occurs due to the integer number of subchannels.

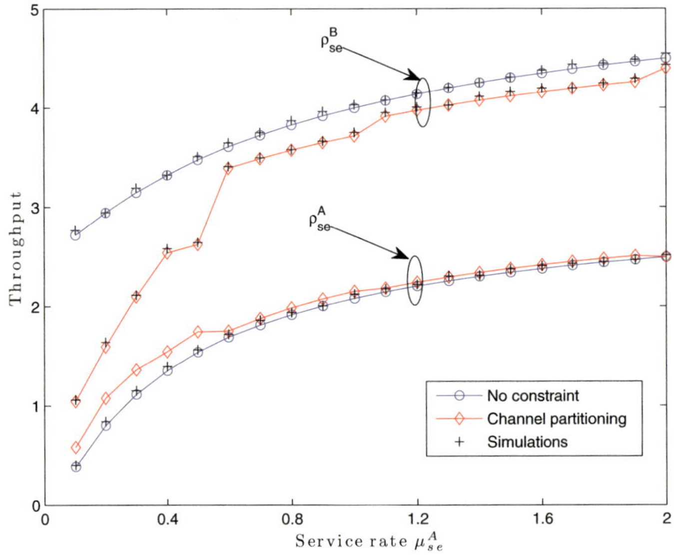

Fig. 15. Throughput ρ versus SU service rate μ_{se}^{A} given $\theta_{se}^{A} = 4$, $\theta_{se}^{B} = 7$, $\mu_{se}^{B} = \mu_{se}^{A} + 0.3$, $\mu_p = 1$ and $\lambda_p = 1$.

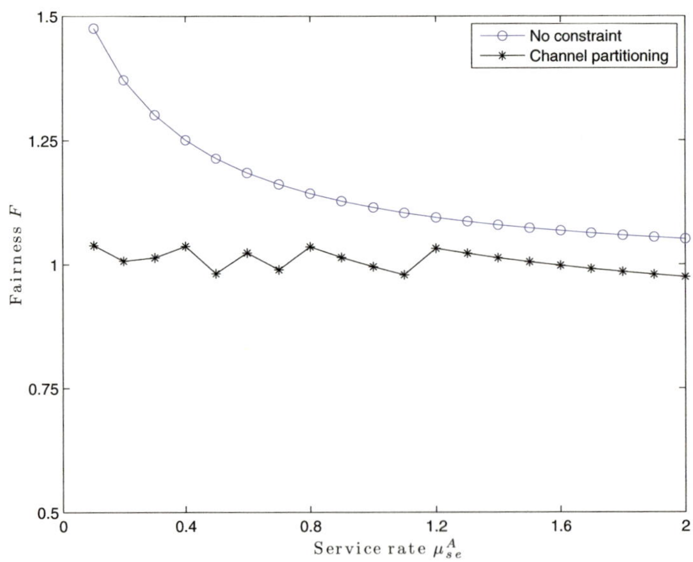

Fig. 16. Fairness F versus SU service rate μ_{se}^{A} given $\theta_{se}^{A}(\theta_{se}^{B}) = 6$, $\mu_{se}^{B} = \mu_{se}^{A} + 0.3$, $\mu_p = 1$ and $\lambda_p = 1$.

5.3 Scenario 2 $\theta_{se}^{A} = \theta_{se}^{B}$, $\mu_{se}^{A} \neq \mu_{se}^{B}$ and $\lambda_{s}^{A} \neq \lambda_{s}^{B}$

Fig. 16 shows the fairness F curves plotted against SU service rate μ_{se}^{A}, given $\mu_{se}^{B} = \mu_{se}^{A} + 0.3$, $\theta_{se}^{A} = 6$ and $\theta_{se}^{B} = 6$. The arrival rate of S^A is higher than S^B because of the offered traffic is constant for both SU groups. The curves are fairly similar to the previous figure and indicate that the smaller durations achieve better fairness.

6. Summary

In this Chapter, using a CTMC we have presented the exact solutions to determine the forced termination and blocking probabilities, and aggregate throughput of a SU group as well as two SU groups. Specifically,

- For multiple PU channels (with more than one subchannel) and two SU groups, three types of systems with were considered; without spectrum handoff (basic system): with spectrum handoff and channel reservation. For all these systems the single SU group was treated as a special case. In the former case, the results show that the blocking and forced termination probabilities increase with PU traffic intensity, as expected. However, SU traffic intensity has a different impact on the forced termination probability. The forced termination probability decreases slightly as SU traffic intensity increases for the basic system. For the systems with spectrum handoff, forced termination probability is always less than the basic system. However, there exists an optimal arrival rate that minimizes the forced termination probability. Spectrum handoff ($r = 0$) is more effective than channel reservation $r > 0$ in reducing forced termination probability for a given increase in blocking probability (tradeoff gain). The tradeoff is much more beneficial when the service rate is either very high or very low compared with the PU service rate.
- For two SU groups, we found that spectrum sharing can result in unfair channel occupancy. The problem is most prevalent when the difference between the two service rates is large. Channel partitioning (where a limit is placed on the maximum number of active connections of each S^A and S^B group) forces fairness but the throughput penalty might not be worth it.

7. References

Mitola III, J. (2000). Cognitive radio: An integrated agent architecture for software defined radio, *PhD. Thesis, KTH Royal Institute of Technology, Sweden*, May 2000.

Haykin, S. (2005). Cognitive radio: Brain-empowered wireless communications. *IEEE Journal on Selected Areas in Communication*, pp. 201-220, vol. 23, Feb. 2005.

Weiss, T. & Jondral, F. (2004). Spectrum pooling: an innovative strategy for the enhancement of spectrum efficiency, *IEEE Communication Magazine*, vol. 42, pp. 8-14, Mar. 2004.

Mehdi, J. (1991). *Stochastic Models in Queueing Theory, Academic Press, San Diego, CA*, 1991.

Raspopovic, M.; Thompson, C.; Chandra K. (2005). Performance Models for Wireless Spectrum Shared by Wideband and Narrowband Sources, *Proceedings of IEEE Military Communication Conference*, pp. 1-6. Oct. 2005.

Raspopovic, M. & Thompson, C. (2007). Finite Population Model for Performance Evaluation Between Narrowband and Wideband Users in the Shared Radio Spectrum, *Proceedings of IEEE Dynamic Spectrum Access Networks*, pp. 340-346, Apr. 2007.

Xing, Y.; Chandramouli, R.; Mangold, S.; N, S. S. (2006). Dynamic Spectrum Access in Open Spectrum Wireless Networks. *IEEE Journal on Selected Areas in Communication*, pp. 626-637, vol. 24(3), Mar. 2006.

Chou, C.-T.; Shankar, S.; Kim, H.; Shin, K. G. (2007). What and How Much to Gain by Spectrum Agility, *IEEE Journal on Selected Areas in Communication*, vol. 25(3), Apr. 2007.

Capar, F.; Martoyo, I.; Weiss, T.; Jondral, F. (2002). Comparison of bandwidth utilization for controlled and uncontrolled channel assignment in a spectrum spooling system, *Proceedings of IEEE Vehicular Technology Conference*, pp. 1069-1073, ISBN 0-7803-7484-3, 2002.

Tang, P. K.; Chew, Y. H.; Ong, L. C.; Haldar, M. K. (2006). Performance of Secondary Radios in Spectrum Sharing with Prioritized Primary Access, *Proceedings of IEEE Military Communication Conference*, pp. 1-7, Oct. 2006.

Tang, P. K. & Chew, Y. H. (2010). On the Modeling and Performance of Three Opportunistic Spectrum Access Schemes, *IEEE Transactions on Vehicular Technology*, vol. 59(8), Oct. 2010.

Huang S.; Liu, X.; Ding, Z. (2008). Opportunistic Spectrum Access in Cognitive Radio Networks, *Proceedings of International Conference on Computer Communications*, pp. 1427-1435, Apr. 2008.

Yuan, G.; Grammenos, R. C.; Yang, Y.; Wang, W. (2010). Performance Analysis of Selective Opportunistic Spectrum Access With Traffic Prediction, *IEEE Transactions on Vehicular Technology*, vol. 59 (4), May. 2010.

Park, J. et al. (2010). To Buffer or to Switch: Design of Multichannel MAC for OSA Ad Hoc Networks, *Proceedings of IEEE Dynamic Spectrum Access Networks*, pp. 1-10, 2010.

Song Y., & Xie, J. (2010). Proactive Spectrum Handoff in Cognitive Radio Ad Hoc Networks based on Common Hopping Coordination, *Proceedings of IEEE International Conference on Computer Communications*, pp. 1-2, 2010.

Yoon S.-U. & Ekici, E. (2010). Voluntary Spectrum Handoff: A Novel Approach to Spectrum Management in CRNs, *Proceedings of IEEE International Conference on Communication*, pp. 1-5. 2010.

Zhu, X.; Shen, L. & Yum, T.-S. P. (2007). Analysis of cognitive radio spectrum access with optimal channel reservation, *IEEE Communication Letters*, vol. 11, pp. 304-306.

Al-Mahdi, H.; Kalil, M. A.; Liers, F.; Mitschele-Thiel, A. (2009). Increasing Spectrum Capacity for Ad Hoc Networks using Cognitive Radios: An Analytical Model, *IEEE Communication Letters*, vol. 13 (9), Sept. 2009.

Ahmed, W.; Gao, J.; Suraweera, H.; Faulkner, M. (2009). Comments on "Analysis of Cognitive Radio Spectrum Access with Optimal Channel Reservation", *IEEE Transactions on Wireless Communications*, pp. 4488-4491, vol. 8(9), Sept. 2009.

Zhang, Y. (2008). Dynamic Spectrum Access in Cognitive Radio Wireless Networks, *Proceedings of International Conference on Communication ICC*, pp. 4927-4932, May 2008.

Bauset, J. M.-; Pla, V.; Param, D. P.- (2009). Comments on "Analysis of Cognitive Radio Spectrum Access with Optimal Channel Reservation", *IEEE Communication Letters*, vol. 13(10), Oct. 2009.

Tzeng, S.-S. (2009). Call admission control policies in cellular wireless networks with spectrum renting, *Computer Communication*, no. 32, pp. 1905-1913, 2009.

Bauset, J. M.-; Pla, V.; Param, D. P.- (2010). Admission Control and Interference Management in Dynamic Spectrum Access Networks, *EURASIP Journal on Wireless Communication and Networks*, no. 708029, 2010.

Kannappa, S. M. & Saquib, M. (2010). Performance Analysis of a Cognitive Network with Dynamic Spectrum Assignment to Secondary Users, *Proceedings of IEEE International Conference on Communication ICC*, pp. 1-5, May 2010.

Heo, J. et al. (2008). Mathematical Analysis of Secondary User Traffic in Cognitive Radio System, *Proceedings of IEEE Vehicular TEchnology Conference*, pp. 1-5, Sept. 2008.

Xue, D.; Yu, H.; Wang, X.; Chen, H.-H. (2009). Adoption of Cognitive Radio Scheme to Class-based Call Admission Control, *Proceedings of IEEE International Conference on Communication*, pp. 1-5, Jun. 2009.

Ramjee, R.; Nagarajan, R.; Towsley, D. (1996). On optimal call admission control in cellular networks, *Proceedings of IEEE International Conference on Computer Communications*, 1996.

Zhang, Y. (2009). Spectrum Handoff in Cognitive Radio Networks: Opportunistic and Negotiated Situations, *Proceedings of IEEE International Conference on Computer Communications*, pp. 1-5, 2009.

Tang, S. & Mark, B. L. Modeling and Analysis of Opportunistic Spectrum Sharing with Unreliable Spectrum Sensing, *IEEE Transactions on Wireless Communications*, vol. 8(4), Apr. 2009.

Wang, L.-C. & Anderson, C. (2008). On the performance of spectrum handoff for link maintenance in cognitive radio," in *Proceedings of IEEE International Conference on Wireless and Pervasive Computing*, pp. 670-674. May 2008.

Willkomm, D.; Gross, J.; Wolisz, A. (2005). Reliable link maintenance in cognitive radio systems, *Proceedings of IEEE Dynamic Spectrum Access Networks*, pp. 371-378, Nov. 2005.

Khalil, M. A.; Al-Mahdi, H.; Mitschele-Thiel, A. (2010). Spectrum handoff reduction for cognitive radio ad hoc networks, *Proceedings of IEEE International Symposium on Wireless Communication Systems*, pp. 1036-1040, Sept. 2010.

Lin, P.; Lin, T. & Wang, C. (2009). Performance Analysis of a Cross-Layer Handoff Ordering Scheme in Wireless Networks, *IEEE Transactions on Wireless Communications*, vol. 7, no. 12, Dec. 2009.

Chung, J.-M.; Kim, C. H.; Park, J.-H.; Shin, J.; Kim, D. (2010). Optimal channel allocation for cognitive radios," *IET Letters*, vol. 46(11), Mar. 2010.

Wang, B.; Ji, Z.; Lui, K. J. R. & Clancy, T.C. (2007). Primary-Prioritized Markov Approach for Dynamic Spectrum Access. *IEEE Transactions on Wireless Communication*, pp. 1854-1865, vol. 8(4), April 2009.

Luo, L. & Roy, S. (2009). Analysis of Dynamic Spectrum Access with Heterogeneous Networks: Benefits of Channel Packing Scheme, *Proc. IEEE GLOBECOM*, pp.1-7, Dec. 2009.

Ma, Z.; Cao, Z. & Chen, W. (2008). A Fair Opportunistic Spectrum Access (FOSA) Scheme in Distributed Cognitive Radio Networks,*Proceedings of IEEE International Conference on Computer Communications*, pp. 4054-4058, May 2008.

Wiggins, U.; Kannan, R.; Chakravarthy, V.; Vasilakos, A. V. (2008). Data- centric prioritization in a cognitive radio network: A quality-of-service based design and integration, in *Proceedings of IEEE Dynamic Spectrum Access Networks*, pp. 1âĂŞ11, Oct. 2008.

Gosh, C., Chen, S., Agarwal D. P. & Wyglinski, A. M. (2009). Priority-based spectrum allocation for cognitive radio networks employing NC-OFDM transmission, *Proceedings of IEEE Military Communication Conference*, pp. 1-5, Oct. 2009.

Tumuluru, V. K.; Wang, P. & Niyato, D. (2011). Performance Analysis of Cognitive Radio Spectrum Access with Prioritized Traffic, *To Appear in Proceedings of IEEE International Conference on Computer Communications*. 2011

Ahmed, W.; Gao, J. & Faulkner, M. (2010). Channel Allocation for Fairness in Opportunistic Spectrum Access Networks *Proceedings of IEEE Wireless Communications and Networking Conference*, pp. 1-6, 18-21 Apr. 2010.

Ahmed, W.; Gao, J.; Zhou, H. & Faulkner, M. (2010) . Throughput and Proportional Fairness in Cognitive Radio Network, *IEEE International Conferences on Advanced Technologies for Communications*, pp. 248-252, 12-14 Oct. 2009.

Part 4

Resource Allocation and Beyond

10

Primary Outage-Based Resource Allocation Strategies

Bassem Zayen and Aawatif Hayar
EURECOM
France

1. Introduction

Cognitive radio (CR) is an emerging technology in wireless technology that uses software-defined radio to aim to the efficient use of the spectrum by exploiting the unused frequency bands at the time and space (J. Mitola, 1999). A look on the state of the art shows that CR research area is very open. A particular problem in the context of CR, when we seek to optimize the secondary system capacity, is to guarantee a quality of service (QoS) to primary users (PUs) and a certain QoS to secondary users (SUs). There is a large number of proposals for all communication layers treating the increase of restrictions to spectrum utilization (Peha, 2005), but the QoS issue still has not been clearly defined. In addition, it is unclear how secondary system opportunism is compatible with the support of QoS for both, CR systems and primary systems. The U.S. Federal Communications Commission (FCC) proposed the concept of "*interference temperature*" as a way to have unlicensed transmitters sharing licensed bands without causing harmful interference (FCC, 2003). Rather than merely regulate transmitter power at fixed levels, as it has been done in the past, the scheme would have governed transmitter power on a variable basis calculated to limit the energy at victim receivers, where interference actually occurs. As a practical matter, however, the FCC abandoned the interference temperature concept recently (FCC, 2003) due to the fact that it is not a workable concept. While offering attractive promises, CRs face various challenges, starting from defining the fundamental performance limits of this radio technology, in order to achieve the capability of using the spectrum in an opportunistic manner. Specifically, CR is required to detect spectrum holes in the spectrum band and to determine if the spectrum allocation meets the QoS requirements of different users. This decision can be made by assessing the channel capacity, which is the most important factor for spectrum characterization.

The purpose of this chapter is to present an analysis of the QoS problem along with a proposed solution, while maintaining a limited scope to provide coherency and depth. The QoS problem will be tackled in this work by proposing three resource management strategies based on outage probability. The motivation behind doing so is that, in any case, the PU will not necessarily need all system rate. In fact, the PU will experience the SU's interference, and as long as all his target rate (depending on his QoS) is achieved, he does not care about if he leaves any channel for SUs. In what follows, we adopt this setting and consider a CR network (CRN) in which primary and secondary users both attempt to communicate in a distributed way, subject to mutual interference. We propose a CR coordination algorithm that maximizes the CRN secondary rate while keeping the interference to the PU acceptable. Our goal is to realize PU-SU spectrum sharing by optimally allocating SU transmit powers, in order to maximize the total SU throughput under interference and noise impairments, and short term (minimum and peak) power constraints, while preserving the QoS of the primary system. In particular, it is of interest to determine the maximum number of SUs allowed to transmit threshold above which SUs can decide to transmit without affecting the PU's QoS. In such approaches, each user individually makes its decision on its transmit power so as to optimize its contribution to the system throughput. At the core of the concept lies the idea that the interference is more predictable when the network is dense, and consequently the resource allocation problem of a given user becomes more dependent to the average behavior, thus facilitating optimization.

Therefore, we will present and develop in this chapter three resource management strategies based on outage probability. We will derive in a first step a distributed user selection algorithm under a cognitive capacity maximization and outage probability constraints (B. Zayen, M. Haddad, A. Hayar and G.E. Øien, 2008). Specifically, we allow SUs to transmit simultaneously with the PU as long as the interference from the SUs to the PU that transmits on the same band remains within an acceptable range. We impose that SUs may transmit simultaneously with the PU as long as the PU in question does not have his QoS affected in terms of outage probability. The second algorithm investigates multiuser multi-antenna channels using a beamforming strategy (B. Zayen, A. Hayar and G.E. Øien, 2009). The proposed strategy tries to maximize the system throughput and to satisfy the signal-to-interference plus-noise ratio (SINR) constraint, as well as to limit interference to the PU. In the proposed algorithm, SUs are first pre-selected to maximize the per-user sum capacity subject to minimizing the mutual interference. Then, the CR system verifies the outage probability constraint to guarantee QoS for the PU. The third algorithm is based on a game theory tools with the objective to maximize a defined utility function with protection for PUs (B. Zayen, A. Hayar and G. Noubir, 2011). Particularly, we formulate a utility function to reflect the needs of PUs by verifying the outage probability constraint, and the per-user capacity by satisfying the SNIR constraint, as well as to limit interference to PUs. Furthermore, the existence of the Nash equilibrium of the proposed game is established, as well as its uniqueness under some sufficient conditions. Both theoretical and simulation results based on a realistic network setting, for the three presented strategies, will be provided in this chapter and a comparison in terms of CR network (CRN) deployment while maintaining QoS for the PU will be presented.

The chapter is organized as follows. Section 2 will introduce a number of theoretical concepts of importance. It will describe the CRN that will be used throughout this chapter. Section 3 will present the performance metrics used to evaluate the proposed resource allocation algorithms. The following two metrics are considered: PU performance metrics including the primary capacity, the outage probability and the interference outage, and, SU's performance metrics including SU's capacity, SU's sum capacity, the interference power and fairness. In Section 4 and Section 5, we will provide a rather straightforward classification of resource allocation strategies attempting to show the diversity and advantages of these techniques. Two types of resource allocation strategies, centralized strategies, and distributed strategies, are discussed in these two sections, respectively. Section 6 will describe the first distributed resource allocation strategy based on outage probability. Section 7 and Section 8 will present

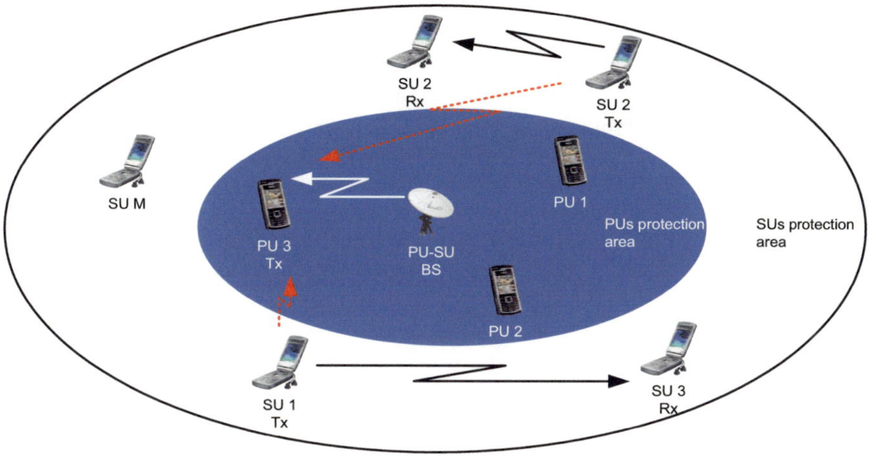

Fig. 1. The cognitive radio network with N primary users and M secondary users attempting to communicate with their respective pairs in an ad-hoc manner during a primary system transmission in downlink mode, subject to mutual interference.

the centralized beamforming-based resource allocation strategy and the one based on game theory, respectively. Section 9 is split in two main subsections. The first subsection will introduce the propagation model that will be used to evaluate the performance of the presented strategy. The second subsection presents simulation results and a comparison of the three user selection strategies presented in this chapter. Section 10 concludes the chapter.

2. Resource allocation goal

In order to facilitate the deployment of CR technologies for the secondary usage of spectrum, it is crucial to prove the reliable detection of PUs by SUs. In fact, primary and secondary users can coexist without a degradation of the PU transmission in order to convince regulatory authorities as well as PUs to enable such technologies. Particulary, PU will not necessarily need all that multi-rate system. In fact, the PU will experience the SU's interference, and as long as all his target rate (depending on his QoS) to be achieved, he does not care about if he leaves any channel for SUs. In what follows, we adopt this setting and consider a CRN in which primary and secondary users both attempt to communicate, subject to mutual interference. This is the main goal of the study in the second part of this chapter.

We consider here a wireless CRN with a collection of users randomly distributed over the geographical area considered. Users can be both transmitters and receivers. By virtue of a scheduling protocol, N PUs and M pairs of SUs are simultaneously selected from these users to communicate at a given time instant, while others remain silent. We will consider in our analysis both downlink and uplink scenarios. We show in Figure 1 the downlink scenario as an example. In this scenario, we assume that a BS transmits to its user which has the highest channel power gain. In the uplink scenario, its users transmit to the network's BS. We introduce also in the presented figure the interference channel gain between SUs transmitters and PU transmitter. Provided that no significant scatterers are presented in the area, the channel gains between any pair of users are assumed i.i.d. in the two proposed scenarios and they depend on the position of the users in the two-dimensional plane. Each PU is allocated with a unique resource slot so that it transmits in an orthogonal manner with respect to other PUs within its coverage area, i.e. no interference between different PUs like in the Orthogonal Frequency-Division Multiple Access-based (OFDMA-based systems). All details about the CRN parameters and the propagation model in the downlink and the uplink mode will be given in Section 9. In this CRN, we will consider only the case when we have one PU and M pairs of SUs.

In order to facilitate the problem formulation of the resource allocation problem, we state the following notations:

- the PU is indexed by pu,
- the index of SU m lies between 1 and M,
- $h_{l,m}$ denotes the channel gain from SU l to the desired user m,
- the data destined from SU m is transmitted with power p_m and a maximum power P_{max},
- $h_{pu,m}$ denotes the channel gain from the PU indexed by pu to the desired user m,
- $h_{pu,pu}$ denotes the channel gain between the BS and the PU,
- the data destined from the primary system is transmitted with power p_{pu}.

In the coverage area of the primary system, there is an *interference boundary* within which no SUs can communicate in an ad-hoc manner. Thus, as can be seen in Figures 1, for the impairment experienced by the primary system to be as small as possible, a SU must be able to detect very reliably whether it is far enough away from a primary base station, i.e., in the area of possible CR operation. Under these schemes, we allow SUs to transmit simultaneously to the PU as long as the interference from the SUs to the PU that transmits on the same band remains within an acceptable range. Specifically, we impose that SUs may transmit simultaneously with the PU as long as the PU in question does not have its QoS affected. Based on PU channel statistics, we determine a QoS bound to ensure a protection to the PU. To compute this bound, we will use the outage probability.

For system design purposes we will need to define metrics that will guide our development. These metrics need to be sufficiently broad, so that a realistic system can be designed through an optimization of all the metrics that we define. This would therefore need to include metrics to measure the performance of the CRN. These metrics will be given in the following section.

3. Resource allocation metrics

Algorithms that aid reliable resource management need to be verified and their performance has to be quantified by some metrics. In this section, we will present the performance metrics used to evaluate the proposed resource allocation algorithms.

3.1 Primary users performance metrics

Primary Capacity In most of resource allocation strategies, algorithms must ensure that the transmission rate of PU is no bigger than the primary capacity for most of the time. The PU instantaneous capacity is give by

$$C_{pu} = \log_2 \left(1 + \frac{p_{pu} \mid h_{pu,pu} \mid^2}{\sum_{m=1}^{M} p_m \mid h_{pu,m} \mid^2 + \sigma^2} \right) \tag{1}$$

where σ^2 is the ambient noise variance. Clearly, the primary capacity is directly related to the PU transmission as well as the SUs transmission.

Outage Probability The notion of *information outage probability*, defined as the probability that the instantaneous mutual information of the channel is below the transmitted code rate, was introduced in (L. H. Ozarow, S. Shamai and A.D. Wyner, 1994). Accordingly, the outage probability can be written as:

$$P_{out}(R) = P\{I(\mathbf{x};\mathbf{y}) \leq R\} \tag{2}$$

where $I(\mathbf{x};\mathbf{y})$ is the mutual information of the channel between the transmitted vector $\mathbf{x}$ and the received vector $\mathbf{y}$, and R is the target data rate in bits/s/Hz. Reliable communication can therefore be achieved when the mutual information of the channel is strong enough to support the target rate R. Thus, a $m-$th cognitive transmitter can adapt its transmit power p_m within the range of $[0; P_{max}]$ to fulfill the following two basic goals:

- *Self-goal*: Trying to transmit as much information for itself as possible,
- *Moral-goal*: Maintaining the PU's outage probability unaffected.

The outage probability can be rewritten as:

$$P_{out} = Prob\left\{C_{pu} \leq R_{pu}\right\} \tag{3}$$

where R_{pu} is the PU transmitted data rate. The information about the outage failure can be carried out by a band manager that mediates between the primary and secondary users, or can be directly fed back from the PU to the secondary transmitters through collaboration and exchange of the CSI between the primary and secondary users.

Interference Outage The CR specific metrics relate to how well the CR is able to avoid PU and the efficiency in using available spectrum. This will require a model for PU dynamics, such as disappearance and reappearance time intervals, the amount of spectrum being used and the strength and location of the PU. In addition to the primary capacity and the outage probability, we define the interference outage meaning when the power of interference at a receiver PU exceeds a pre-defined absolute limit. Let q be this absolute limit (i.e. the maximum outage probability).

3.2 Secondary users performance metrics

Secondary User's Capacity By making SUs access the primary system spectrum, the m-th SU experiences interference from the PU and all neighboring co-channel SU links that transmit on the same band. Accordingly, the m-th SU instantaneous capacity is given by:

$$C_m = \log_2\left(1 + \mathrm{SINR}_m\right) \tag{4}$$

where

$$\text{SINR}_m = \frac{p_m |h_{m,m}|^2}{\sum\limits_{\substack{l=1 \\ l \neq m}}^{M} p_l |h_{l,m}|^2 + p_{pu} |h_{pu,m}|^2 + \sigma^2} \quad (5)$$

Secondary User's Sum Capacity SUs need to recognize their communication environment and to adjust the parameters of their communication scheme in order to maximize the per-user cognitive capacity, expressed as

$$C_{su} = \sum_{m=1}^{M} C_m \quad (6)$$

while minimizing the interference to the PUs, in a *distributed* fashion. The sum here is made over the M SUs allowed to transmit. Moreover, we assume that the coherence time is sufficiently large so that the channel stays constant over each scheduling period length. We also assume that SUs know the channel state information (CSI) of their own links, but have no information on the channel conditions of other SUs.

Interference Power No interference cancelation capability is considered in our study. Power control is used for SUs both in an effort to preserve power and to limit interference and fading effects. The interference power (Intf) is given by:

$$\text{Intf}_m = \sum_{\substack{l=1 \\ l \neq m}}^{M} p_l |h_{l,m}|^2 + p_{pu} |h_{pu,m}|^2 + \sigma^2 \quad (7)$$

Combining (5) and (7), we define the SINR as a function of Intf:

$$\text{SINR}_m = \frac{p_m |h_{m,m}|^2}{\text{Intf}_m} \quad (8)$$

and

$$p_m = \frac{\text{SINR}_m \text{Intf}_m}{|h_{m,m}|^2} \quad (9)$$

4. Centralized resource allocation strategies

In the centralized mode, the resource allocation system would require a central controller and information about all users and channels. The centralized resource allocation have been the main focus of some research efforts in CRNs. We will provide in this section some solutions to this issue that have been proposed in the literature.

The authors in (L. Qian, X. Li, J. Attia and Z. Gajic, 2007) derived a centralized power control method for the CRN to maximize the energy efficiency of the SUs and guarantee the QoS of both the PUs and the SUs. The feasibility condition was derived in (L. Qian, X. Li, J. Attia and Z. Gajic, 2007) and a joint power control and admission control procedure was suggested such that the priority of the PUs is ensured all the time. However, in (L. Qian, X. Li, J. Attia and Z. Gajic, 2007) only one CRN was considered.

In (Y. Xing, C.N. Mathur, M.A. Haleem, R. Chandramouli and K.P. Subbalakshmi, 2007), the authors considered spectrum sharing among a group of spread spectrum users with a constraint on the total interference temperature at a particular measurement point, and a QoS constraint for each secondary link. An optimization solution of this problem was proposed in (Y. Xing, C.N. Mathur, M.A. Haleem, R. Chandramouli and K.P. Subbalakshmi, 2007) by using a game theory method. Specifically, the authors defined the secondary spectrum sharing problem as a potential game which takes different priority classes into consideration. Firstly, this game is solved through sequential play. Then a learning automata algorithm is introduced which only requires a feedback of the utility value. The same idea was proposed in (J. Huang, R.A. Berry and M.L. Honig, 2006), where the authors study a centralized auction mechanisms to allocate the received powers. They consider an objective function of maximizing utility which is a function of SINR. In (E. Jorswieck and R. Mochaourab, 2010) the authors tried to solve the centralized resource allocation problem by including a beamforming strategy. In this work, the primary systems are assumed to tolerate an amount of interference originating from secondary systems. This amount of interference is controlled by a pricing mechanism that penalizes the secondary systems in proportion to the interference they produce on the PUs.

Two centralized optimization frameworks were proposed in (L. Akter and B. Natarajan, 2009) in order to solve for the optimal resource management strategies. In the first framework, authors determine the minimum transmit power that SUs should employ in order to maintain a certain SINR and use that result to calculate the optimal rate allocation strategy across channels. In the second framework, both transmit power and rate per channel are simultaneously optimized with the help of a bi-objective problem formulation.

Though there have been ample research efforts on centralized resource management in CRNs, there is still a lack of a complete framework that considers QoS for SUs as well as resource management in a fair manner. One of the objective in this chapter is to take a step towards such a solution.

In a realistic network, centralized system coordination is hard to implement, especially in fast fading environments and in particular if there is no fixed infrastructure for SUs, i.e., no back-haul network over which overhead can be transmitted between users. In fact, centralized channel state information for a dense network involves immense signaling overhead and will not allow the extraction of diversity gains in fast-fading channel components. To alleviate this problem, distributed methods were proposed in the literature where SUs can communicate without PU knowledge.

5. Distributed resource allocation strategies

In the centralized case, the need may exist, as mentioned above, for the perfect knowledge of all channel and interference state conditions for all nodes in the network. To circumvent this problem, the design of so-called distributed resource allocation techniques is crucial. Distributed optimization refers to the ability for each user to manage its local resources (e.g. rate and power control, user scheduling) based only on locally observable channel conditions such as the channel gain between the access point and a chosen user, and possibly locally measured noise and interference.

A number of distributed resource allocation strategies for CRNs have been proposed in literature. In addition to the two centralized frameworks presented in last section, the authors in (L. Akter and B. Natarajan, 2009) designed a distributed suboptimal joint coordination and power control mechanism to allocate transmit powers to SUs. A lower bound on SINR is used as a QoS constraint for SUs. In (N. Nie and C. Comaniciu, 2005), the authors propose a game theoretic framework to analyze the behavior of CRs for distributed adaptive channel allocation. They define two different objective functions for the spectrum sharing games, which capture the utility of selfish users and cooperative users, respectively. The channel allocation problem is modeled in (N. Nie and C. Comaniciu, 2005) to a potential game which converges to a deterministic Nash equilibrium channel allocation point. Game theory was applied in (F. Wang, M. Krunz and S. Cui, 2008) to develop a distributed power allocation algorithm. In this work, each user maximizes its own utility function (which includes a pricing term) by performing a single-user price-based water-filling. However, in (F. Wang, M. Krunz and S. Cui, 2008), coexistence of multiple SUs in a channel has not been considered. Also, the QoS requirement of SUs has been ignored. In (Y. Wu and D.H.K. Tsang, 2009), the authors studied the distributed multi-channel power allocation for spectrum sharing CRNs with QoS guarantee. They formulate the problem as a noncooperative game with coupled strategy space to address both the co-channel interference among SUs and the interference temperature regulation imposed by primary systems.

The authors in (P. Cheng, Z. Zhang, H. Huang and P. Qiu, 2008) presented a general analytical framework, in which SU's rate, frequency, and power resource can be jointly optimized under the interference temperature constraints. This framework was used to design an optimal distributed resource allocation algorithm with low polynomial time complexities in multiuser broadband CRNs. In (J. Huang, Z. Han, M. Chiang and V. Poor, 2007), the authors focus on designing distributed resource allocation algorithms for cooperative networks. They proposed two share auction mechanisms, the SNR auction and the power auction, to distributively coordinate the relay power allocation among users. The authors in (J. Huang, Z. Han, M. Chiang and V. Poor, 2007) demonstrate that the SNR auction achieves the fair allocation, while the power auction achieves the efficient allocation.

In (D. Schmidt, C. Shi, R. Berry, M. Honig and W. Utschick, 2009), the authors propose a distributed resource allocation scheme where SUs are penalized for interfering on the primary systems. The penalty is proportional to the interference rate produced from the secondary transmitter to each PU. This mechanism is referred to as pricing and is interpreted as introducing the effect of disturbance created from a user as a penalty measure in his utility function. In this means, the secondary transmitters can be controlled to choose their transmission strategies satisfying soft interference constraints on the PUs. In (J. S. Pang, G. Scutari, D. Palomar and F. Facchinei, 2010), this model of exogenous prices is used to analyze a noncooperative game between the SUs. Distributed algorithms are provided that iteratively modify the prices weights and eventually reach the Nash equilibrium that satisfies the interference temperature constraints.

6. Distributed strategy

In the current study, we adopt a QoS guarantee to the PU by means of an outage constraint. This knowledge can be obtained with a centralized mode where the resource allocation system would require information from a third party (i.e. central database maintained by regulator or another authorized entity) to schedule SUs coming. This is the case of the user selection strategy presented in (M. Haddad, A. Hayar and G.E. Øien, 2008). In fact, to compute the P_{out}, the CR system requires knowledge of the PU and SUs channels. To alleviate this problem, we propose in this chapter a distributed method where SUs can get rid of PU knowledge. In this distributed framework, the information about the outage failure can be computed without exchange of information between the primary and secondary users. In this section, we will present in a first step a reformulation of the outage probability that will be used throughout the development of the proposed user selection strategy. Then, we will present the optimization problem of this strategy.

6.1 Binary power control policy

One basic assumption throughout this part is that a SU can vary its transmit power, under short term (minimum and peak) power constraints, in order to maximize the cognitive capacity, while maintaining a QoS guarantee to the PU. For the first proposed distributed resource allocation algorithm, we will use a binary power control (nodes transmitting at maximum power P_{max} or being silent).

The idea of the binary "on"/"off" power control is simple, as well as yielding quasi-optimal results in a number of cases (A. Gjendemsjø, D. Gesbert, G. E. Øien and S.G. Kiani, 2007). As such, it constitutes a promising tool for making spectrum sharing a reality. Besides complexity reduction, an important additional benefit of binary power control is to allow distributed optimization. With binary power constraints, power control reduces to deciding if links should be "on" or "off". The power p_m of the m-th SU transmitter is selected from the binary set $\{0, P_{max}\}$. It is intuitively clear that if the cross-gain is sufficiently low, then all links should be "on".

The key idea within the iterative algorithm used in the development of the proposed distributed user selection algorithm is, as in (Kiani & Gesbert, 2006), to subsequently limit p_m to $\{0, P_{max}\}$, i.e., to switch "off" transmission in SUs' links which do not contribute enough capacity to outweigh the interference degradation caused by them to the rest of the network. The authors in (M. Haddad, A. Hayar and G.E. Øien, 2008) propose an adaptation of the distributed algorithm which allows a subset of controlled size $\tilde{M}$ of the total number of SUs M to transmit simultaneously on the same sub-band. We will give in this section a summary of the presented method in (M. Haddad, A. Hayar and G.E. Øien, 2008).

At High SINR Regime Assuming all SUs to be in "on" condition for the mentioned CRN, at high SINR regime, we have dense environment with more users within smaller geometrical area and hence a SU requires higher threshold to be active. Assuming that 1+ SINR = SINR holds, the signal-to-interference ratio (SIR) threshold for high region comes out to be,

$$\mathrm{SIR}_m = \frac{p_m \mid h_{m,m} \mid^2}{p_{pu} \mid h_{pu,m} \mid^2 + \sum\limits_{\substack{k \in \Psi \\ k \neq m}} p_k \mid h_{k,m} \mid^2} > e = 2.718281... \tag{10}$$

At Low SINR Regime By definition in the low SINR region $\ln(1+x) \simeq x$ holds with good accuracy, and binary power control is always optimal. The active user threshold at low SINR region is expressed as,

$$\text{SIR}_m > 1 \tag{11}$$

Detailed derivations of the two threshold at high and low SINR are given in (M. Haddad, A. Hayar and G.E. Øien, 2008). Results given in (10) and (11) confirm, as intuition would expect, that SUs under better SINR conditions would transmit only above a higher threshold than in the low SINR regime.

6.2 Outage probability constraint

To proceed further with the analysis of the distributed strategy and for the sake of emphasis, we introduce the PU average channel gain estimate G_{pu} based on the following decomposition:

$$h_{pu,pu} \triangleq G_{pu} * h'_{pu,pu} \tag{12}$$

where $h'_{pu,pu}$ is the random component of channel gain and represents the *normalized* channel impulse response tap (A. Gjendemsjø, D. Gesbert, G. E. Øien and S.G. Kiani, 2007). This gives us the following PU outage probability expression:

$$P_{out} = Prob\left\{ \log_2\left(1 + \frac{p_{pu} G_{pu}^2 \mid h'_{pu,pu} \mid^2}{\sum_{m=1}^{M} p_m \mid h_{pu,m} \mid^2 + \sigma^2}\right) \leq R_{pu} \right\} \tag{13}$$

Let $\tilde{M}_d$ be the maximum number of SUs allowed to transmit using the distributed method and G_{su} be the SU average channel gain estimate. If we insert these two parameters in (13), we obtain

$$\begin{aligned} P_{out} &\simeq Prob\left\{ \frac{p_{pu} G_{pu}^2 \mid h'_{pu,pu} \mid^2}{G_{su}^2 \sum_{m=1}^{\tilde{M}_d} p_m + \sigma^2} \leq 2^{R_{pu}} - 1 \right\} \leq q \\ &\simeq Prob\left\{ \mid h'_{pu,pu} \mid^2 \leq \left(2^{R_{pu}} - 1\right)\left(\frac{\tilde{M}_d G_{su}^2 P_{max} + \sigma^2}{G_{pu}^2 p_{pu}}\right)\right\} \leq q \end{aligned} \tag{14}$$

From now on we assume for simplicity of analysis that the channel gains are i.i.d Rayleigh distributed. However, the results can be immediately translated into results for any other channel model by replacing the appropriate probability distribution function. Continuing from (14), we have:

$$P_{out} \simeq \int_0^{\left(2^{R_{pu}} - 1\right)\left(\frac{\tilde{M}_d G_{su}^2 P_{max} + \sigma^2}{G_{pu}^2 p_{pu}}\right)} \exp(-t)dt \leq q \tag{15}$$

Finally, we get the following outage constraint:

$$P_{out} \simeq 1 - \exp\left[-\left(2^{R_{pu}} - 1\right)\left(\frac{\tilde{M}_d G_{su}^2 P_{max} + \sigma^2}{G_{pu}^2 p_{pu}}\right)\right] \leq q \tag{16}$$

and the maximum number $\tilde{M}_d$ of active "on" SUs that transmit with P_{max} is given by

$$0 \leq \tilde{M}_d \leq \frac{-\log(1-q)}{\left(2^{R_{pu}} - 1\right)} \cdot \frac{G_{pu}^2 p_{pu}}{G_{su}^2 P_{max}} - \frac{\sigma^2}{G_{su}^2 P_{max}} \tag{17}$$

By writing $\text{SNR} = \frac{G_{su}^2 P_{max}}{\sigma^2}$, equation (17) can be expressed as:

$$0 \leq \tilde{M}_d \leq \frac{-\log(1-q)}{\left(2^{R_{pu}} - 1\right)} \cdot \frac{G_{pu}^2 p_{pu}}{G_{su}^2 P_{max}} - \frac{1}{\text{SNR}} = \tilde{M}_{theory} \tag{18}$$

where $\tilde{M}_{theory}$ is the theoretic maximum number of SUs allowed to transmit. The LHS in (18) prevents from obtaining a negative number of users when the SNR decreases significantly. The formula in (18) points out that the number of SUs allowed to transmit increases as their SNR increases.

6.3 Optimization problem

The SUs offer the opportunity to improve the system throughput by detecting the PU activity and adapting their transmissions accordingly while avoiding the interference to the PU by satisfying the QoS constraint on outage. We present in this subsection a distributed user selection strategy using the binary power allocation policy given in Section 6.1. The proposed strategy tries to limit the number of SUs interfering with the PU so as to guarantee the QoS for the primary system. Specifically, a SU will be deactivated if its action results in an increase in the cognitive capacity of SUs or if its transmission violates the PU outage constraint. The optimization problem can therefore be expressed as follows:

$$\text{Find } p_m|_{m=1,\ldots,M} = \arg\max_{p_m} C_{su} \tag{19}$$

subject to:

$$\begin{cases} p_m \in \{0, P_{max}\}, \quad \text{for} \quad m = 1, \ldots, M \\ 0 \leq \tilde{M}_d \leq \frac{-\log(1-q)}{\left(2^{R_{pu}} - 1\right)} \cdot \frac{G_{pu}^2 p_{pu}}{G_{su}^2 P_{max}} - \frac{1}{\text{SNR}} \end{cases} \tag{20}$$

where $\tilde{M}_d$ is the maximum number of SUs allowed to transmit using the distributed algorithm and q the maximum outage probability. As we can see from (20), the CR system does not require knowledge about the PU and SUs channels in the sense that it decides *distributively* to either SU transmit data or stay silent over the channel coherence time depending on the specified P_{out} threshold (q). On the other hand, the optimization problem presented in (M.

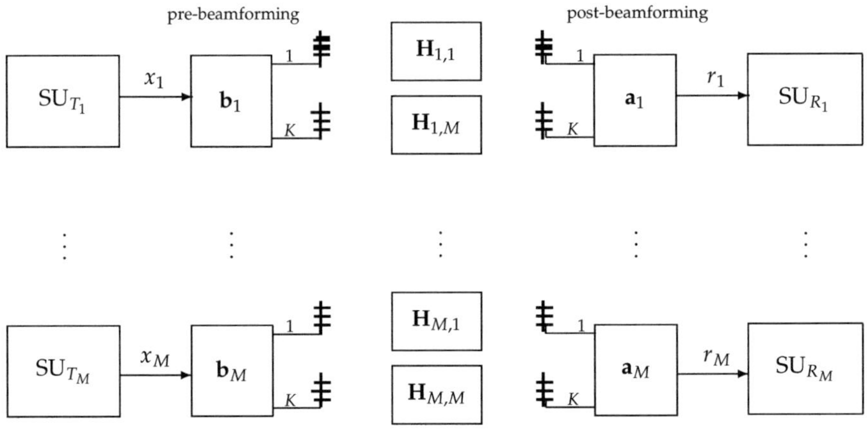

Fig. 2. Multiple transmit and receive secondary users system structure.

Haddad, A. Hayar and G.E. Øien, 2008) requires all $h_{m,pu}$ and $h_{pu,pu}$ data to compute the outage probability and to select then the SUs able to transmit without affecting the PUs' QoS.

An iterative approach is adopted throughout this algorithm. The algorithm is first initialized with a zero power allocation vector. Each SU simultaneously measures his SIR; depending on whether the SU is on high or low average SINR, he remains active or inactive during the next time slot based on (10) and (11), respectively. Similarly, at every iteration of the Monte Carlo simulation, $\tilde{M}_d$ is evaluated for the SU in question based on the average channel gains G_{su} and G_{pu} estimation. The algorithm is run until the outage probability stabilizes. The last SU entering in the system is removed from the transmission.

7. Centralized beamforming-based resource allocation strategy

In this section, we will present the design of the transmit and receive beamvectors. In fact, beamvector associated with each SU is determined by optimizing a certain criterion to reach a specific purpose such as maximizing the throughput or minimizing the interference. In the literature, depending on the objective function and the constraints, the beamforming optimization problems can be divided into two classes. One is the SINR balancing problem (M. Schubert and H. Boche, 2004), i.e., maximizing the sum SINR among all the users. The other one is the power minimization problem with SINR constraints (M. Schubert and H. Boche, 2004), i.e., minimizing some power function with SINR constraints. In this work, we adopt the first class combined with an outage probability constraint, i. e., we will maximize the per-user sum capacity subject to minimize the mutual interference. The goal here is to choose for each user who has the best equivalent channel created by the multiple antennas and beamforming where resources are allocated. In this section, we introduce the power constraints to compute beamvectors. Then, we present the outage probability constrain. Finally, we present the optimization problem of the proposed strategy.

7.1 Secondary users MIMO system

In this subsection, we will describe the SU MIMO system model and multiuser diversity scheme that we are considering in this chapter, and discuss the primary and secondary users metrics when we use a SU MIMO system. We will reformulate the resource allocations metrics given in Section 3 when we use a SU single-input-single-output (SISO) system.

The proposed system in this chapter consists of multiple transmit/receive SU links randomly distributed over the geographical area considered. MIMO systems have great potential to enhance the capacity in the framework of wireless cellular networks (E.A. Jorswieck and H. Boche, 2004). Multiple antennas can for example be deployed at a cognitive BS. Many wireless network standards provision the use of transmit antenna arrays. Using baseband beamforming, it is possible to steer energy in the direction of the intended users, whose channels can often be accurately estimated (A. Tarighat, M. Sadek and A.H. Sayed, 2005; E.A. Jorswieck and H. Boche, 2004). Beamforming has been also exploited as a strategy that can serve many users at similar throughput. Moreover, beamforming has the advantage of limiting interference. Thus, we are interested in transmit beamforming schemes for cognitive transmission. For this purpose, we utilize joint beamforming that implies an extension to the transmitter side of classical receive beamforming (P. Viswanath, D. Tse and R. Laroia, 2004).

The SU system structure is based on beamforming at both the transmitter (K antennas) and the receiver (K antennas) for each SU link as given in Fig. 2. The number of secondary transmitters (SU_T) is equal to M, and is equal to the number of secondary receivers (SU_R). Assuming that many scatterers are located around the transmitter and receivers, the channel coefficient matrix $\mathbf{H}_{rt}$ (the channel between the t-th transmit SU and the r-th receive SU) exhibits flat fading. The channel gain vector $\mathbf{h}_{pu,m}$ from the PU indexed by pu to a desired SU m (m between 1 and M) is given by:

$$\mathbf{h}_{pu,m} = [h_{pu,m_1} ... h_{pu,m_K}]^T \tag{21}$$

where the channel gains are assumed i.i.d. random variables. We consider that the channels between different users are independent. We then set the received signal of the m-th user as follows (the index of SUs m lies between 1 and M):

$$\mathbf{y}_m = \mathbf{H}_{m,m}\mathbf{s}_m + \sum_{l=1, l \neq m}^{M} \mathbf{H}_{m,l}\mathbf{s}_l + \mathbf{h}_{pu,m}x_{pu} + \mathbf{n}_m, \quad m = 1, ..., M \tag{22}$$

with $\mathbf{n}_m$ of size $K \times 1$ being zero-mean i.i.d. Gaussian noise with power σ_m^2, and K being the number of antennas. $\mathbf{s}_m$ is the transmit vector of size $K \times 1$ for the m-th SU and x_{pu} being the transmit sample sent from PU. $\mathbf{y}_m$ is the receive vector of size $K \times 1$. $\mathbf{H}_{m,m}$ ($K \times K$ matrix) is the channel between the m-th SU_T and the m-th SU_R and $\mathbf{H}_{m,l}(l = 1, ..., m-1, m+1, ..., M)$ are channel matrices between the other SUs, referred to as the *interference channel matrices.*

Here, a joint beamforming approach is proposed for the SU system, that is, all the transmitters and receivers exploit a beamforming architecture (E.A. Jorswieck and H. Boche, 2004). The transmission scheme is characterized by the power allocation (eigenvalues of the transmit covariance matrix) and the orientation (eigenvectors of the transmit covariance matrix) (Tse, 2004). This yields

$$\mathbf{s}_m = \mathbf{b}_m x_m, \quad m = 1, ..., M \tag{23}$$

where $\mathbf{b}_m$ is the pre-beamforming vector and x_m is the transmit sample for m between 1 and M. The output of the m-th receiver beamformer is:

$$
\begin{aligned}
r_m &= \mathbf{a}_m^H \mathbf{y}_m \\
&= \mathbf{a}_m^H \mathbf{H}_{m,m} \mathbf{b}_m x_m + \mathbf{a}_m^H \sum_{l=1,l\neq m}^{M} \mathbf{H}_{m,l} \mathbf{b}_l x_l + \mathbf{a}_m^H \mathbf{h}_{pu,m} x_{pu} + \mathbf{a}_m^H \mathbf{n}_m
\end{aligned} \tag{24}
$$

where $\mathbf{a}_m$ is the post-beamforming vector at the receive SUs. $\Phi_m = E\{\mathbf{n}_m \mathbf{n}_m^H\}$ is the associated covariance matrix. The SINR defined in (5) at the m-th SU can be rewritten as:

$$
\begin{aligned}
\mathrm{SINR}_m &= \frac{E\{|\mathbf{a}_m^H \mathbf{H}_{m,m} \mathbf{b}_m x_m|^2\}}{E\left\{\sum_{l=1,l\neq m}^{M} |\mathbf{a}_m^H \mathbf{H}_{m,l} \mathbf{b}_l x_l|^2\right\} + E\{|\mathbf{a}_m^H \mathbf{h}_{pu,m} x_{pu}|^2\} + E\{|\mathbf{a}_m^H \mathbf{n}_m|^2\}} \\
&= \frac{|\mathbf{a}_m^H \mathbf{H}_{mm} \mathbf{b}_m|^2}{|\mathbf{a}_m^H \mathbf{h}_{pu,m}|^2 + \sum_{l=1,l\neq m}^{M} |\mathbf{a}_m^H \mathbf{H}_{m,l} \mathbf{b}_l|^2 + \mathbf{a}_m^H \mathbf{R}_m \mathbf{a}_m}
\end{aligned} \tag{25}
$$

and the capacity of PU is given in this context by:

$$
C_{pu} = \log_2\left(1 + \frac{p_{pu}|h_{pu,pu}|^2}{\sum_{m=1}^{M} |\mathbf{h}_{pu,m}\mathbf{h}_{pu,m}{}^H|||\mathbf{b}_m||^2 + \sigma^2}\right) \tag{26}
$$

An efficient transmit beamforming technique combined with user selection will be proposed in the following section by optimizing a certain problem.

7.2 Power constraints

To compute the beamvectors, we consider just the SU MIMO system. The reason for this is that the interference among PU is nulled in SINR equation (i.e. $\mathbf{a}_m^H \mathbf{h}_{pu,m} = 0$). In fact, we propose an algorithm that can minimize the interference between cognitive users. SUs are first pre-selected so as to maximize the per-user sum capacity, and then, the PU verifies the outage probability constraint and a number of SUs are selected from those pre-selected SUs. Specifically, beamvectors are selected such that they satisfy the interference free condition $\mathbf{a}_m^H \mathbf{h}_{pu,m} = 0$. If we consider this condition, the SINR at the m-th SU can then be written as:

$$
\mathrm{SINR}_m = \frac{\left(\mathbf{a}_m^H \mathbf{H}_{m,m} \mathbf{b}_m\right)^H \left(\mathbf{a}_m^H \mathbf{H}_{m,m} \mathbf{b}_m\right)}{\mathbf{a}_m^H \left(\Phi_m + \sum_{l=1,l\neq m}^{M} \mathbf{H}_{m,l} \mathbf{b}_l \mathbf{b}_l^H \mathbf{H}_{m,l}^H\right) \mathbf{a}_m} \tag{27}
$$

We define the total interference plus noise covariance matrix at the m-th SU as:

$$
\mathbf{R}_m = \Phi_m + \sum_{l=1,l\neq m}^{M} \mathbf{H}_{m,l} \mathbf{b}_l \mathbf{b}_l^H \mathbf{H}_{m,l}^H \tag{28}
$$

Therefore, the SINR at the m-th SU can be formulated as follows:

$$\begin{aligned}\text{SINR}_m &= \frac{\left(\mathbf{a}_m^H \mathbf{H}_{m,m} \mathbf{b}_m\right)^H \left(\mathbf{a}_m^H \mathbf{H}_{m,m} \mathbf{b}_m\right)}{\mathbf{a}_m^H \mathbf{R}_m \mathbf{a}_m} \\ &= \mathbf{b}_m^H \mathbf{H}_{m,m} \mathbf{R}_m^{-1} \mathbf{H}_{m,m}^H \mathbf{b}_m\end{aligned} \tag{29}$$

From (29), the post-beamforming vector can be expressed as follows:

$$\mathbf{a}_m = \mathbf{R}_m^{-1} \mathbf{H}_{m,m} \mathbf{b}_m \tag{30}$$

This gives us the following maximization of SINR at the m-th SU:

$$\mathbf{b}_m^H \mathbf{H}_{m,m}^H \mathbf{R}_m^{-1} \mathbf{H}_{m,m} \mathbf{b}_m \leq \lambda_{max}(m) p_m = \text{SINR}_m|_{max} \tag{31}$$

where $\lambda_{max}(m)$ is the maximum eigenvalue of $\mathbf{H}_{m,m}^H \mathbf{R}_m^{-1} \mathbf{H}_{m,m}$ and $p_m = \mathbf{b}_m^H \mathbf{b}_m$. For beamforming, the transmitted power through all the SUs for the m-th SU is proportional to $||\mathbf{b}_m||^2$. The design goal is to find the optimum transmit weight vector subject to a carrier power constraint. We consider the power allocation problem corresponding to the distribution of all the available power at the transmitter among all SUs, when the data destined from SU m is transmitted with a maximum power P_{max}. This per-user power constraint is given by:

$$||\mathbf{b}_m||^2 = p_m \leq P_{max}, \quad \forall m = 1, ..., M \tag{32}$$

and the global power constraint is formulated as follows:

$$\sum_{m=1}^{M} ||\mathbf{b}_m||^2 = \sum_{m=1}^{M} p_m \leq M P_{max} \tag{33}$$

7.3 Outage probability constraint

The outage probability is given by (51). In this subsection, we will reformulate this equation using the beamforming strategy. Proceeding in the same manner as in Subsection 6.2, the outage probability can be written as:

$$P_{out} = Prob\left\{\log_2\left(1 + \frac{p_{pu} G_{pu}^2 \mid h'_{pu,pu} \mid^2}{\sum_{m=1}^{M} |\mathbf{h}_{pu,m} \mathbf{h}_{pu,m}{}^H| ||\mathbf{b}_m||^2 + \sigma^2}\right) \leq R_{pu}\right\} \tag{34}$$

As in the development of the distributed user selection strategy in Section 6, we introduce here the PU and SU average channel gain estimate G_{pu} and G_{su}, respectively, defined in the Subsection 6.2. These assumptions give the same PU outage probability expression given by (14).

7.4 Optimization problem

Concluding that the maximum eigenvalue $\lambda_{max}(m)$ must be chosen so as to maximize the capacity of SUs given a fixed transmit power. In the first step of the proposed beamforming user selection strategy, SUs are first pre-selected so as to maximize the per-user sum capacity. In the second step of the user selection strategy, the PU verifies the outage probability constraint and a number of SUs are selected from those pre-selected SUs. If we maximize the per-user sum capacity (C_{su}): i.e. the sum of the SINR averaged over all SUs under the constraints of maintaining the global power lower than MP_{max} and of satisfying the QoS constraint on outage, the problem can be written as:

$$\begin{cases} \text{maximize } f(p_1, ..., p_M) = \dfrac{1}{\ln 2} \sum\limits_{m=1}^{M} \ln\left(1 + \lambda_{max}(m) p_m\right) \\ \text{subject to } \sum\limits_{m=1}^{M} p_m \leq MP_{max} \\ \qquad\qquad P_{out} \leq q \end{cases} \tag{35}$$

To compute the transmitted power through all SUs, we define the Lagrangian expression for this maximization problem as follows:

$$J = \frac{1}{\ln 2} \sum_{i=1}^{M} \ln\left(1 + \lambda_{max}(i) p_i\right) - \mu \left(\sum_{i=1}^{M} p_i - MP_{max} \right) - \nu \left(P_{out} - q\right) \tag{36}$$

We introduce in (36) two variables, μ and ν, called Lagrange multipliers. The solution of all the system is found by calculating the derivatives of J with respect to the power allocation parameters $p_m|_{m=1..M}$ and Lagrange multipliers μ and ν. By calculating the derivatives of J with respect to the power allocation parameters p_m, we obtain:

$$\frac{\partial J}{\partial p_m} = \frac{(\ln 2)^{-1} \lambda_{max}(m)}{1 + \lambda_{max}(m) p_m} - \mu - \nu \frac{\left(2^{R_{pu}} - 1\right) G_{su}^2}{G_{pu}^2 P_{pu}} \exp\left[-\left(2^{R_{pu}} - 1\right) \left(\frac{G_{su}^2 \sum\limits_{i=1}^{M} p_i + \sigma^2}{G_{pu}^2 P_{pu}} \right) \right] = 0 \tag{37}$$

Let $g(p_i) = \frac{(2^{R_{pu}}-1) G_{su}^2}{G_{pu}^2 P_{pu}} \exp\left[-\left(2^{R_{pu}} - 1\right) \left(\frac{G_{su}^2 \sum\limits_{i=1}^{M} p_i + \sigma^2}{G_{pu}^2 P_{pu}} \right) \right]$, we can express the solution of (37) as:

$$\frac{1}{(\mu + \nu g(p_i)) \ln 2} \lambda_{max}(m) = 1 + \lambda_{max}(m) p_m \tag{38}$$

The solution of this problem is formulated as follows:

$$p_m = \frac{1}{(\mu + \nu g(p_i)) \ln 2} - \frac{1}{\lambda_{max}(m)} \tag{39}$$

The derivatives of J with respect to the power allocation parameters $p_i|_{i=1..M}$:

$$\begin{cases} p_1 = \frac{1}{(\mu+\nu g(p_i)) \ln 2} - \frac{1}{\lambda_{max}(1)} \\ \cdot \\ \cdot \\ p_M = \frac{1}{(\mu+\nu g(p_i)) \ln 2} - \frac{1}{\lambda_{max}(M)} \end{cases} \tag{40}$$

The sum of all equations in (55) gives:

$$\begin{aligned} \sum_{i=1}^{M} p_i &= \frac{M}{(\mu + \nu g(p_i)) \ln 2} - \sum_{i=1}^{M} \frac{1}{\lambda_{max}(i)} \\ &= M\left(p_m + \frac{1}{\lambda_{max}(m)}\right) - \sum_{i=1}^{M} \frac{1}{\lambda_{max}(i)} \\ &= MP_{max} \end{aligned} \tag{41}$$

Finally, we obtain the following set of equalities:

$$p_m = P_{max} - \frac{1}{\lambda_{max}(m)} + \frac{1}{M} \sum_{i=1}^{M} \frac{1}{\lambda_{max}(i)} \quad \text{for} \quad m = 1, ..., M \tag{42}$$

This equation gives the power allocation solution using the global power constraint given by (33). Firstly, the per-user power constraint given in (32) has been utilized to solve the problem, i.e. maximizing the per-user sum capacity under the constraint of maintaining the per-user power constraint lower than P_{max} for all users. In this case, the Lagrangian expression is given by:

$$J = \frac{1}{\ln 2} \sum_{m=1}^{M} \ln\left(1 + \lambda_{max}(i) p_i\right) - \sum_{i=1}^{M} \mu_i \left(p_i - P_{max}\right) \tag{43}$$

and the transmitted power through all SUs is:

$$p_m = P_{max}, \qquad m = 1, ..., M \tag{44}$$

but it is not the optimal solution. Besides, from (42), p_m can have values higher than P_{max} which contradicts condition (32). To optimally solve this problem, one should adopt this solution:

$$\begin{array}{ll} p_m = P_{max} & \text{if} \quad \frac{1}{\lambda_{max}(m)} < \frac{1}{M} \sum_{i=1}^{M} \frac{1}{\lambda_{max}(i)} \\ p_m = P_{max} - \frac{1}{\lambda_{max}(m)} + \frac{1}{M} \sum_{i=1}^{M} \frac{1}{\lambda_{max}(i)} & \text{else} \end{array} \tag{45}$$

Therefore, it will be shown later from simulation results that (45) can approximate very well the per-user sum capacity with optimal power allocation.

8. Distributed game theory-based resource allocation strategy

8.1 Game theory tools

Game theory was at first a mathematical tool used for economics, political and business studies. It helps understand situations in which decision-makers interact in a complex environment according to a set of rule (M. J. Osborne, 2003). Many different types of game exist which are used to analyze different situations: potential games, repeated game, cooperative or non-cooperative games. In the cognitive radio network (CRN), the formal game model for the power control can be defined as follows:

- Players: are the cognitive users (secondary users (SUs)).
- Actions: called also as the decisions, and are defined by the transmission power allocation strategy.
- Utility function: represents the value of the observed quality-of-service (QoS) for a player, and is defined later in this section.

The central idea in game theory is how the decision from one player will affects the decision-making process from all other players and how to reach a state of equilibrium that would satisfy most of the players. A well known contributor in the field is Nash for the Nash equilibrium (R. W. Rosenthal, 1973). The theory shows that you can reach a state equilibrium for your system where all decisions are set, unchanging and is the best possible situation for the players.

CR need to perform sophisticated adaptation and dynamically learn from the environment. This situation makes the learning process a very complicated one comparable to situation found in economics. Game theory is already used in other field of communication to better understand congestion control, routing, power control, topology control and trust management (A. B. MacKenzie, L. Dasilva, and W. Tranter, 2006). Our interests rest in its use for power control as it can be considered a game with fixed number of players where each tries to optimize their power levels. There are a number of properties that makes this problem appropriate for a cognitive radio game model:

- The player's payoff is a function of her own transmit power level and her signal-to-noise and interference ratio (SINR). The player's SINR is a function of her own transmit power and the transmit powers of the other players in the cell.
- When a player increases her power level, this will increase her own SINR, but will decrease the SINRs of all other players.
- For a fixed SINR, the players prefer lower power levels to higher ones. That is, players wish to conserve power and extend their battery life when possible.
- For a fixed power level, players prefer higher SINR to lower one. That is, players want the best possible channel conditions for a given expenditure of power.

There are many ways to cope with these issues such as to add restriction to the use of the power resource by charging it to users. This is done by adding a cost component to the *payoff function* to add fairness to the network. Another idea is to model the scenario as a repeated game (A. B. MacKenzie, L. Dasilva, and W. Tranter, 2006).

In this part we formulate the problem of resource allocation in the context of a CRN to reflect the needs of PUs and SUs. We consider the primary uplink of a single CRN, where cognitive transmitters transmit signals to a number of SUs, while the primary BS receives its desired signal from a primary transmitter and interference from all the cognitive transmitters.

To resolve the problem of resource allocation, we propose a *utility function* that meets the objective to maximize the SUs capacity, and the protection for PUs. Specifically, we define a *payoff function* that represents the SNIR constraint, and a *price function* specifies the outage probability constraint. The *utility function* is defined as:

$$\text{utility function} = \text{payoff function} - \text{price function}$$

We introduce a *payoff* to express the capacity need of SU m, and a *price function* to represent the protection for PUs by means of the outage probability. And each SU adjusts its transmitted power to maximize its *utility function*. Therefore, we will present in this part a power allocation algorithm that maximize the defined *utility function* to compute the transmitted power of each SU.

8.2 Power allocation algorithm

We derive in this section the *utility function*: we define a *payoff function* specifies the SU capacity constraint and a *price function* that represents the interference constraint as a function of the outage probability constraint. Therefore, the *price function* is given by:

$$\mathrm{L}_{\mathrm{Intf}m} = \frac{p_m|h_{pu,m}|^2}{P_T - \sum\limits_{\substack{l=1 \\ l\neq m}}^{M} p_l|h_{pu,l}|^2} \tag{46}$$

which is a normalized value. As long as this ratio $\in [0,1]$, the protection for PU is met. The margin of $P_T - \sum_{\substack{l=1 \\ l\neq m}}^{M} p_l|h_{pu,l}|^2$ is the maximum interference that SU m could generate. We will derive in this section the equation of the interference constraint P_T as a function of the outage probability.

To proceed further with the analysis and for the sake of emphasis, we introduce the PU average channel gain estimate G_{pu} based on the following decomposition:

$$h_{pu,pu} \equiv G_{pu} * h'_{pu,pu} \tag{47}$$

where h'_{pupu} is the random component of channel gain and represents the *normalized* channel impulse response tap. This gives us the following PU outage probability expression in an interference-limited context:

$$P_{out} \simeq Prob\left\{ |h'_{pu,pu}|^2 \leq \left(2^{R_{pu}} - 1\right)\left(\frac{\sum\limits_{m=1}^{M} p_m|h_{m,pu}|^2}{G_{pu}^2 p_{pu}}\right)\right\} \tag{48}$$

From now on we assume for simplicity of analysis that the channel gains are i.i.d rayleigh distributed. However, the results can be immediately translated into results for any other channel model by replacing by the appropriate probability distribution function. Continuing from (48), we have:

$$P_{out} \simeq \int_0^{\left(2^{R_{pu}}-1\right)\left(\frac{\sum\limits_{m=1}^{M} p_m |h_{m,pu}|^2}{G_{pu}^2 p_{pu}}\right)} \exp(-t)dt \tag{49}$$

Finally, we get the following outage constraint:

$$P_{out} \simeq 1 - \exp\left[-\left(2^{R_{pu}}-1\right)\left(\frac{\sum\limits_{m=1}^{M} p_m |h_{m,pu}|^2}{G_{pu}^2 p_{pu}}\right)\right] \tag{50}$$

Replacing the interference constraint equation in (50), we can express the probability outage as:

$$P_{out} \simeq 1 - \exp\left[-\left(2^{R_{pu}}-1\right)\frac{P_T}{G_{pu}^2 p_{pu}}\right] \tag{51}$$

Then, the corresponding interference constraint is:

$$P_T \simeq \frac{p_{pu} G_{pu}^2}{1-2^{R_{pu}}} \ln\left(1-P_{out}\right) \tag{52}$$

We introduce now a *utility function* for which each SU adjusts its transmitted power in order to maximize it. It is composed of a *payoff function* expressed as the capacity C_m of the SU, and of a *price function* composed of the interference level to the PU and the power consumption.

Then, the *utility function* is expressed as follow:

$$U_m = C_m - \left(\frac{p_m |h_{pu,m}|^2}{P_T - \sum\limits_{\substack{l=1\\ l\neq m}}^{M} p_l |h_{l,m}|^2}\right)^{a_m} \tag{53}$$

The parameter a_m is adjustable to have a comparable values, i.e. the *payoff function* value and the *price function* value. This parameter gives the flexibility needed to adjust the SU capacity over the interference to the PU. We choose $a_m < 0$. It could be easily obtained that the *price function* decreases as the ratio $\mathrm{L}_{\mathrm{Intf}m}$ increases. This fact is caused by the negative property of a_m.

Mathematically, the game G can be expressed as:

$$\text{Find } p_m|_{m=1,\ldots,M} = \arg\max_{p_m} U_m(p_m, \mathbf{P}_{-m}) \tag{54}$$

subject to:

$$\begin{cases} \sum_{m=1}^{M} p_m |h_{pu,m}|^2 \leq P_T \\ P_{out} \leq q \\ 0 \leq p_m \leq P_{max} \end{cases} \tag{55}$$

Recall that p_m denotes the strategy adopted by SU m and $\mathbf{P}_{-m} = (p_l)_{l \neq m, l \in \{1,\ldots,M\}}$ denotes the strategy adopted by the other SUs. We replace the capacity by expression given by (4) and use (9) to obtain the following equation:

$$U_m = \log_2(1 + \text{SINR}_m) - \left(\frac{|h_{pu,m}|^2}{P_T - \sum\limits_{\substack{l=1 \\ l \neq m}}^{M} p_l |h_{l,m}|^2} \right)^{a_m} \times \left(\frac{\text{SINR}_m \text{Intf}_m}{|h_{m,m}|^2} \right)^{a_m} \tag{56}$$

We are going to maximize the *utility function* in terms of the SINR. The solution of the system is found by calculating the derivatives of U_m with respect to the signal-to-noise and interference ratio parameters SINR_m:

$$\frac{\partial U_m}{\partial \text{SINR}_m} = \frac{1}{(1 + \text{SINR}_m)\ln 2} - \left(\frac{|h_{pu,m}|^2}{P_T - \sum\limits_{\substack{l=1 \\ l \neq m}}^{M} p_l h_{l,m}} \right)^{a_m} \times a_m \left(\frac{\text{SINR}_m \text{Intf}_m}{|h_{m,m}|^2} \right)^{a_m - 1} \frac{\text{Intf}_m}{|h_{m,m}|^2} \tag{57}$$

We can express the solution of (57) as:

$$(1 + \text{SINR}_m)\, \text{SINR}_m^{a_m - 1} = \frac{1}{a_m \beta_m \ln 2} \tag{58}$$

where:

$$\beta_m = \left(\frac{|h_{pu,m}|^2}{P_T - \sum\limits_{\substack{l=1 \\ l \neq m}}^{M} p_l |h_{l,m}|^2} \right)^{a_m} \left(\frac{\text{Intf}_m}{|h_{m,m}|^2} \right)^{a_m} \tag{59}$$

denoting the slope of the *price function*. Let $f(\text{SINR}_m) = (1 + \text{SINR}_m)\,\text{SINR}_m^{a_m - 1}$. Finally, we obtain the following set of equalities:

$$\text{SINR}_m = f^{-1}\left(\frac{1}{a_m \beta_m \ln 2}\right) \tag{60}$$

The maximization problem is dependent on a_m which is defined in the *utility function* as an adjustment parameter to the *price function*. For simulation results $a_m = -0.2$. It was chosen to stay with this value after different simulations to show its influence on the obtained results.

8.3 Existence and uniqueness of the Nash equilibrium

In the proposed game, each SU chooses an appropriate power to maximize its *utility function*. In this context, it is important to ensure the stability of the system. A concept which relates to this issue is the Nash equilibrium. As definition in (R. W. Rosenthal, 1973), a pure strategy profile $\{p_l^*\}_{l \neq m, l \in \{1,\ldots,M\}}$ is a Nash equilibrium of the proposed game if, for every player m (i.e. SU m):

$$U_m(p_m^*, \mathbf{P}_{-m}^*) \geq U_m(p_m, \mathbf{P}_{-m}^*), \quad \forall m \in \{1, \ldots, M\} \tag{61}$$

A Nash equilibrium can be regraded as a stable solution, at which none of the users has the incentive to change its power p_m.

8.3.1 Existence of the Nash equilibrium

Theorem 1: Game G admits at least one Nash equilibrium.

proof: The conditions for the existence of Nash equilibrium in a strategic game are given in (Fudenberg & Tirole, 1991):

1. The set P_m is a nonempty, convex, and compact subset of some Euclidean space for all m.
2. The *utility function* $U_m(p_m, \mathbf{P}_{-m})$ is continuous on P and quasi-concave on P_m.

According to the above description of the strategy space, it is straightforward to see that P_m is nonempty, convex and compact. Notice that $U_m(p_m, \mathbf{P}_{-m})$ is a linear function of either p_m, which means the second condition is satisfied. Hence, game G admits at least one Nash equilibrium.

8.3.2 Uniqueness of the Nash equilibrium

Theorem 2: Game G always possesses a unique Nash equilibrium under the sufficient conditions.

proof: It's established in (Yates, 1995) that if the *utility function* $U_m(p_m) : (p_m)_{m \in \{1,\ldots,M\}}$ is a standard function, then the Nash equilibrium in this game will be unique. A function $f(x)$ is said to be a standard function if it satisfies the following three properties (Yates, 1995):

1. Positivity: $f(x) > 0$.
2. Monotonicity: If $x \geq x'$, then $f(x) \geq f(x')$.
3. Scalability: For all $\mu > 1$, $\mu f(x) \geq f(\mu x)$.

The positivity is obviously satisfied by adjusting parameter a_m.

Considering $p_m \geq p'_m$, we have

$$C_m(p_m) \geq C_m(p'_m) \tag{62}$$

Using the propriety that $a_m < 0$, we can obtain that

$$\left(\frac{p_m |h_{pu,m}|^2}{P_T - \sum_{\substack{l=1 \\ l \neq m}}^{M} p_l |h_{l,m}|^2} \right)^{a_m} \leq \left(\frac{p'_m |h_{pu,m}|^2}{P_T - \sum_{\substack{l=1 \\ l \neq m}}^{M} p_l |h_{l,m}|^2} \right)^{a_m} \tag{63}$$

According to (62) and (63), the monotonicity property is proved $\forall m \in \{1, ..., M\}$.

For all $\mu > 1$, it's got that:

$$\mu C_m(p_m) = \mu \log_2 (1 + \text{SINR}_m) = \log_2 (1 + \text{SINR}_m)^{\mu} \geq \log_2 (1 + \mu \text{SINR}_m) = C_m(\mu p_m) \tag{64}$$

Since $a_m < 0$, we have also:

$$\left(\frac{\mu p_m |h_{pu,m}|^2}{P_T - \sum_{\substack{l=1 \\ l \neq m}}^{M} p_l |h_{l,m}|^2} \right)^{a_m} = \mu^{a_m} \left(\frac{p_m |h_{pu,m}|^2}{P_T - \sum_{\substack{l=1 \\ l \neq m}}^{M} p_l |h_{l,m}|^2} \right)^{a_m} \leq \mu \left(\frac{p_m |h_{pu,m}|^2}{P_T - \sum_{\substack{l=1 \\ l \neq m}}^{M} p_l |h_{l,m}|^2} \right)^{a_m} \tag{65}$$

Finally, according to (64) and (65) the scalability property is proved. Therefore, the proposed game G always possesses a unique Nash equilibrium.

9. Performance evaluation

To go further with the analysis, we resort to realistic network simulations. Specifically, we consider a CRN in the downlink and the uplink mode, respectively, with one PU and M SUs attempting to communicate during a transmission, subject to mutual interference. A hexagonal cellular system functioning at 1.8GHz with a secondary cell of radius R and a primary protection area of radius R_p is considered. Secondary transmitters may communicate with their respective receivers of distances $d < R_p$ from the BS. We assume that the PU and the SUs are randomly distributed in a two-dimensional plane.

The channel gains are based on the COST-231 Hata model (*Urban Transmission Loss Models for Mobile Radio in the 900 and 1800 MHz Bands*, 1991) including log-normal shadowing with standard deviation of 10dB, plus fast-fading assumed to be i.i.d. circularly symmetric with distribution $\mathcal{CN}(0, 1)$.

The performance of the proposed distributed user selection strategy is evaluated by Monte Carlo simulations ($IT_{max} = 10^4$). It is assumed that the maximum outage probability $q = 1\%$

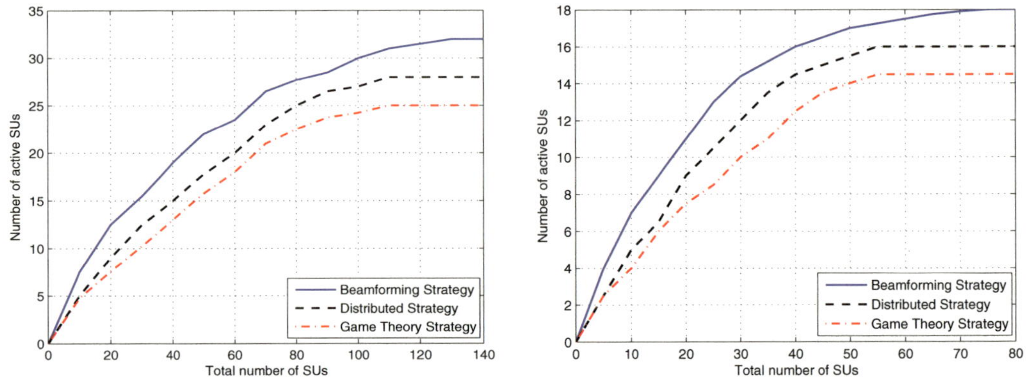

(a) Downlink (b) Uplink

Fig. 3. Performance evaluation of the proposed user selection strategies: Number of active secondary users versus total number of secondary users with rate = 0.1bits/s/Hz and $q = 1\%$ in the downlink and the uplink mode.

for both distributed and centralized algorithms. We considered also that the radius of the secondary cell $R = 1000$ meters and the radius of the primary protection area $R_p = 600$ meters. The derivation of the maximum number of SUs allowed to transmit using the distributed algorithm is based on the average channel gains G_{su} and G_{pu} estimation. From the locations of the users in the two-dimensional plane and the propagation characteristics of the environment, we can estimate the two average channel gains for the downlink and the uplink mode. These values are estimated assuming a wireless ad hoc network affected by a large number of interferers. From simulation results, using $M = 500$ SUs and one PU, we find $G_{pu}^2/G_{su}^2 \simeq 15$ in the downlink mode and $G_{pu}^2/G_{su}^2 \simeq 20$ in the uplink mode.

Fig. 3 shows the behavior of the three strategies, for both downlink and uplink scenarios. This figure presents the number of active SUs versus the total number of SUs ranging between 1 user and a maximum of 140 users, and using a rate equal to 0.1bits/s/Hz). It can be seen from the figure that increasing the number of SUs produces improvements in the number of active SUs. We show also that the beamforming user selection method outperforms the game theory-based and the distributed strategy. We gain almost 4 additional active SUs using the centralized beamforming strategy in comparison with the simple distributed strategy. It is obvious from Fig. 3 that the number of active SUs slopes in the distributed strategy start dropping at a lower number of SUs than in the beamforming strategy case. While the number of active SUs curve has dropped off starting from approximately 40 SUs in the distributed algorithm, the curve of the beamforming strategy starts dropping off after 60 SUs for the beamforming strategy, in the case of downlink scenario. Observing this figure, we get also that the number of active SUs is lower than 20 users for any number of transmitter SUs in the system. This means that the PU outage probability is upper-bounded by the maximum outage probability q. Fig. 4 confirms these results. As reflected in the figure, the required maximum outage probability is respected, since all outage probability values are lower then 1% for any number of SUs. From these results, the P_{out} curves in both uplink and downlink cases can be observed to have very similar slopes as in Fig. 3.

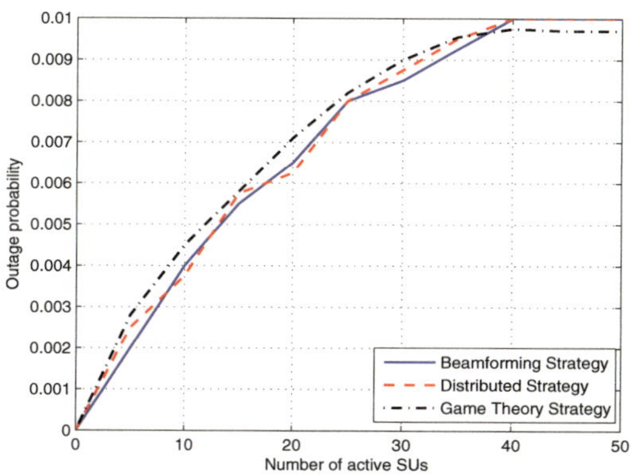

(a) Downlink

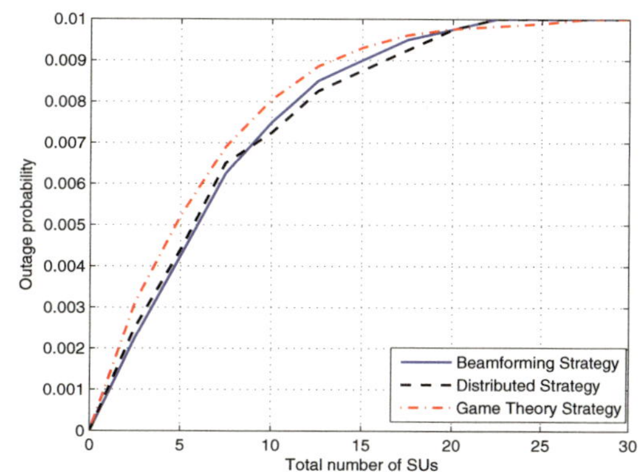

(b) Uplink

Fig. 4. Performance evaluation of the proposed user selection strategies: Outage probability as function of the number of secondary users for a target outage probability = 1% and a rate = 0.3bits/s/Hz in the downlink and the uplink mode.

So far, we verified the first goal of the proposed methods, maintaining the outage probability of the PU not degraded. The second goal in developing these new strategies and specially the beamforming-based one is to reduce the interference from SUs transmitters. However, we must show the impact of the proposed centralized beamforming scheme on the interference power. Fig. 5 depicts the normalized interference power of the beamforming user selection strategy versus the number of SUs in the uplink mode, in comparison with the distributed and the centralized methods. This figure shows that the interference power increases with the increasing number of SUs. It shows as well that the beamforming strategy performs better in terms of interference power. On the other hand, the distributed and the centralized strategies have virtually identical curves. Indeed, the proposed technique reduces interfering power by about 45% in comparison with the distributed and the centralized techniques. Therefore, we conclude that the beamforming strategy is highly efficient in terms of reducing the interference power as well as robust in maintaining a certain QoS to a PU.

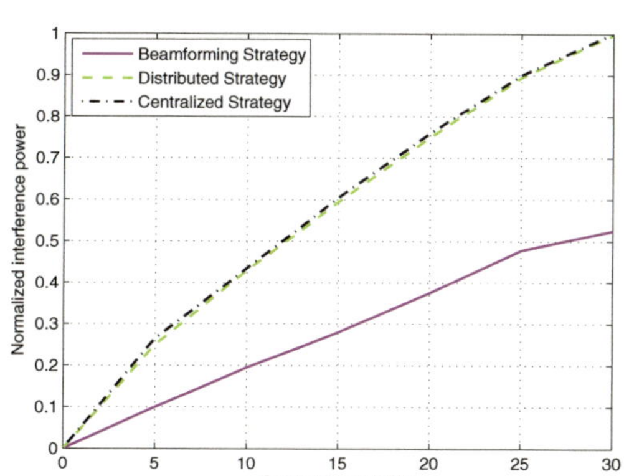

Fig. 5. Performance evaluation of the user selection strategies: Interference power versus number of SUs with $q = 1\%$ and a rate = 0.3bits/s/Hz in the uplink mode.

10. Conclusion

In this chapter, we focused on resource allocation and interference management. Within this setting, we considered different system models in which SUs compete for a chance to transmit simultaneously or orthogonally with the PU. On the basis of these models, we have also defined the specific resource allocation problem and offer insights into how to design such scenario in a CRN environments and we proposed three user selection strategies. One first key idea is based on outage probability to mange the QoS of the CR system. We have derived a distributed user selection algorithm under a cognitive capacity maximization criterion and outage probability constraint. We found out that we should make a tradeoff between cognitive capacity maximization and number of active SUs maximization. Then, we investigated the problem of resource allocation for multiuser multi-antenna channels using a beamforming strategy. The proposed strategy was proved to be the optimal one that achieves the maximum rate for both users under the constraint that the SU guarantees a QoS for the primary system within the outage probability constraint. We have explicitly derived the capacity of the primary as well as the SU. Finally, we explored the idea of combining game theory with resource allocation in CRN. Specifically, we defined a utility/pricing strategy that meets the objective to maximize the SUs capacity and the protection for PUs. Indeed, we discussed the existence of the Nash equilibrium of the proposed game, as well as its uniqueness. We demonstrated that the proposed game admits one and only one Nash equilibrium. Both theoretical and simulation results based on a realistic network setting provide substantial throughput gains, thereby illustrating interesting features in terms of CRN deployment while maintaining QoS for the primary system by means of outage probability for the three strategies. The main contribution within this work is the QoS management of the CR system. The originality in the proposed methods is that we guarantee a QoS to PU by maintaining the PU's outage probability unaffected in addition to a certain QoS to SUs and ensuring the continuity of service even when the spectrum sub-bands change from vacant to occupied. Thus by the outage probability control, if we have a vacant spectrum holes in the PU band, we set the outage probability $P_{out} = 1$ to exploit the available spectrum band by SUs, and if we have occupied sub-bands, the outage probability is set to $P_{out} = q$ depending on the PU's QoS.

11. Acknowledgements

The research leading to these results has received funding from the European Community's Seventh Framework Programme (FP7/2007-2013) under grant agreement SACRA n°249060. This work was supported also by the European projects: ACROPOLIS (advanced coexistence technologies for radio optimization and unlicensed spectrum), and CROWN (cognitive radio oriented wireless networks).

12. References

A. B. MacKenzie, L. Dasilva, and W. Tranter (2006). Game theory for wireless engineers, *Morgan and Claypool Publishers* .

A. Gjendemsjø, D. Gesbert, G. E. Øien and S.G. Kiani (2007). Binary power control for multi-cell capacity maximization, *IEEE Trans. Wireless Comm.* .

A. Tarighat, M. Sadek and A.H. Sayed (2005). A multi user beamforming scheme for downlink mimo channels based on maximizing signal-to-leakage ratios, *IEEE International Conference on Acoustics, Speech, and Signal Processing, ICASSP* .

B. Zayen, A. Hayar and G. Noubir (2011). Utility/pricing-based resource allocation strategy for cognitive radio systems, *ICMCS'11, 2nd International Conference on Multimedia Computing and Systems* .

B. Zayen, A. Hayar and G.E. Øien (2009). Resource allocation for cognitive radio networks with a beamforming user selection strategy, *Asilomar* .

B. Zayen, M. Haddad, A. Hayar and G.E. Øien (2008). Binary power allocation for cognitive radio networks with centralized and distributed user selection strategies, *Elsevier Physical Communication Journal* 1(3): 183–193.

D. Schmidt, C. Shi, R. Berry, M. Honig and W. Utschick (2009). Distributed resource allocation schemes, *IEEE Signal Process. Mag.* 26(5): 53–63.

E. Jorswieck and R. Mochaourab (2010). Beamforming in underlay cognitive radio: Null-shaping design for efficient nash equilibrium, *International Workshop on Cognitive Information Processing*.

E.A. Jorswieck and H. Boche (2004). Channel capacity and capacity-range of beamforming in mimo wireless systems under correlated fading with covariance feedback, *IEEE Transactions on Wireless Communications* 3(5).

F. Wang, M. Krunz and S. Cui (2008). Spectrum sharing in cognitive radio networks, *IEEE Conference on Computer Communications*, pp. 1885–1893.

FCC (2003). Cognitive radio technologies proceeding, Web Page, http://www.fcc.gov/oet/cognitiveradio/.

Fudenberg, D. & Tirole, J. (1991). *Game Theory*, MIT Press, Cambridge,MA.

J. Huang, R.A. Berry and M.L. Honig (2006). Auction-based spectrum sharing, *Mobile Networks and Applications*, Vol. 11, pp. 405–418.

J. Huang, Z. Han, M. Chiang and V. Poor (2007). Auction-based distributed resource allocation for cooperation transmission in wireless networks, *IEEE Global Communications Conference (GLOBECOM)*.

J. Mitola (1999). Cognitive radio for flexible mobile multimedia communications, *Mobile Multimedia Communications (MoMUC)*, New York.

J. S. Pang, G. Scutari, D. Palomar and F. Facchinei (2010). Design of cognitive radio systems under temperature-interference constraints: A variational inequality approach, *IEEE Trans. Signal Process.* 58(6): 3251–3271.

Kiani, S. G. & Gesbert, D. (2006). Optimal and distributed scheduling for multicell capacity maximization. IEEE International Conference on Communications.

L. Akter and B. Natarajan (2009). Qos constrained resource allocation to secondary users in cognitive radio networks, *Elsevier, Computer Communications* .

L. H. Ozarow, S. Shamai and A.D. Wyner (1994). Information theoretic considerations for cellular mobile radio, *IEEE Trans. Veh. Technol.* 43(5): 359–378.

L. Qian, X. Li, J. Attia and Z. Gajic (2007). Power control for cognitive radio ad hoc networks, *IEEE Workshop on Local and Metropolitan Area Networks*, pp. 07–12.

M. Haddad, A. Hayar and G.E. Øien (2008). Uplink distributed binary power allocation for cognitive radio networks, *CrownCom*, Singapore.

M. J. Osborne (2003). An introduction to game theory, *Oxford University Press* .

M. Schubert and H. Boche (2004). Solution of the multiuser downlink beamforming problem with individual sinr constraints, *IEEE Trans. Veh. Technol.* 53(1): 18–28.

N. Nie and C. Comaniciu (2005). Adaptive channel allocation spectrum etiquette for cognitive radio networks, *New Frontiers in Dynamic Spectrum Access Networks*.

P. Cheng, Z. Zhang, H. Huang and P. Qiu (2008). A distributed algorithm for optimal resource allocation in cognitive ofdma systems, *IEEE International Conference on Communications*.

P. Viswanath, D. Tse and R. Laroia (2004). Random beamforming for mimo systems with multiuser diversity, *IEEE Personal, Indoor and Mobile Radio Commun. Conf.* 1: 290–294.

Peha, J. M. (2005). Approaches to spectrum sharing, *IEEE Communications Magazine* 43(2): 10–12.

R. W. Rosenthal (1973). A class of games possessing pure-strategy nash equilibria, *International Journal of Game Theory* 2 pp. 65–67.

Tse, D. (2004). *Fundamentals of Wireless Communication*, University of California, Berkeley Pramod Viswanath, University of Illinois, Urbana-Champaign.

Urban Transmission Loss Models for Mobile Radio in the 900 and 1800 MHz Bands (1991).

Y. Wu and D.H.K. Tsang (2009). Distributed power allocation algorithm for spectrum sharing cognitive radio networks with qos guarantee, *IEEE INFOCOM*.

Y. Xing, C.N. Mathur, M.A. Haleem, R. Chandramouli and K.P. Subbalakshmi (2007). Dynamic spectrum access with qos and interference temperature constraints, *IEEE Transactions on Mobile Computing* 6: 423–433.

Yates, R. D. (1995). A framework for uplink power control in cellular radio systems, 13: 1341–1347.

11

Power Control for Cognitive Radios: A Mixed-Strategy Game-Theoretic Framework

Chungang Yang and Jiandong Li
State Key Lab. of ISN, Xidian University
China

1. Introduction

Motivated by the spectrum drought and explosive growth of increasing quality of service requirements, cognitive radio as a promising technique has attracted a significant attention in wireless community. In this chapter, we investigate and summarize the following contents:

- After a simple introduction of cognitive radio development in recent years, we focus on the issue of how to implement interference mitigation by power control techniques amongst multiple cognitive radios. An overview of concurrent power control schemes is provided first, then we point out the existing problems and new challenges of power control in cognitive radio networks, which leads us to concentrate on a novel mathematical model-game theory.
- Game theory, which captures the dynamic decision-making behavior of selfish and rational players have attracted a wide attention from cognitive radio community, specifically for the game theory-based power control in cognitive radio networks. Several specific game models which suit cognitive radios well are explored and introduced, and these models are with typical good properties, for instance, they can well guarantee the existence and uniqueness of the celebrated Nash equilibrium solution.
- There are many impractical assumptions in these existing literature, for example, complete information and rationality and so on. In cognitive radio networks/systems, the complexities of mobility and traffic models, coupled with the dynamic topology and the unpredictability of link quality make these convectional game models meet with limited success. So that we employ mixed-based power control game (MPCG) to deal with the discrete power control issue. MPCG provides a novel point of view to investigate other resource management problems in the uncertain environment including cognitive radio context.
- We also discuss several related open problems, such as the lack of proper models for dynamic and incomplete information games. We use the application prospect of game theory to conclude this chapter.

We will relax the full nature of the information requirements, and investigate the effective power control from a very creative perspective termed as mix-strategy based matrix power control game (MPCG) model. The typical max-min fairness criteria is chosen as the fair and optimal criteria for the mixed-strategy based power control algorithm, so that, the scheme proposed in this chapter can greatly improve the fairness of the multi-user located at different distance and with diversities of channel state information. The contributions in this chapter are summarized as:

- An efficient and fair (max-min fairness) discrete power control scheme is proposed in this chapter. Our scheme is based on the mixed-strategy of the matrix power control game model. The convergence and uniqueness properties are well guaranteed as long as available strategy space, e.g., the available power level is in a finite countable number.
- Additionally, mixed-strategy provides much larger strategy space for each player, so that the opportunity for each player to achieve the Nash Equilibrium Solution can well satisfy the Pareto optimality (effectiveness criteria). Because the pure strategy of the traditional game-theoretic model is a special case of the mixed-strategy with the determinate distribution.
- Last but not the least, the power control algorithm is greatly simplified by employing an amazing transformation from the mathematical point of view. With the conventional simplex method, the reformulated system model can be efficiently solved.

2. Background

The wireless industry is witnessing an explosive growth due to the increase in the number of the mobile users, paralleled by the widespread deployment of heterogeneous wireless networks. The requirements of the high transmit rate is becoming serious, and the high wide-band data service urgently requires more spectrums. Unfortunately, the available spectrum has been allocated completely. Meanwhile, recent measurement studies suggest that radio spectrum is gradually becoming an under-utilized resource that should be better explored. According to FCC, 15% to 85% assigned spectrum is used with large temporal and geographical variations (1). By now, it has been recognized that the scarcity of radio spectrum is mainly due to inefficiency of traditional static spectrum-allocation policies (1; 2). Motivated by the promising cognitive radio (CR) technology, both academic and industrial communities have shifted attention to dynamic spectrum access to alleviate alleviate spectrum scarcity and improve spectrum efficiency. Dynamic spectrum access represented as cognitive radio technology attracts wide attention to improve the spectrum-hunger situation. An introduction to CR basics, different spectrum sharing models, and challenges and issues in designing dynamic spectrum access networks can be found in (1–4).

While the cognitive radio community has had significant success popularizing the concept of cognitive radio and developing prototypes, applications, and critical components, the community has had a surprisingly difficult time agreeing upon exactly what is and is not a cognitive radio beyond. Some commonalities have developed different definitions of cognitive radios. However, as the original cognition cycle shown in Figure 1, the basic characteristics can be summarized as follows: First, all of these definitions assume that cognition will be implemented as a control process, presumably as part of a software defined radio. Second, all of the definitions at least imply some capability of autonomous operation. In detail, Observation: whether directly or indirectly, the radio is capable of acquiring information about its operating environment. Adaptability: the radio is capable of changing

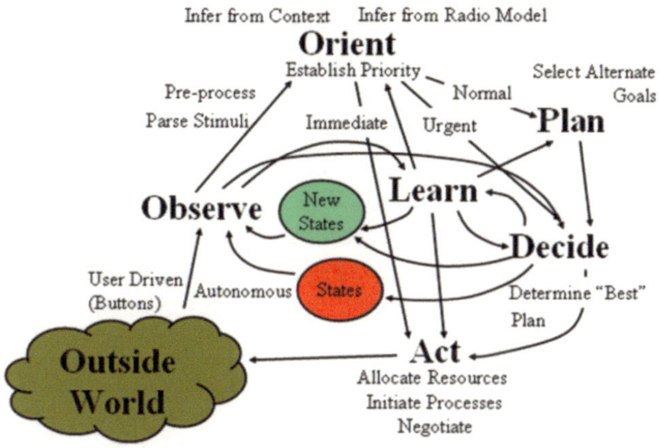

Fig. 1. Cognitive radio and cognition cycle (1; 2).

its waveform. Intelligence: the radio is capable of applying information towards a purposeful goal (6). The interference avoidance problem between the primary user and the secondary user is a critical issue for the cognitive radio networks.

2.1 Background of game theory for wireless communications

In recent years, game theory has found an increasingly important role, especially for the issues of radio resource management (17; 18). There are many prior game theory literature which investigate various issues in wireless communications, especially, in the context of Cognitive Radios (CRs). Game theory is a powerful tool to analyze the interactions among decision-makers with conflicting interests and finds a rich extent of application in communication systems including network routing, load balancing, resource allocation, flow control and power control. There is an extensive power control strategies based on game theoretic and utility theory (2). Meanwhile, they did achieve certain progress and better results, especially for resource management issue for cognitive radios. But most of them are based on the Nash game (5), which is essentially a non-cooperative game model.

Based on game theory, there are extensive research on the radio resource management (RRM) issues, we cite some here, including the power control (4), spectrum sharing (2), spectrum access (13), channel selection (12) and congestion control (19). However, the concurrent research on the basis of the game theory almost all focuses on characterization description and identification of the feasible equilibrium operating point, e.g., the typical Nash equilibrium (6–9) and Nash bargaining solution (5), also including some other extensive equilibrium solutions , e.g., Stackelberg equilibrium solution and correlated equilibrium solution(23). Some others concentrate the existence and uniqueness of the equilibrium solution. For example, the investigations in the potential game (18) and the super-modular game (6), which are all game models with some nice properties guaranteeing the existence and uniqueness. Actually, this is guaranteed by the specific utility function design in the game model (6–9; 18).

A great number of resource allocation and management problems in communication networks can be formulated as game models, which are summarized in (17). There are also lots of works on dealing with a diversity of new issues in current wireless networks. (1; 2) investigate the non-cooperative selfish behavior and spectrum sharing games of multiple WiFi access point in the open/unlicensed spectrum, considers the interference management problem in the ad hoc networks using the super-modular games (3). For example, (8; 9) investigate the pricing function design for improving the Pareto optimality of the Nash equilibrium solution in the power control games in CDMA systems, and others study the bandwidth allocation in broadband networks, channel allocation in OFDMA networks, and the resource management in the multi-media transmission networks, respectively from the cooperative game-theoretical perspective, for example, the Nash bargaining game, and coalition formulation games. On the other hand, some drawbacks and disadvantages have been found and encountered of the traditional mathematical tools, which are unprecedentedly faced before. For example, the convex optimization can not well formulate the dynamic decision making problem of the multiple CRs. In addition, the decision making process is interactive, coupled among each other, inter-dependently. Meanwhile, the dynamic topology and changing radio spectrum holes, the opportunistic spectrum access and various service characteristics cause people to find new mathematical molding tool in CR context. How to devise an adaptive QoS measurement for the cognitive radios is really full of absolute challenge in CRNs. In addition, from the concurrent research, we have seen that the game theory is really suitable for analysis of cognitive radios, which is shown in Figure 2.

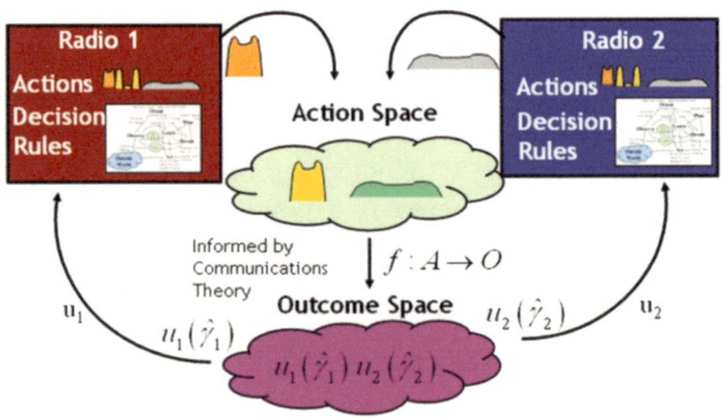

Fig. 2. Mapping of cognitive radio to game model (6).

2.2 Mixed-strategy considerations

John von Neumann's (1928) theoretical formulation and analysis of such strategic situations is generally regarded as the birth of game theory. von Neumann introduced the concept of a mixed strategy: a mixed strategy is a probability distribution one uses to randomly choose among available actions in order to avoid being predictable. In a mixed strategy equilibrium each player in a game is using a mixed strategy, one that is best for him against the strategies the other players are using. John Nash (1950) introduced the powerful notion of equilibrium in games (including non-zero-sum games and games with an arbitrary number of players): an equilibrium is a combination of strategies (one for each player) in which each player's strategy is a best strategy for him against the strategies all the other players are using. An equilibrium is thus a sustainable combination of strategies, in the sense that no player has an incentive to change unilaterally to a different strategy. A mixed-strategy equilibrium (MSE) is one in which each player is using a mixed strategy; if a game's only equilibria are mixed, we say it is an MSE game. In two-person zero-sum games there is an equivalence between minimax and equilibrium: it is an equilibrium for each player to use a minimax strategy, and an equilibrium can consist only of minimax strategies. Non-zero-sum games and games with more than two players often have mixed strategy equilibria as well. Important examples are decisions whether to enter a competition (such as an industry, a tournament, or an auction), 'wars of attrition' (decisions about whether and when to exit a competition), and models of price dispersion (which explain how the same good may sell at different prices), as well as many others. Every finite n-person strategic game has a mixed Nash Equilibrium.

To the best of our knowledge, first, there is always a impractical complete information assumption, which is players know choices of strategies and corresponding payoffs of other players [1] (but not their actions [2]). Second, the previous mentioned work mostly focused on research of continuous power control scheme, since under the continuous assumption, it is easy to deal with from the mathematical perspective. Traditional discrete power control is based on the continuous power space, which is adaptive to the practical scenario and the traditional method "discretizing "the continuous value that will not always guarantee the convergence and uniqueness of continuous power control.

We assume that there exists only one time step, which means that the players have only one move as a strategy. In game-theoretic terms, this is called a single stage or static game. Please note that the definition of a static game means that the players have only one move as a strategy, but this does not necessarily correspond to the time slot of an underlying networking protocol. In many strategic situations a player's success depends upon his actions being unpredictable. Competitive sports are replete with examples. One of the simplest occurs repeatedly in soccer (football): if a kicker knows which side of the goal the goaltender has chosen to defend, he will kick to the opposite side; and if the goaltender knows to which side the kicker will direct his kick, he will choose that side to defend. In the language of game theory, this is a simple 2x2 game which has no pure strategy equilibrium. So that, mixed strategy-based game theoretical formulation with nice existence of equilibrium solutions has received a great attention.

2.3 Power control in cognitive radio systems

In a cognitive radio network, proper power control is of importance to ensure efficient operation of both primary and secondary users. Even without the presence of primary users, power control is still an issue among secondary users since the signal of one user may cause interference to the transmissions of others. Thus, how to develop an efficient power allocation scheme that is able to jointly optimize the performance of multiple cognitive radios in the presence of mutual interference is of interest to such a system.

2.3.1 Power control

Power control mitigates unnecessary interference, and it can save the battery life of the mobile devices, hence, increasing the network capacity and prolong battery's life. Centralized power control requires extensive information interaction between the base station and the mobile

[1] In this chapter, we use player, secondary user and CR interchangeably.

[2] We also use strategy, action and power level interchangeably throughout this chapter.

terminal, which is applied inefficiently in practice. The distributive versions only depend on local information, e.g., the received interference power or the Signal-to-interference and noise ratio (SINR) to adaptively adjust the power level until each user obtains the objective SINR threshold.

Recently, the max-min fairness criterion is widely accepted as the fairness criteria. The max-min fairness is regarded as the one that the player can not increase the utility without decreasing the utility of his components. It is a standardized fairness concept in the ATM networks, and now it is widely accepted as the fairness criterion of the resource allocation technique in the wireless communication networks. E.g. (9) addresses the joint transmit power control and beam-forming technology if the multi-antenna systems with the aid of two different objective function design. The same as (10), the authors of the (11) investigate the max-min fairness for the MISO downlink systems. Based on the max-min fairness framework, a distributed power control algorithm is proposed for the Ad-hoc networks.

2.3.2 Game theoretical consideration for power control

Autonomously dynamic behavior and performance analysis is of great importance in dynamic spectrum sharing scenario, especially, when context information perceived by multiple secondary users (SUs) of different levels of cognition is asymmetric, which is definitely necessary in cognitive radio networks (CRNs) (1; 2). Game theory, which captures the dynamic decision-making behavior of selfish and rational players have attracted a wide attention from wireless community (6–8). Meanwhile, its excellent predictability of next action employed by the player, along with well established equilibrium solution concepts, lends itself well to the design and analysis of CRNs. A survey of game theory for wireless engineers is provided in (5), and its increasing use for spectrum management is exemplified in CRNs (6–10).

2.4 Special game models

Only when the game has certain special structure, the gaming iteration algorithm can be converged and lead to equilibrium solution, especially, in the distributed decision making context, for instance, the cognitive radio networks, since there is only local information support. There are several special cases of utility function design besides the above mentioned several design criteria.

2.4.1 Potential games

A potential game is a special type of game where $\mathcal{U}$ are such that the change in value seen by a unilaterally deviating player is reflected in the potential function $\mathcal{V}$. A game $\mathcal{G} = \{\mathcal{N}, \mathcal{S}, \mathcal{U}\}$ is a potential game if there is a potential function $\mathcal{V} : S \rightarrow \mathbb{R}$ such that one of the following conditions holds.

- $\mathcal{U}_i(s_i, s_{-i}) - \mathcal{U}_i(s_i', s_{-i}) = \mathcal{V}_i(s_i, s_{-i}) - \mathcal{V}_i(s_i', s_{-i})$, where for any $i \in N$, $s \in S$, and $s_i' \in S_i$;
- $sgn\{\mathcal{U}_i(s_i, s_{-i}) - \mathcal{U}_i(s_i', s_{-i})\} = sgn\{\mathcal{V}_i(s_i, s_{-i}) - \mathcal{V}_i(s_i', s_{-i})\}$, where sgn is the signal function.

It is an exact potential game, if and only if the first condition above denoted is satisfied. In addition, the necessary and sufficient condition for a game to be an exact potential game is

$$\frac{\partial^2 u_i(s_i, s_{-i})}{\partial s_i \partial s_j} = \frac{\partial^2 u_i(s_i, s_{-i})}{\partial s_j \partial s_i}, \forall j \neq i \in N. \tag{1}$$

There are also some other special potential game, which requests different properties owned by the utility function. For example, the coordination-dummy games, self-motivated games, and bilateral symmetric interaction games. Detailed about this can be found in (1).

2.4.2 S-modular games

An S-modular game restricts $\{u_i\}$ such that for $i \in N$, either the following two equations (2) or (3) is satisfied.

$$\frac{\partial^2 u_i(s_i, s_{-i})}{\partial s_i \partial s_j} \geq 0, \forall j \neq i \in N \tag{2}$$

$$\frac{\partial^2 u_i(s_i, s_{-i})}{\partial s_i \partial s_j} \leq 0, \forall j \neq i \in N \tag{3}$$

When (2) is satisfied, the game is said to be super-modular; when (3) is satisfied, the game is said to be sub-modular. Myopic games whose stages are S-modular games and potential games with a unique Nash equilibrium solution(NES) and follow a best response dynamic converge to the NES when the NES is unique.

3. System model and problem formulation

In this chapter, a distributed scenario is considered as Figure 3, multiple secondary users (SUs) opportunistically access in the spectrum holes of the GSM system by sensing technology who works as the primary users (PUs), we don't care about how the SUs access and how to obtain such access opportunities in this chapter, but focus on how the multiple SUs choose the optimal power control strategy to mostly improve performance of the secondary system and maximize the payoff function of the individual SU.

In Figure 3, the rectangles represent the transmitters of the cognitive radio, and the circles represent the respective receiver, the communication link is tagged as the solid lines with the arrow. The lines depict the interference links of the CR-transmitter to the base station (BS) of the primary system, e.g. the GSM system; and also including the mutual interference between the multiple CR-transmitter and the specific CR-receiver. Here, we assume that each CR-transmitter can well obtain the necessary information, e.g. the channel state information and the interference situation of the considering scenario with the help of the BS. Consider the heterogeneous networks, and the GSM coexists with the secondary network composed by the multiple CRs who will employ the same available power levels as the GSM users.

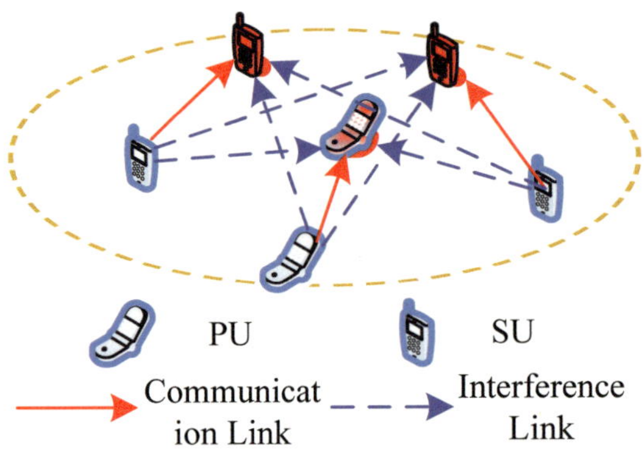

Fig. 3. Secondary system model: where GSM as primary user who is the provider of the spectrum holes; and multiple secondary form a ad hoc cognitive scenario to implement an opportunistically spectrum accessing fashion.

A typical strategic power control game model is a three-tuple defined as $\mathcal{G} = \{\mathcal{N}, \mathcal{A}, \mathcal{U}\}$, where $\mathcal{N}$ is the player set, $\mathcal{A} = \prod_i \mathcal{A}_i$ is the action set, and $\mathcal{U}$ denotes the utility function which depicts the preference relationship of the various players in the game model. Here, we summarize a general joint rate and power control game model, which means that the action set is (R_i, p_i). There exists a tradeoff relationship among large SINR, low power consumption and high transmit rate, which are shown in Figure 4. As Figure 4 shown, we have some intrinsic characterizes summarized as follows.

- when the SINR λ_i and the transmit rate R_i are fixed, the utility function U_i won't increase with the increasing power level. This is partially due to the more power introduced into the game process, the more mutual interference power to the other players in the same gaming situation. That means when a player achieve the available SINR threshold, then increasing more power will not do good to the performance improvement, but damage it.
- Meanwhile, if one player has obtained the required QoS, that means more power consumption will shorten battery life of the equipment. So that the power control is necessary.
- If the consuming power is fixed and one player is transmit in the fixed transmit rate, the utility perceived by the player will increase with the SINR, which is illustrated in Figure 4. This tells us that when the higher SINR is guaranteed, the spectrum efficiency will be higher too.
- In addition, we capture the case of the fixed power level, when the SINR is also maintained on some fixed level, we can see that the utility leads a proportional relationship with respect to the transmit rate as Figure 4 shown.

The utility functions denoted in the (6–9; 17; 18) are all satisfied above mentioned these observations. From the typical power control game, we have some conclusions on the concept of utility function. The design or selection of a suitable utility function form in the extension of game theory for communications networks is always the bottleneck factor.

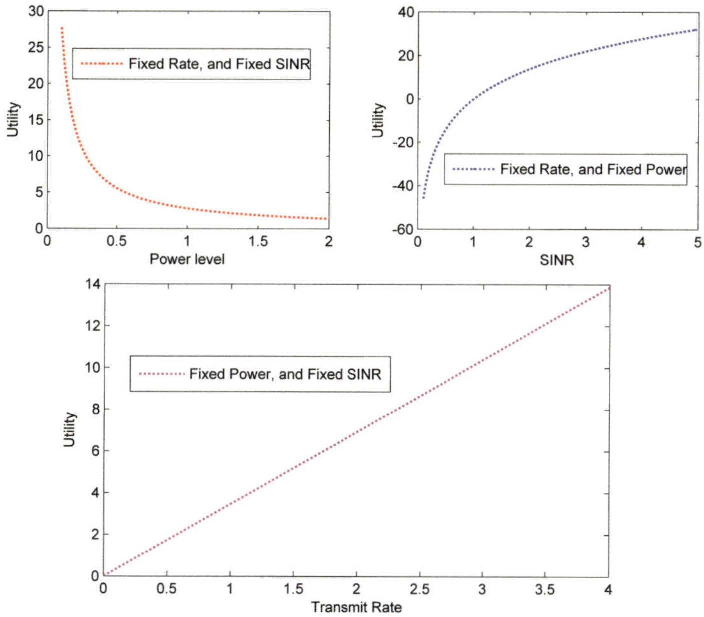

Fig. 4. Characteristics description of the utility function in the power control games, with the power level p_i, transmission rate R_i and the obtained SINR λ_i into consideration. The typical properties of the utility function (here, we choose the utility function as $u_i(p_i, R_i, \lambda_i) = \frac{R_i log(\lambda_i)}{p_i}$ in the similar form of (7; 8)) are reflected among these impacting parameters.

Lemma 1. *Utility function development for the investigated resource management issue, for example, power control, must satisfy two basic criteria: 1) utility function can be with physical meaning of the formulated problem as described in Figure 4. 2) utility function should well capture the characteristics of the preference of the users/players in the resource management game, for example, the relationship of resource consumption and the QoS satisfaction perceived by users.*

Definition 1. *Utility Function: Without loss of generality, in this chapter, we employ the Shannon channel capacity as the utility function $U_i(p_i, \lambda_i)$, which is shown as*

$$U_i(p_i, \lambda_i) = \log(1 + \lambda_i), \tag{4}$$

The terms p_i, g_i and N represent the transmit power, the channel gain of CR_i, and the CR transceiver pairs number, where the signal-to-interference and noise ratio λ_i is defined as

$$\lambda_i = \frac{p_i g_i}{\sum\limits_{j=1, j\neq i}^{N} p_j g_j + \sigma^2}, \tag{5}$$

where σ^2 is the power density of background noise, and $\sum_{j=1,j\neq i}^{N} p_j g_j$ represents the total interference power perceived by the SU i, which is introduced by the other players who are sharing the same spectrum hole.

In this chapter, we apply the max-min fairness as the fair and optimal power control criterion, which is shown to be Pareto optimal and fair. As the primary user in the scenario considered is the listened user, the CR opportunistic access can not damage the performance of the PU, and the interference temperature constraints of the PU firstly satisfied, the discrete power control problem can be formulated as

$$\max_{p_i \in P_i} \min_i \; U_i(p_i, \lambda_i) = \log(1 + \lambda_i)$$

$$\text{subject to} \sum_{i=1}^{N} p_i h_{i,k} \le T_k, k \in K, \tag{6a}$$

$$p_i \in [p_{i,\min} : p_{i,step} : p_{i,\max}], \tag{6b}$$

where $\sum_{i=1}^{N} p_i h_{i,k} \le T_k, k = 1, ..., K$ is the interference temperature constraint of the GSM BS. Each user selfishly chooses the optimal power level to max-minimize the utility function. Basically speaking, the problem is still a non-cooperative game model, for the cognitive radio the interference temperature is introduced to constraint the power control of the secondary spectrum utilization in order that the data transportation of the primary user is guaranteed. In this chapter, we assume that the available discrete power level for each secondary user is well satisfied the ITL constraints of the primary users.

Notice: In this chapter, we assume that the various SUs can observe the possible policy (actions, e.g., the available power level) and further to derive the possible utility function, which forms the payoff matrix, but they cannot determine what is the exact strategy during the gaming process. But to exploit the mixed-strategy to guess. And the biggest probability of some specific power level for each SU deserves to most exact strategy in this decision-making step.

3.1 MPCG: Matrix discrete power control game

The decision-making flow chart of the matrix-game model-based discrete power control approach can be shown as Figure 5.

Each CR-transmitter can achieve the context information including the channel state information (CSI) and the interference temperature constraint thresholds of the primary user for some specific spectrum by using sensing techniques. For instance, the signal sensing in Gaussian noise environment can be carried out using higher order statistics (HOS). Then CR-transmitters determine suitable discrete power level according to the context information sensed/reasoned and the available power policy set. For instance, each CR-transmitter has 13 power level, which can be used accordingly in set $P = [5dBm : 2dBm : 33dBm]$.

The most important thing in the proposed MPCG model is to compute the utility matrix due to based on which, we can determine the useful power in line with mixed policy design. Therefore, the utility value achieved by each CR when it selects the available power level can be predicted according to the power strategy interactions and combinations among multiple CR-transmitters. After each CR gets the final utility value, then the utility matrix is made up. Here, it deserves to pay attention to that not all the power level in the strategy space can be utilized because the interference power level of the primary user must be satisfied first. By now, we know it is of great interest to find the threshold of the available power level, which is not the main objective in this chapter. However, in this chapter we assume that every power strategy in the strategy space can be employed during the next analysis. In the following step, the adaptively mixed-strategy selecting algorithm that will be implemented by each CR-transmitter.

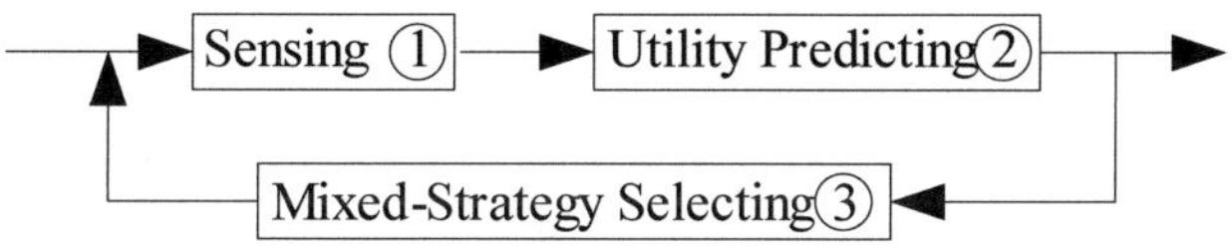

Fig. 5. Stream architecture

On the strategy space design, for instance, we choose the available power levels for each CR from $5dBm$ to $33dBm$, which is in accordance with the GSM system with the updating step $2dBm$. The channel state information of each CR is $G = [g_{=}0.5632, g_2 = 1.2321]$. Therefore, the power level of CR1 and CR2 can select is defined as $P = [5dBm : 2dBm : 33dBm]$.

According to definition of the utility function, we can obtain the following utility matrix when different CRs choose multi-power levels. For CR1, the transmit power level is denoted as $P = [5dBm : 2dBm : 33dBm]$, and the channel gain between the CR1 transmitter to the receiver is g_1 and the channel gain between the CR2 transmitter to the CR1 receiver is termed as g_{21}, so the interference power of CR2 introducing to the CR1 is $p_2 g_{21}$, and the utility function of CR1 takes the form of

$$U_1(p_1, p_2) = \log(1 + p_1 g_1 \big/ (p_2 g_{21} + \sigma^2)). \tag{7}$$

Similarly, the utility function of CR2 takes the form of

$$U_2(p_1, p_2) = \log(1 + p_2 g_2 \big/ (p_2 g_{12} + \sigma^2)). \tag{8}$$

From above utility function design, we can see that the optimal utility value achieved by each CR depends not only its own power selection but also the others in the wireless environment. And the strategy space of CR2 is also $P_2 = [5dBm : 2dBm : 33dBm]$.

3.2 Basics of matrix game definition

Consider the complex interference radio environment, we assume that each CR will adjust the power level entirely, e.g. when CR1 chooses the power level of 5dBm, the utility of the CR1 maybe of 12 different cases, since CR2 has 12 different power level to choose in the set of $P_2 = [5dBm : 2dBm : 33dBm]$. So that, we get a utility matrix A_{12*12} for a two-player MPCG model, where $A = \{a_{ij}, i = 1, ..., M, j = 1, ..., M\}$ represents utility achieved when the CR1 selects any power strategy p_i in set $P_1 = [5dBm : 2dBm : 33dBm]$, and the CR2 selects any power strategy p_j in set $P_2 = [5dBm : 2dBm : 33dBm]$.

Definition 2. *Mixed-strategy definition: A Mixed-strategy for CR1 $S_1 = \{x_i, i = 1, .., M\}$ is denoted as the distribution function of the pure strategy $P_1 = [5dBm : 2dBm : 33dBm]$. That is to say, the*

CR1 chooses the pure power level p_i at a probability of x_i. These vectors form a novel strategy space which is referred as the mixed-strategy.

Therefore, we can conclude that the pure strategy can be considered as a special case of that the CR1 choose the power level p_i with a probability of "1". The vector x_i satisfies $0 \leq x_i \leq 1$, and $\sum_{i=1}^{M} x_i = 1$. In fact, the mixed-strategy provides more opportunity for each user, in other words, the strategy space for each user is becoming larger so they all achieve more available strategy to choose, the optimal strategy can more easily achieved. For CR2, we can also similarly denote the mixed-strategy is $S_2 = \{y_i, i = 1, ..., M\}$.

Throughout this chapter, we assume that the available power level for each user is with the same dimension, and the Matrix Power Control Game (MPCG) model can be defined as

Definition 3. *MPCG: a three-element $G = \langle S_1, S_2, A \rangle$ is called as matrix game, where $S_1 = \{x_i, i = 1, ..., M\}$ and $S_2 = \{y_i, i = 1, ..., M\}$ is the mixed-strategy of CR1 and CR2, respectively. The term $A_{M\times M}$ is the utility matrix as the above section described which is also where the concept comes out.*

3.3 How to solve matrix power control game

First, the expected utility function must be clearly described for the matrix power control game. As the utility matrix definition, the term represents the utility obtained when the CR1 selects the power strategy x_i and the CR2 selects the power strategy y_j. For the CR1, the matrix power control game model can be formulated as

$$U_{CR_1} = \max_{X\in S_1} \min_{1\leq j\leq M} \sum_{i=1}^{M} a_{i,j} x_i$$

$$\text{subject to } 0 \leq x_i \leq 1, \tag{9a}$$

$$\sum_{i=1}^{M} x_i = 1, \tag{9b}$$

where X is the mixed power vector selected by CR1. For simplicity, we utilize the term $L(X)$ represent $\min_{1\leq j\leq m} \sum_{i=1}^{M} a_{i,j} x_i$, which means $L(X) = \min_{1\leq j\leq m} \sum_{i=1}^{M} a_{i,j} x_i$. Then the matrix power control game model with the mixed strategy design can be reformulated as

$$U_{CR_1} = \max_{X\in S_1} L(X)$$

$$\text{subject to } L(X) \leq \sum_{i=1}^{M} a_{i,j} x_i, j = 1, 2, ..., M, \tag{10a}$$

$$0 \leq x_i \leq 1, \tag{10b}$$

$$\sum_{i=1}^{M} x_i = 1. \tag{10c}$$

Let $x_i = x_i/L(X)$, and the model of (10) can be further represented as

$$U_{CR_1} = \min_{X \in S_1} \sum_{i=1}^{M} x_i'$$

$$\text{subject to } \sum_{i=1}^{M} a_{i,j} x_i' \geq 1, j = 1, 2, ..., M, \tag{11a}$$

$$0 \leq x_i' \leq 1. \tag{11b}$$

Theorem 1. *(10) is equivalent to (11).*

Proof. The optimal utility of $\tilde{G} = \langle S_1, S_2, \tilde{A} \rangle$ is termed as $\tilde{U}_1^*$ and U_1^* for $G = \langle S_1, S_2, A \rangle$. Using $x_i' = x_i/L(X)$, that is $x_i = x_i' L(X)$ to take place of the x_i in the Eq. (10), the objective function is temporally unchanged, and next we focus on the constraint conditions. We assume that the term $L(X) > 0$ always holds, if it initially can not guaranteed, luckily, we further prove during the sequel section provides us a powerful technique to satisfy the assumption. First, we see the constraint condition (3) in the Eq. (10) $\sum_{i=1}^{M} x_i' L(X) = 1$, that is $\sum_{i=1}^{M} x_i' = 1/L(X)$. So observe the objective function again, we can conclude that the objective function takes the form of $U_{CR_1} = \max_{X \in S_1} L(X)$, which can represent as $U_{CR_1} = \min_{X \in S_1} 1/L(X)$, and it finally appears as $U_{CR_1} = \min_{X \in S_1} 1/L(X) = \min_{X \in S_1} \sum_{i=1}^{M} x_i'$, as the Eq. (11) shown. Because we assume that $L(X) > 0$ always holds, and the constraint condition $x_i \geq 0, i = 1, 2, ..., M$, when $x_i = x_i' L(X)$, and we can conclude that $x_i' L(X) = x_i \geq 0$, that is $x_i' \geq 0, i = 1, 2, ..., M$. Further, we only need to represent the term x_i as x_i', $L(X) \leq \sum_{i=1}^{M} a_{i,j} x_i = \sum_{i=1}^{M} a_{i,j} x_i' L(X), j = 1, 2, ..., M$, and the $L(X) > 0$ always holds, and it can be missed, so the constraint condition transforms into $\sum_{i=1}^{M} a_{i,j} x_i' \geq 1, j = 1, 2, ..., M$. We can conclude that Eq. (10) is equivalent to Eq. (11). □

3.4 Simplex algorithm for linear programming problem

Finally, the problem is transformed into the simple linear programming as the Eq. (11) shown, which can be easily solved using the simplex method. Meanwhile, for the power level of each CR is limited, the matrix power control game model is absolutely guaranteed to have at least one mixed-strategy. But in the context of a more practical scenario when CR2 employs a so large power level that introduces more interference power to CR1, the utility achieved by CR1 maybe zero even negative. We find the matrix game will be hard to be solved from the mathematical perspective. Fortunately, we find a powerful tool to deal with this situation.

If we directly to employ the simplex method to search the mixed-strategy optimal power level, the computation complexity will be very high, now to simply the problem an equivalent mathematical model is introduced. If the original matrix power control game model takes the form of $G = \langle S_1, S_2, A \rangle$, and $A = \{a_{ij}, i = 1, ..., M, j = 1, ..., M\}$. Meanwhile, we define the novel matrix game $\tilde{G} = \langle S_1, S_2, \tilde{A} \rangle$, where $\tilde{A} = \{\tilde{a}_{ij} = a_{ij} + d, i = 1, ..., M, j = 1, ..., M\}$.

Lemma 2. *The optimal mixed-strategy of original model is the same as the newly-designed matrix power control model. The optimal utility $\tilde{G} = \langle S_1, S_2, \tilde{A} \rangle$ of is termed as $\tilde{U}_1^*$ and U_1^* for $G = \langle S_1, S_2, A \rangle$. Meanwhile, we can conclude that $\tilde{U}_1^* = U_1^* + d$.*

Proof. For $G = \langle S_1, S_2, A \rangle$, we assume that the optimal mixed-strategy for CR1 and CR2 is X_1^* and Y_2^* respectively. So that, we know

$$U_1^* = E(X_1^*, Y_2^*) = \max_{X \in S_1} \min E(X, Y_2^*). \tag{12}$$

Adding a suitable parameter d to the both sides of the Eq. (12), and we get

$$U_1^* + d = E(X_1^*, Y_2^*) + d = \max_{X \in S_1} \min E(X, Y_2^*) + d, \tag{13}$$

further, mathematically we get

$$\begin{aligned}
&\tilde{U}_1^* = E(\tilde{X}_1^*, \tilde{Y}_2^*) \\
&= \max_{X \in S_1, 1 \le j \le M} \min \sum_{i=1}^{M} \tilde{a}_{i,j} x_i \\
&= \max_{X \in S_1, 1 \le j \le M} \min \sum_{i=1}^{M} (a_{i,j} + d) x_i \\
&= \max_{X \in S_1, 1 \le j \le M} \min \sum_{i=1}^{M} a_{i,j} x_i + d \\
&= E(\tilde{X}_1^*, \tilde{Y}_2^*) + d \\
&= U_1^* + d.
\end{aligned} \tag{14}$$

That is to say, for any X, and any optimal mixed-strategy of X_1^* and Y_2^*, the Eq. (14) always holds. So the conclusion always holds. □

Lemma 2 tells us that the objective function $L(X) > 0$, if we choose the suitable constant parameter of d. So that, we conclude thta the proposed simplex method based power control approach always works well.

3.5 Proposed algorithm

In this subsection, the proposed discrete power control algorithm is depicted as the pseudo-code:

- Each CR collects the channel information (g_i) and the accumulated interference temperature (T) of the primary user from the sensing block of the CR system;
- In line with the necessary information achieved in the last step, each CR determines minimum available discrete power level (p_i) from the strategy space;
- The utility matrix (U_i) is predicted in this step, meanwhile, according to the method introduced in section IV the matrix is simplified to easily deal with;
- Using the simplex method to find the optimal discrete power level (P_i^*) and obtain the optimal utility (U_i^*) of each CR.

4. Numerical results

We consider a time-varying channel model which obeys the Raleigh distribution, and the channel characteristics change with the time, each CR must dynamically and adaptively the power level according to the wireless environment, e.g., the channel state information and the interference temperature of the requirement of the PU which are all considered in this chapter. The background noise is in accordance with the Gauss distribution, and the power density is 1e-5 w.

4.1 Analysis of proposed algorithm

Figure 6 is the utility obtained of CR1 and CR2 after a limited iterations, because the interaction of the strategy (e.g. the power level) and the interference between them, the strategy choosing process is harshly hard. Fortunately based on the basics of the non-cooperative game theory, the mixed-strategy optimal power level for each CR is always existed. From the figures, we can see that in the context of the max-min fairness criterion we finally achieved the optimal power level for CR1 and CR2, respectively, which are shown as the peak value of the two three-dimensions pictures. The max-min utilities obtained of the CR1 and CR2 are 7.3584 and 7.5888, respectively. From the numerical results, we can conclude that our algorithm is fair, for the two CRs obtain the similar utility. Simultaneously, the existence of the proposed algorithm is investigated as well.

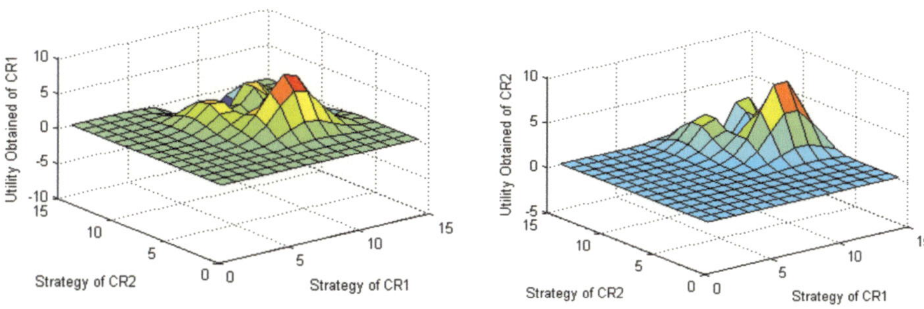

Fig. 6. Utility of CR1 and CR2 (max-min) (27).

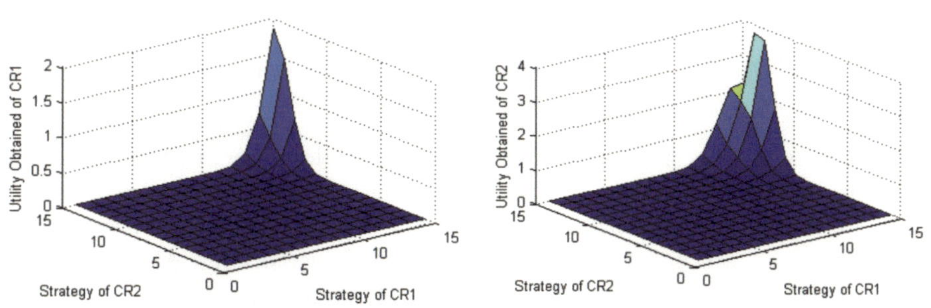

Fig. 7. Utility of CR1 and CR2 (max-max) (27).

4.2 Performance comparison

To show the effectiveness of our algorithm, here we compare with the greedy scheme which is based on the max-max principle. When other factors are all the same as the proposed algorithm, we obtain Figure 7. Though the fairness criterion changed into the max-max principle, the optimal power level can also achieved, which tells that the proposed algorithm is robust. The most important issue reflected in the pictures that when the user become greedily to pursue the maximum utility, first the fairness between them is damaged entirely, and the entire performance also declined absolutely as a result of the selfish behavior. The utilities obtained of the CR1 and CR2 using the max-max based proposed algorithm are 1.8512 and 3.7225, respectively. Observing from the perspectives of the two numerical results, we can conclude that our algorithm can guarantee the existence, the fairness and effectiveness.

5. Conclusion and prospect of game theory for wireless communications

Due to the typical dynamic behavior of PU, various categories requirement for service, limited resource constraints and complicated interactive context of the cognitive radio context, the traditional mathematical tools encounter the unprecedented drawback in the multiple users who interact with each other including the context/user information. Meanwhile, the behavior of a given wireless device may affect the communication capabilities of a neighboring device, notably because the radio communication channel is usually shared in wireless networks. Game theory is a discipline aimed at modeling situations in which decision makers have to make specific actions that have mutual, possibly conflicting, consequences. It has been used primarily in economics, in order to model competition between companies: for example, should a given company enter a new market, considering that its competitors could make similar (or different) moves? Game theory has also been applied to other areas, including politics and biology. So that, from the engineering point of view, researchers capture the similar characteristics between the issues from communication networks and the game theory. Therefore, game-theoretic framework have been widely developed in this cognitive radio scenarios.

The proposed discrete power control approach highlights a practical approach for the cognitive system designer. The proposed mixed-strategy based scheme can entirely avoid the convergence issue of the "discretizing scheme". The max-min fairness improves a lot of the fairness and spectrum efficiency. Meanwhile, the Matrix game based scheme can easily extend to multiple secondary users cognitive context but with huge computation complexity. This is our next focus research topic as how implementation our proposed scheme in a multiple user case, but is still represents a potential point in this chapter.

Game theory is a fascinating field of study. Due to the development of game theory, there are always novel game model suits a great number of issue the wireless communication networks, for instance, the conventional radio resource management, access control and the media access control policy design. For example, potential game, super-modular game and Markovian game have found a lot of use in wireless communication networks.

6. Acknowledgement

This work was supported in part by the National Science Fund for Distinguished Young Scholars under Grant 60725105, by the National Basic Research Program of China under Grant 2009CB320404, by the Program for Changjiang Scholars and Innovative Research Team in University, by the 863 projects under Grant 2007AA01Z288, by the National Nature Science Foundation of China under Grant 60572146, 60902032, 60902033, by the Research Fund for the Doctoral Program of Higher Education under Grant 20050701007, by the Teaching and

Research Award Program for Outstanding Young Teachers in Higher Education Institutions of Ministry of Education, China, by the Key Project of Chinese Ministry of Education under Grant 107103, by ISN02080001, 2009ZX03007-004, 2010ZX03003-001, and by the 111 Project under Grant B08038.

7. References

[1] FCC, *Spectrum policy task force report, FCC,* Nov. 2002.

[2] J. Mitola III, J. Maguire, Q. Gerald, *Cognitive radio: making software radios more personal,* IEEE Personal Communications, 6(4): 13-18, 1999.

[3] J. Mitola III, *Cognitive Radio for Flexible Mobile Multimedia Communications,* Mobile Networks and Applications, 6(5): 435-441, 2001.

[4] S. Haykin, *Cognitive Radio: Brain-Empowered Wireless Communications,* IEEE journal on Selected Areas in Communications, 2005, 23(2):201-220.

[5] C.W. Sung, *A Non-cooperative Power Control Game for Multi-rate CDMA Data Networks,* IEEE Transactions on Wireless Communications, 2003, 2(1): 186-194.

[6] J. Neel, J. Reed, R. Gilles, *Convergence for cognitive radio networks,* Wireless Communications and Networking Conference, 2004: 2250-2255.

[7] N. Dusit, H. Zhu, *Dynamics of multiple-seller and multiple-buyer spectrum trading in cognitive radio networks: a game-theoretic modeling approach,* IEEE Transactions on Mobile Computing, 8(8): 1009-1022, August 2009.

[8] H. Zhu; C. Pandana, K.J. Kay Liu, *Distributed opportunistic spectrum access for cognitive radio using correlated equilibrium and no-regret learning,* IEEE Wireless Communications and Networking Conference, 11-15, 2007.

[9] Y. Su, M. Van Der Schaar, *A new perspective on multi-user power control games in interference channels,* IEEE Transactions on Wireless Communications, 8(6): 2910-2919, 2009.

[10] C.G. YANG, J.D. Li, *A Game-Theoretic Approach to Adaptive Utility-Based Power Control in Cognitive Radio Networks,* VTC-Fall, 2009, PP:1-6, Sept.,Anchorage, Alaska, USA.

[11] C.G. YANG, J.D. Li, *Joint Rate and Power Control Based on Game Theory in Cognitive Radio Networks,* ChinaCom 2009,Xi'an,China.

[12] E. Altman, T. Boulogne, R.E. Azouzi, T. Jimenez, and L. Wynter, *A survey on networking games in telecommunications,* Computers and Operations Research, 33(2): 286-311, 2006.

[13] M.F. Elegyhazi and J.P. Hubaux, *Game theory in wireless networks: A tutorial,* EFL, Tech. Rep. LCA-REPORT-2006-002, Feb. 2006.

[14] A.B. MacKenzie, S.B. Wicker, *Game theory in communications: motivation, explanation, and application to power control,* IEEE Global Telecommunications Conference, 2001, 25(2):821 - 826.

[15] B.M. Allen, A.D. Luiz, *Game theory for wireless engineer,* Morgan and Claypool Publishers.

[16] B. Wang, Y. Wu, K.J. Ray Liu, *Game Theory for Cognitive Radio Networks: A Tutorial Survey,* Computer Networks, 54(14): 2537-2561, 2010.

[17] J.W. Huang, V. Krishnamurthy, *Game theoretic issues in cognitive radio systems,* Journal of Communications, 4(10):790-802, 2009.

[18] A.B. MacKenzie, S.B. Wicker, *Game theory in communications: motivation, explanation, and application to power control,* IEEE Global Telecommunications Conference, 25(2): 821-826, 2001.

[19] V. Srivastava, J. Neel, A.B. MacKenzie, R. Menon, L. A. DaSilva, J. E. Hicks, J. H. Reed, and R. P. Gilles, *Using game theory to analyze wireless ad hoc networks*, IEEE Communication Surveys and Tutorials, 2005.

[20] J. Huang, D.P. Palomar, N. Mandayam, J. Walrand, S.B. Wicker, and T. Basar, *IEEE J. Select. Areas Commun. (Special Issue on Game Theory in Communication Systems)*, vol. 26, no. 7, Sept. 2008.

[21] E.A. Jorswieck, E.G. Larsson, M. Luise, and H.V. Poor, *IEEE Signal Processing Mag. (Special Issue on Game Theory in Signal Processing and Communications)*, vol. 26, no. 5, Sept. 2009.

[22] W. Saad, Z. Han, M. Debbah, A. Hjorungnes, T. Basar, *Coalitional game theory for communication networks*, IEEE Signal Processing Magazine, 26(5): 77-97, 2009.

[23] B. Holger, S.Martin, *Resource allocation in multiantenna systems - Achieving max-min fairness by Optimizing a sum of inverse SIR*, IEEE Transactions on Signal Processing, 2006,54(6I):1990-1997.

[24] B.Y. Song, Y.H. Li, *Weighted max-min fair beamforming, power control, and scheduling for a MISO downlink*, IEEE Transactions on Wireless Communications, 2008,7(2):464-469.

[25] W. Marcin, S. Slawomir, B. Holger, *Quadratically converging decentralized power allocation algorithm for wireless ad-hoc networks - The max-min framework*, ICASSP-2006, 4:245-248.

[26] Y. Xing, and R. Chandramouli, *Stochastic Learning Solution for Distributed Discrete Power Control Game in Wireless Data Networks*, IEEE/ACM Transactions on networking, 2008,16(4):932-944.

[27] C.G. YANG, J.D. Li, *Mixed-Strategy Based Discrete Power Control Approach for Cognitive Radios: A Matrix Game-Theoretic Framework*, ICFCC 2010, pp: V3-806-810.

12

Joint Spectrum Sensing and Resource Scheduling for Cognitive Radio Networks Via Duality Optimization

Guoru Ding, Qihui Wu, Jinlong Wang, Fei Song and Yuping Gong
Institute of Communications Engineering
PLA University of Science and Technology
China

1. Introduction

With the increasing growth of wireless spectrum demand and the corresponding weak supply, wireless bandwidth has been reported as a new black gold of the new decade by Time News. To resolve the well known dilemma between wireless bandwidth scarcity and underutilization, cognitive radio (CR) (J. Mitola, 2000) is a promising technology in which secondary user (SU) can be allowed to opportunistically access a spectrum hole (S. Haykin, 2005) without interrupting primary user (PU)'s communication. During the past decade, worldwide researchers have created a mass of interesting results to promote CR technology and a comprehensive survey on the recent advances can be found in (Q. Zhao, et al., 2007). However, many challenges are still ahead.

Spectrum sensing and resource scheduling are two critical components of enabling CR technology. Generally, the former aims at finding the spectrum hole quickly and reliably (P. Cheng, et al., 2008, Y. Chen, et al., 2008), and the latter's objective is to obtain a network performance (e.g., system throughput) gain as much as possible (R. Wang, et al., 2008). Due to receiver uncertainty, wireless channel fading and shadowing effect, sensing capability of single spectrum sensor is limited (Z. Chair & P. K. Varshney, 1986). Previous work (Z. Quan, et al., 2008) has shown that cooperation among SUs can significantly improve the sensing reliability by exploiting multi-user spatial diversity. On the other hand, adaptive scheduling spectrum resource (spectrum band, power, etc.) among SUs with different channel conditions can obtain a higher network performance by exploiting multi-user frequency diversity (J. Ma, et al., 2008).

In the most existing designs of CR systems, spectrum sensing and resource scheduling are always implemented separately or sequentially (P. Cheng, et al., 2008, Y. Chen, et al., 2008). In these related works, two main steps are included as follows,

Step 1. A spectrum sensing module makes a one-bit hard decision (i.e., idle or busy) based on the soft sensing information collecting from all local spectrum sensors.

Step 2. A resource scheduling module implements spectrum assignment and/or power allocation based on the one-bit hard decision sensing information (HSI).

However, intuitively, the separated/isolated design of spectrum sensing and resource scheduling might not be the best choice to improve wireless spectrum efficiency. For one thing, any process (e.g., the hard decision-making) of passive information processing can bring information loss according to classical information theory (G.M. Antonio, et al., 2009). For another, a joint design or cross-layer design can often outperform a separated/isolated design by exploiting correlation between different components/layers. Moreover, (Y. Chen, et al., 2008) shows that a joint PHY sensing and MAC access strategy outperforms a separation design by considering a scenario with single SU.

On the shoulder of previous valuable works, our contributions in this chapter will be twofold:

- First of all, we will design a joint PHY layer cooperative spectrum sensing and MAC layer resource scheduling scheme. In the joint design, the soft sensing information (SSI) collecting from all local spectrum sensors will be directly fused at a base station for resource scheduling, without any hard decision-making process, to decrease the information loss and further exploit the spectrum opportunity.
- Secondly, we will formulate the joint design as a non-convex optimization problem and find the asymptotic optimum solution based on recent advances in duality optimization theory.

2. System model

In this paper, we consider a centralized CRN consists of a secondary base station (SBS) and K SUs (Fig.1). In each frame, every SU independently senses the PU's activity and sends its sensing results to the SBS. Based on the results collecting from all SUs, the SBS performs resource scheduling decision-making, i.e., spectrum assignment and/or power allocation.

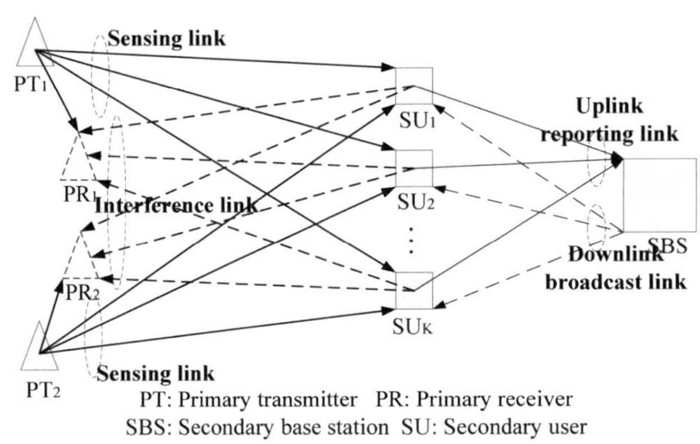

Fig. 1. System model of CRN

2.1 Cooperative spectrum sensing and sensing confidence level

In general, CSS can be divided into two major steps: Local processing and Global processing. For subsequent comparison, we further classify CSS into three classes:

1. Local 1 bit hard decision and Global 1 bit hard decision (LHGH) (Z. Chair & P. K. Varshney, 1986).
2. Local 1 bit hard decision and Global non-hard fusion (LHGN) (R. Wang et al., 2008).
3. Local soft sensing and Global soft fusion (LSGS) (Z. Quan, et al., 2008).

Let S_m denotes the real instantaneous state of channel m in a frame where $S_m = 0$ if the m-th channel is idle and $S_m = 1$ otherwise. We denote the occupy probability of the channel as Q_p. Note that the aim of spectrum sensing is to determine whether a channel is occupied. Due to capability limit of the spectrum sensor, noise uncertainty and something else, sensing error is inevitable. Therefore, we shall concern that how much confidence we should put on the sensing results, which can be formulated as follows:

$$\begin{aligned}\phi_m &= E[S_m \mid \hat{\mathbf{S}}_m] \\ &= \frac{Q_p \Pr[\hat{\mathbf{S}}_m \mid S_m = 1]}{Q_p \Pr[\hat{\mathbf{S}}_m \mid S_m = 1] + (1-Q_p)\Pr[\hat{\mathbf{S}}_m \mid S_m = 0]}\end{aligned} \tag{1}$$

where $\phi_m \in [0,1]$ is defined as sensing confidence level that the SBS believes the m-th channel is busy. $\hat{\mathbf{S}}_m$ denotes the sensing results (vector) of the m-th subcarrier collected by the SBS. For different CSS classes mentioned above, sensing results (vector) shall have different forms and the corresponding sensing confidence levels are given by:

$$\begin{aligned}&\phi_{m\_LHGH} \\ &= \frac{Q_{p,m} \Pr(\hat{\mathbf{S}}_m \mid S_m = 1)}{Q_{p,m} \Pr(\hat{\mathbf{S}}_m \mid S_m = 1) + (1-Q_{p,m})\Pr(\hat{\mathbf{S}}_m \mid S_m = 0)}\Bigg|_{\hat{\mathbf{S}}_m \in \{0,1\}} \\ &= \begin{cases} \dfrac{Q_{p,m}(1-Q_m^{cd})}{Q_{p,m}(1-Q_m^{cd})+(1-Q_{p,m})(1-Q_m^{cf})}, & \hat{\mathbf{S}}_m = 0 \\[2ex] \dfrac{Q_{p,m}Q_m^{cd}}{Q_{p,m}Q_m^{cd}+(1-Q_{p,m})Q_m^{cf}}, & \hat{\mathbf{S}}_m = 1 \end{cases}\end{aligned} \tag{2}$$

where $\hat{S}_m$ denotes the global 1 bit HSI. Q_m^{cf} and Q_m^{cd} represent the probability of false alarm and detection after LHGH, respectively.

$$\begin{aligned}&\phi_{m\_LHGS} \\ &= \frac{Q_{p,m} \Pr(\hat{\mathbf{S}}_m \mid S_m = 1)}{Q_{p,m} \Pr(\hat{\mathbf{S}}_m \mid S_m = 1) + (1-Q_{p,m})\Pr(\hat{\mathbf{S}}_m \mid S_m = 0)}\Bigg|_{\hat{\mathbf{S}}_m = [d_{1,m}, \ldots, d_{K,m}]} \\ &= \frac{Q_{p,m}\binom{K}{N}\prod_{i=1}^{N} Q_{i,m}^d \cdot \prod_{j=1}^{K-N}(1-Q_{j,m}^d)}{Q_{p,m}\binom{K}{N}\prod_{i=1}^{N} Q_{i,m}^d \cdot \prod_{j=1}^{K-N}(1-Q_{j,m}^d) + (1-Q_{p,m})\binom{K}{N}\prod_{i=1}^{N} Q_{i,m}^f \cdot \prod_{j=1}^{K-N}(1-Q_{j,m}^f)}\end{aligned} \tag{3}$$

where $Q_{i,m}^f$ and $Q_{i,m}^d$ represent the probability of false alarm and detection of the i-th SU after local 1 bit hard decision. N is the number of users whose local 1 bit HSI is 0.

$$\phi_{m\_LSGS}$$

$$= \frac{Q_{p,m}\Pr(\hat{\mathbf{S}}_m \mid S_m = 1)}{Q_{p,m}\Pr(\hat{\mathbf{S}}_m \mid S_m = 1) + (1 - Q_{p,m})\Pr(\hat{\mathbf{S}}_m \mid S_m = 0)}\Big|_{\hat{\mathbf{S}}_m = [E_{1,m}, \ldots, E_{K,m}]} \quad (4)$$

$$= \frac{Q_{p,m}\prod_{k=1}^{K} f(E_{k,m} \mid S_m = 1)}{Q_{p,m}\prod_{k=1}^{K} f(E_{k,m} \mid S_m = 1) + (1 - Q_{p,m})\prod_{k=1}^{K} f(E_{k,m} \mid S_m = 0)} \quad (4)$$

where $f(\hat{S}_{k,m} \mid S_m = 0)$ and $f(\hat{S}_{k,m} \mid S_m = 1)$ represent the probability density of the instantaneous SSI of the *k*-th SU under given hypothesis, respectively. For detail expressions of them, one can refer to (J. Ma, et al., 2008).

2.2 System throughput and interference constraints

In this paper, we consider the downlink of the centralized CRN with an OFDMA air interface. We assume that the SBS could transmit only when PUs are inactive and it has perfect channel state information (CSI) knowledge.

With the imperfect sensing information and perfect CSI, the SBS make decision of resource scheduling based on the following definitions:

Definition 1 (Exclusive subcarrier assignment policy A): For any feasible realization of CSI **H** and sensing information $\hat{\mathbf{S}}$, we characterize an exclusive subcarrier assignment policy by a subcarrier assignment indicator matrix **A**, whose element $a_{k,m} = 1$ means that the *m*-th subcarrier is assigned to the *k*-th SU and $a_{k,m} = 0$ otherwise.

Definition 2 (Power allocation policy P): For any feasible realization of CSI **H** and sensing information $\hat{\mathbf{S}}$, we characterize the power allocation policy with a matrix **P**, whose element $p_{k,m}$ denotes the power that the SBS allocates to the *k*-th SU over the *m*-th subcarrier. As the peak power budget of the SBS should not be exceeded, we have $\sum_{k,m} p_{k,m} \le P_{max}$.

After the above definitions, system throughput is given by:

$$\begin{aligned} &R(A,P) \\ &= E_s[\sum_{m=1}^{M}(1 - S_m)\sum_{k=1}^{K} a_{k,m}\log_2(1 + p_{k,m}r_{k,m}) \mid \hat{S}, \mathbf{H}] \\ &= \sum_{m=1}^{M}(1 - \phi_m)\sum_{k=1}^{K} a_{k,m}\log_2(1 + p_{k,m}r_{k,m}) \end{aligned} \quad (5)$$

where $r_{k,m}$ denotes the squared channel gain between the SBS and the *k*-th SU over the *m*-th subcarrier, which is assumed normalized by the receiver noise variance.

Note that imperfect spectrum sensing implies aggressive access to busy subcarriers and thus potential interference to the PU. The interference constraint over the *m*-th subcarrier is:

$$\begin{aligned}\bar{I}_m &= E[\sum_{k=1}^{K} a_{k,m} p_{k,m} \sigma_{p,m}^2 S_m \mid \hat{\mathbf{S}}_m] \\ &= \sum_{k=1}^{K} a_{k,m} p_{k,m} \sigma_{p,m}^2 \phi_m \le \bar{I}_{th,m}\end{aligned} \tag{6}$$

where $\sigma_{p,m}^2$ denotes the squared channel gain from the SBS to the p-th primary receiver over the m-th subcarrier and the interference threshold is $\bar{I}_{th,m}$.

3. Joint spectrum sensing and resource scheduling

3.1 Problem formulation

In this section, the proposed joint spectrum sensing and resource scheduling scheme is formulated as follows:

Given a feasible realization of CSI **H** and sensing information $\hat{\mathbf{S}}$, find the optimal exclusive subcarrier assignment policy A and Power allocation policy P such that the system throughput is maximized, while satisfying the peak power budget of the SBS and the interference constraint of the PU on each subcarrier simultaneously. The mixed integer nonlinear programming problem can be modelled by

$$\begin{aligned} g^* &= \max_{P,A} \sum_{m=1}^{M} (1-\phi_m) \sum_{k=1}^{K} a_{k,m} \log_2(1 + p_{k,m} r_{k,m}) \\ \text{s.t.} & \quad \sum_{m=1}^{M} \sum_{k=1}^{K} a_{k,m} p_{k,m} \le P_{max} \\ & \quad \sum_{k=1}^{K} a_{k,m} p_{k,m} \sigma_m^2 \phi_m \le \bar{I}_{th,m} \qquad \forall m \in \{1,\dots,M\} \end{aligned} \tag{7}$$

3.2 Duality optimization

Note that two optimal variables **P** and **A** are coupled in the problem above, so global optimal solution is often difficult to obtain. However, we will propose an approximately global optimal solution based on duality theory (Y. Wei & R. Lui, 2006).

Firstly, by introducing nonnegative dual variables λ and $\boldsymbol{\mu} = [\mu_1, \mu_2, \dots, \mu_m]$, the Lagrange function is:

$$\begin{aligned} & L(A,P,\lambda,\boldsymbol{\mu}) \\ &= \sum_{m=1}^{M} (1-\phi_m) \sum_{k=1}^{K} a_{k,m} \log_2(1 + p_{k,m} r_{k,m}) \\ &+ \lambda(P_{max} - \sum_{m=1}^{M} \sum_{k=1}^{K} a_{k,m} p_{k,m}) + \sum_{m=1}^{M} \mu_m (\bar{I}_{th,m} - \sum_{k=1}^{K} a_{k,m} p_{k,m} \sigma_m^2 \phi_m) \\ &= \sum_{m=1}^{M} \sum_{k=1}^{K} l(a_{k,m}, p_{k,m}, \lambda, \mu_m) + \lambda P_{max} + \sum_{m=1}^{M} \mu_m \bar{I}_{th,m} \end{aligned} \tag{8}$$

where

$$
\begin{aligned}
l(a_{k,m}, p_{k,m}, \lambda, \mu_m) = a_{k,m}[(1-\phi_m)\log_2(1+p_{k,m}r_{k,m}) \\
- \lambda p_{k,m} - \mu_m p_{k,m}\sigma_m^2\phi_m]
\end{aligned}
\tag{9}
$$

Secondly, Lagrange dual function can be obtained by

$$
D(\lambda, \mathbf{\mu}) = \max_{P,A} L(A, P, \lambda, \mathbf{\mu}) \tag{10}
$$

And the dual problem is

$$
d^* = \min_{\lambda \geq 0, \mathbf{\mu} \geq 0} D(\lambda, \mathbf{\mu}) . \tag{11}
$$

which can be decomposed into three sub problems (P. Cheng, et al., 2008, G.M. Antonio, et al., 2009, R. Wang, et al., 2009)：

Sub problem 1 (Power allocation): Given the dual variables λ and $\mathbf{\mu}$, for any $k \in \{1,...,K\}$ and any $m \in \{1,...,M\}$, maximizing (10) will bring the optimized variable as follows:

$$
p_{k,m}^* = \arg\max_{p_{k,m}} l(a_{k,m}, p_{k,m}, \lambda, \mu_m) = \{\frac{1-\phi_m}{[\lambda + \mu_m\sigma_m^2\phi_m]\ln 2} - \frac{1}{r_{k,m}}\}^+ \tag{12}
$$

where $[\bullet]^+ = \max\{\bullet, 0\}$.

Sub problem 2 (Subcarrier assignment): Substituting (12) into (9) will bring $l^*(a_{k,m}, p_{k,m}^*, \lambda, \mu_m)$, and maximizing which will bring the best SU of each subcarrier,

$$
a_{k,m}^* = \begin{cases} 1, \ k = \text{argmax}_{(k)} \ l^*(a_{k,m}, p_{k,m}^*, \lambda, \mu_m) \\ 0, \ \text{otherwise} \end{cases} \tag{13}
$$

and we have $l^{*\lozenge}(a_{k,m}^*, p_{k,m}^*, \lambda, \mu_m) = \max_{a_{k,m}} l^*(a_{k,m}, p_{k,m}^*, \lambda, \mu_m)$.

Sub problem 3 (Dual variables update): The optimal dual variables can be obtained by solving its dual problem:

$$
\begin{aligned}
(\lambda^*, \mathbf{\mu}^*) = \min_{\lambda \geq 0, \mathbf{\mu} \geq 0} \sum_{m=1}^{M}\sum_{k=1}^{K} l^{*\lozenge}(a_{k,m}^*, p_{k,m}^*, \lambda, \mu_m) \\
+ \lambda P_{max} + \sum_{m=1}^{M} \mu_m \overline{I}_{th,m}
\end{aligned}
\tag{14}
$$

Because dual function is always convex (S. Boyd & L. Vandenberghe, 2004), a gradient-type search is guaranteed to converge to the global optimum. However, dual function is not necessarily differentiable and thus it does not always have a gradient. Here we use a sub-gradient (a generalization of gradient) update method,

$$\lambda^{n+1} = [\lambda^n - s^n (P_{max} - \sum_{m=1}^{M} \sum_{k=1}^{K} a^*_{k,m} p^*_{k,m})]^+ \quad (15)$$

$$\mu_m^{n+1} = [\mu_m^n - s^n (\bar{I}_{th,m} - \sum_{k=1}^{K} a^*_{k,m} p^*_{k,m} \sigma_m^2 \phi_m)]^+ \quad \forall m \quad (16)$$

where *n* is the iteration number. The above sub-gradient update is guaranteed to converge to the optimal dual variables as long as the sequence of scalar step s^n is chosen appropriately (Y. Wei & R. Lui, 2006). The duality gap $g^* - d^*$ can be zero as long as the number of subcarrier is sufficiently large (Y. Wei & R. Lui, 2006).

3.3 Distributive implementation

Without loss of optimality, we proposed a distributive iterative algorithm to alleviate the computing overhead of the SBS.

Distributive Algorithm:

Step 1. Each SU performs local spectrum sensing on each subcarrier and sends the sensing information to the SBS.

Step 2. The SBS computes the sensing confidence level according to (2)/(3)/(4) and broadcasts ϕ_m and initial dual variables λ^1 and μ_m^1 to all SUs.

Step 3. Step 3: In each iteration, first each SU solves subproblem 1 and sends $p^*_{k,m}$ and l^* to the SBS. Secondly the SBS solves subproblem 2 and assigns each subcarrier to the corresponding best SU. Then the SBS update the dual variables according to (16) and (17).

Step 4. If $| \mu_m^{n+1} (\bar{I}_{th,m} - \sum_{k=1}^{K} a^*_{k,m} p^*_{k,m} \sigma_m^2 \phi_m) | \le \varepsilon, \ \forall m$ and $| \lambda^{n+1} (P_{max} - \sum_{m=1}^{M} \sum_{k=1}^{K} a^*_{k,m} p^*_{k,m}) | \le \varepsilon$ are satisfied simultaneously, then terminate the algorithm. Otherwise, jump to Step 3.

4. Nurmerical results

4.1 Simulation setup

In this section, we compare our proposed scheme LSGS with several previous designs, i.e., LHGH(Z. Chair & P. K. Varshney, 1986), LHGN (R. Wang et al., 2008).

- LHGH here refers to the design that the SBS fuses the 1-bit HSI from all SUs (using an OR rule) to make a 1-bit hard decision on the availability of each subcarrier.
- LHGN refers to the design that the SBS fuses the 1-bit HSI from all SUs for resource scheduling without centralized hard decision.
- The proposed scheme LSGS means that the SBS directly uses the SSI collecting from all SUs for resource scheduling without any hard decision.

In the following simulation, we use an OFDMA system with $K = 64$ subcarriers and the interference constraint of each subcarrier is 0dB. All sensing links from primary transmitters to SUs and all scheduling links from the SBS to SUs are assumed to be i.i.d. Rayleigh fading, respectively.

4.2 Sensing performances versus number of cooperative users

Fig. 2 presents the comparison of different schemes in terms of sensing error, which equals to the percentage of the number of incorrect decisions among all decisions. The error can be caused by false alarms (i.e., an idle subcarrier is mistaken as a busy one) or miss detection (i.e., a busy subcarrier is mistaken as an idle one). It is shown in Fig. 2 that the sensing errors of all schemes are highly related to the number of cooperative users. Specifically, the sensing errors of both LHGN and LSGS decrease with the number of cooperative users while LHGH performs a much higher sensing error when the number of cooperative users is large. Note that the proposed LSGS always outperforms the other schemes due to less information loss.

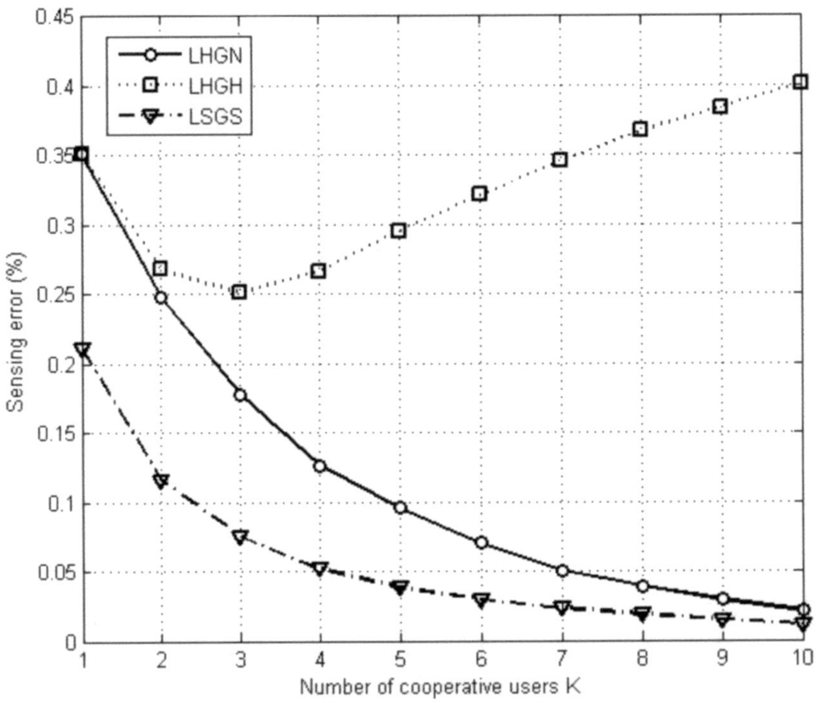

Fig. 2. Sensing performances versus number of cooperative users

4.3 System throughput versus number of cooperative users

In Fig.3, we compare the performance of different schemes in terms of system throughput. It is shown that the system throughput of all schemes increase with the number of cooperative users K. This increment mainly profits from multiuser diversity of CSS. Note that the proposed scheme LSGS always outperforms the other two schemes because of less sensing information loss, which is consistent with the results in Fig. 2. The separated design, LHGH performs the worst and becomes saturated much earlier than the joint designs, LHGN and LSGS. This fact demonstrates that joint design could exploit spectrum opportunity more effectively, and less sensing information loss means better system performance.

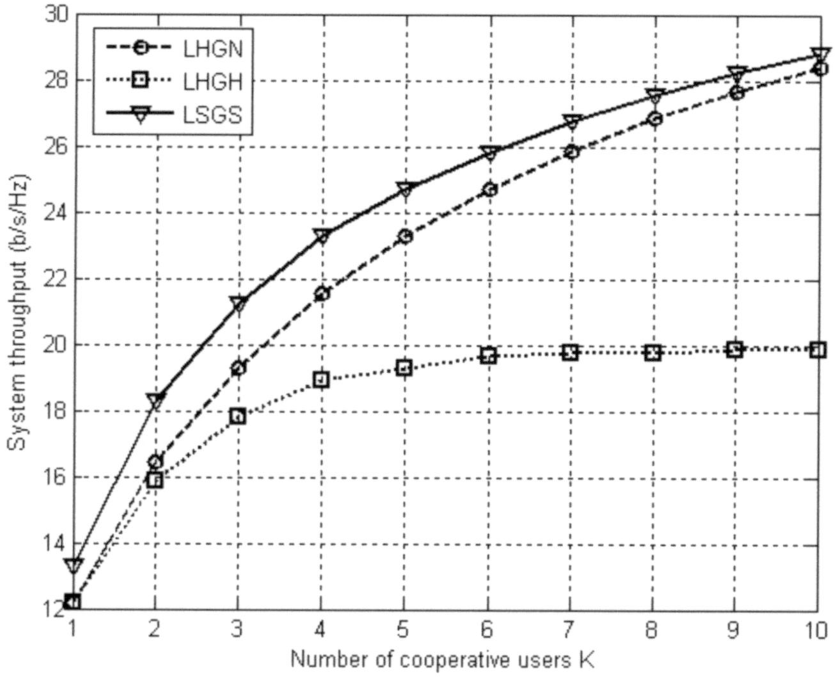

Fig. 3. System throughput versus number of cooperative users at $P_0 = 10\text{dB}$, $Q_p = 0.5$.

4.4 System throughput versus signal-to-noise ratio (SNR)

Signal-to-Noise Ratio (SNR) is an important parameter that affects the system performance. Fig. 4 indicates that the higher the SNR is, the better the system throughput of all schemes are. System throughput of all schemes increases with SNR. The reason behind is that: given the average noise power, a higher SNR means a bigger transmission power the BS can employ, which in turn brings a higher bite rates.

4.5 System throughput versus peak power budget of the SBS

Fig.5 shows that the system throughput of all schemes increase with the peak power budget of the SBS and the proposed LSGS performs the best. Especially, in the scenario of heavy load of the primary user activity $Q_p = 0.8$ (i.e., available spectrum for CRN is rare), the proposed scheme LSGS obtains much higher throughput than LHGN and LHGH, which is valuable for CRN design.

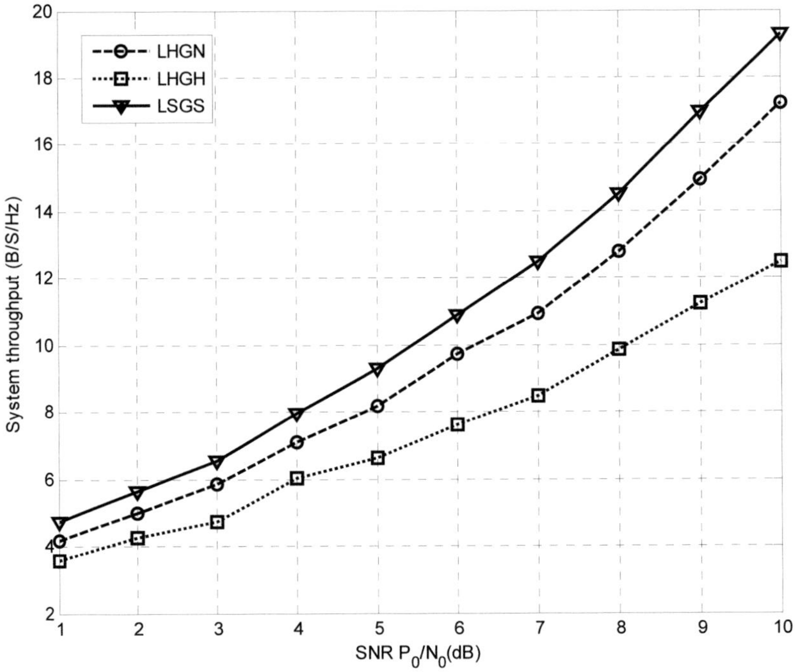

Fig. 4. System throughput versus Signal-to-Noise Ratio (SNR) at K=6

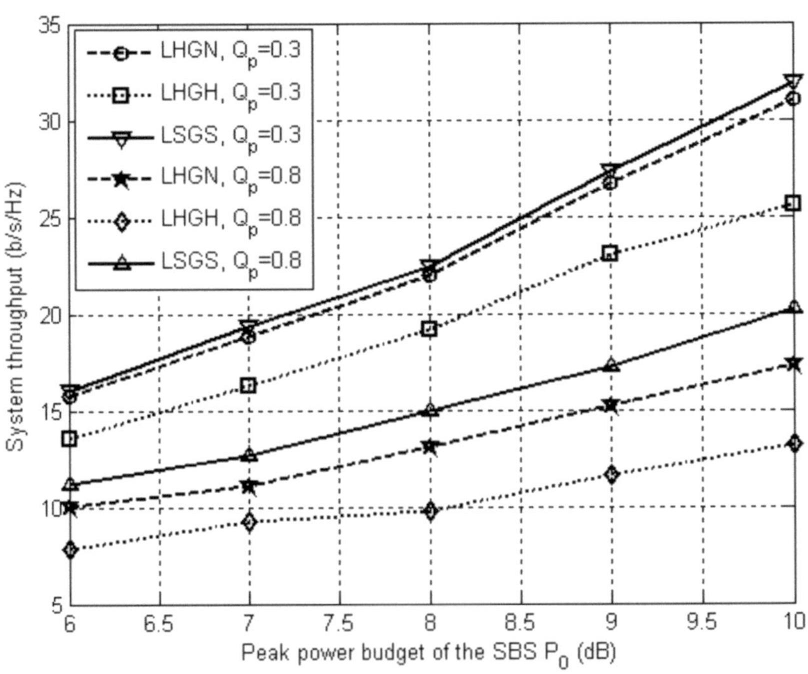

Fig. 5. System throughput versus peak power budget at K=6

5. Conclusion and future work

We have developed a joint cooperative spectrum sensing and resource scheduling scheme for cognitive radio networks in this paper. Numerical results tell that joint design exploits spectrum opportunity more effectively than isolated design; especially much higher system throughput performance can be obtained with soft sensing information feedback to the SBS. The next work will focus on quantizing the SSI or/and clustering the spectrum sensors to achieve a better trade-off between performance and feedback overhead.

6. Acknowledgment

This work is supported by the national basic research program (973) of China under grant no.2009CB320400, the national high-tech research and development program (863) of China under grant no.2009AA01Z243, the national natural science foundation under grant no. 60932002, the national natural science fund of China under grant no. 61172062, and the natural science fund of Jiangsu, China under grant no. BK2011116 .

7. References

G.M. Antonio, X. Wang & B.G. Georgios. (2009). Dynaimic resource management for cognitive radios using limited-rate feedback, *IEEE Trans. Signal Processing*, vol. 57, no. 9, pp. 3651–3666.

J. Ma, G. Zhao & Y. Li. (2008). Soft combination and detection for cooperative spectrum sensing in cognitive radio networks, *IEEE Trans. Wireless Commun.*, vol. 7, no. 11, pp. 4502–4507.

J. Mitola. (2000). Cognitive radio: An integrated agent architecture for software defined radio, *Doctor of technology*, Royal Institute Technology (KTH) Stockholm., Sweden.

P. Cheng, Z. Zhang, H. H. Chen & P. Qiu. (2008). Optimal distributed joint frequency, rate and power allocation in cognitive OFDMA systems, *IET Commun.*, vol. 2, no. 6, pp. 815–826.

Q. Zhao, L. Tong, A. Swami & Y. Chen. (2007). Decentralized cognitive MAC for opportunistic spectrum access in ad hoc networks: A POMDP framework, *IEEE J. Sel. Areas. Commun.*, vol. 25, pp. 589–600.

R. Wang, Vincent K.N.Lau, L. Lv & B. Chen. (2009). Joint cross-layer scheduling and spectrum sensing for OFDMA cognitive radio systems," *IEEE Trans. Wireless Commun.*, vol. 8, no. 5, pp. 2410–2416.

S. Haykin. (2005). Cognitive radio: Brain-empowered wireless communications, *IEEE J. Sel. Areas. Commun.*, vol. 23, no. 2, pp. 201–220.

S. Boyd & L. Vandenberghe. (2004). *Convex Optimization*, Cambridge University Press, ISBN 0521833787.

Y. Chen, Q. Zhao & A. Swami. (2008). Joint design and separation principle for opportunistic spectrum access in the presence of sensing errors, *IEEE Trans. Info. Theory*, vol. 54, no. 5, pp. 2053–2071.

Y. Wei & R. Lui. (2006). Dual methods for nonconvex spectrum optimization of multicarrier systems, *IEEE Transactions on Communications*, vol. 54, pp. 1310–1322.

Z. Chair & P. K. Varshney. (1986). Optimal data fusion in multiple sensor detection systems, *IEEE Trans. Aerosp. Electron. Syst.*, vol. AES-22, no. 1, pp. 98–101.

Z. Quan, S. Cui & A. H. Sayed. (2008). Optimal linear cooperation for spectrum sensing in cognitive radio networks," *IEEE J. Sel. Topics Signal Process,* vol. 2, no. 1, pp. 28–40.

13

A Roadmap to International Standards Development for Cognitive Radio Systems and Dynamic Spectrum Access

Jim Hoffmeyer
Western Telecom Consultants, Inc.
United States

1. Introduction

This chapter provides a summary of standards activities related to cognitive radio systems (CRS) and dynamic spectrum access (DSA) systems. Because of the vast amount of standardization activity in this arena, this chapter can provide only a high-level snapshot of current CRS/DSA standardization activities and a projection of activities expected in the near future. Sufficient references including URL references are included so that the interested reader can review in as much detail as desired the standards that have and are being developed by numerous international standards development organizations (SDOs). These standards will be evolving in time as will be described later in this chapter in the descriptions of the various SDOs. It is expected that standardization in this arena will continue by these SDOs for at least the next decade.

Section 2 sets the stage by describing the terminology that is used by researchers and practitioners because there are many terms closely related to CRS and DSA. As noted in Section 2, there is not total agreement on definitions between the various standards development organizations. Section 3 describes how standards development for CRS/DSA is but one piece of the puzzle that must be solved in order for CRS/DSA systems to achieve their full potential. In that sense, Section 3 can be viewed as the statement of the problem. In order to solve the standardization portion of the puzzle, numerous international SDOs are working on different pieces of the standardization puzzle. The work of these SDOs is described in Sections 4 through 6. A brief summary and a description of future standards work are both provided in Section 7.

2. Terminology

The term "cognitive radio" was first coined by Dr. J. Mitola (Mitola, 2000). Since that time the term "cognitive radio" and the acronym "CR" have been used in countless books, journal articles, and telecommunication standards documents. More recently, the term "cognitive radio system" and the acronym "CRS" frequently have been used. The term "cognitive radio systems" has gained international recognition as a result of its adoption by the International Telecommunication Union Radiocommunication Sector. (ITU-R, 2009). It has also been adopted by the US National Telecommunication and Information Administration. (NTIA, 2011).

Organization	Focus of CRS and DSA Activities	Section
ITU-R	Develops Radio Regulations which is a treaty-level international agreement. Develops ITU-R Reports and Recommendations which support the agreements formally accepted within the Radio Regulations.	Section 4.
IEEE DySPAN-SC	The IEEE Dynamic Spectrum Access Networks Standards Committee develops basic cognitive radio system/dynamic spectrum access (CRS/DSA) technology standards with the focus on the improved use of spectrum in any spectrum band.	Section 5.1
IEEE 802	The IEEE 802 Standards Committee develops standards for local and regional networks. This organization has focused on CRS/DSA for TV white space bands.	Section 5.2
Other	There are many organizations that either directly or indirectly develop standards related to CRS/DSA or standards which incorporate CRS/DSA technology as part of the overall standard. A partial list of these organizations includes ETSI, Ecma, IEEE ComSoc Technical Committees, Wireless Innovation Forum WRAN Alliance and the Wireless Research Forum. These and other organizations such as GSMA and NGMN provide input to the standards development process from the research, operator, and other communities. Some of these organizations develop specifications based on the standards published by accredited Standards Development Organizations. An example is the WiMax Forum which certifies products based on the IEEE 802.16 Standard.	Section 6

Some of the difficulty in the terminology is related to the fact that some of the functionality typically associated with cognitive radios is outside the functionality of what is historically associated with a "radio." As will be described in Section 5.1.1, the Institute of Electronics and Electrical Engineers (IEEE) Standard IEEE 1900.1™ addresses the terminology issue and discusses basic concepts associated with several technologies related to CRS.

There are many terms that are being used in the literature being used that are closely related to CRS technology including dynamic spectrum access, software defined radio, hardware radio, adaptive radio, policy-based radio, reconfigurable radio, and software-controlled radio. In this chapter we focus on standards for CRS and DSA which have been defined as:

- Cognitive Radio System: "*A radio system employing technology that allows the system to obtain knowledge of its operational and geographical environment, established policies and its internal state; to dynamically and autonomously adjust its operational parameters and protocols according to its obtained knowledge in order to achieve predefined objectives; and to learn from the results obtained.*" (ITU-R, 2009).
- Dynamic Spectrum Access: "*The real-time adjustment of spectrum utilization in response to changing circumstances and objectives.*

NOTE—Changing circumstances and objectives include (and are not limited to) energy-conservation, changes of the radio's state (operational mode, battery life, location, etc.), interference-avoidance (either suffered or inflicted), changes in environmental/external constraints (spectrum, propagation, operational policies, etc.), spectrum-usage efficiency targets, quality of service (QoS), graceful degradation guidelines, and maximization of radio lifetime." (IEEE 2008a).

3. Standards – A critical part of the evolution to CRS

As conceptualized in Figure 1, the full potential of cognitive radio systems is dependent upon activities in the following areas:

- Economic feasibility based on business case scenarios:
  - User requirements which drive new business cases which in turn drive the need for more spectrum (e.g., 4th generation commercial wireless).
  - New radio system technologies (e.g., ultra wideband, policy-based DSA radio systems, cognitive radio systems) and networks which utilize these radio system technologies - these new technologies open up new business opportunities.
- Technology Development:
  - Responsive to new users requirements which frequently require significant new spectrum.
  - Responsive to the need for more efficient use of the spectrum.
  - Responsive to, and drives, new regulatory proceedings (e.g., DSA radios and networks, policy-based radio, software defined radio, ultra wideband radio (UWB), cognitive radio systems, etc.) and new questions within the ITU-R.
  - Regulatory and policy changes:
  - Regulatory changes that recognizes new requirements and new technologies.
  - Regulatory changes for CRS and DSA are driven by need for a new paradigm for spectrum management to accommodate requirements for additional spectrum (e.g., the need for more spectrum for cellular systems to support wideband applications such as audio and video steaming).

As depicted in Figure 1, technical standards are a central part needed to bring the above components together to enable the fielding of CRS and DSA-based systems. One example of this is the development of IEEE 802 standards for TV white space as will be described later in this chapter.

Another example is the development of standards relative to advanced mobile telecommunications systems over the last several years. Clearly there is a business case as can be seen for the broadband applications. This required the consideration by the ITU-R and by national regulatory agencies for the identification of additional spectrum to support these new broadband applications. Technologies were developed to support these applications - these technologies were then standardized by the 3rd Generation Partnership Project (3GPP), IEEE, the ITU-R, and other standards organizations. Although this example does not specifically include CRS/DSA, these technologies are envisioned as being a part of the solution in future years for the continued evolution of mobile telecommunications requirements and the identification of spectrum to satisfy these ever increasing requirements. As will be described in later sections, the IEEE and the ITU-R have both been working on standards and reports relative to CRS/DSA for mobile telecommunication applications.

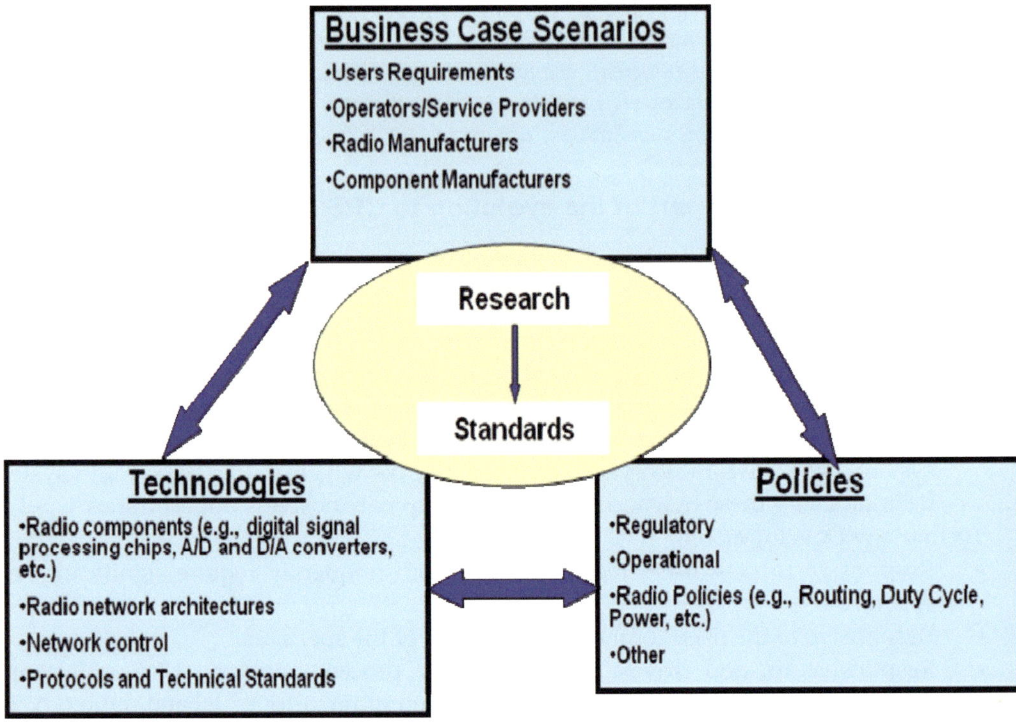

Fig. 1. Relationships between standards, technologies, policies and business cases for CRS and DSA

4. International Telecommunication Union Radiocommunication Sector

In our discussion of standards for CSA/DSA, a logical starting place is the International Telecommunication Union Radiocommunication Sector (ITU-R). The ITU-R is the sector of the ITU of primary relevance to this chapter because of its role as a primary global forum on the evolution of radio technologies and because of its decisions on spectrum allocations necessary for this evolution. Information on other sectors of the ITU as well as ITU-R may be found at www.itu.int.

The *Radio Regulations* (RR) is one of the most significant products of the ITU-R and is published in the 6 official languages of the ITU. (ITU-R, 2008) The *Radio Regulations* document is the international treaty governing the use of the radio-frequency spectrum and the geostationary-satellite and non-geostationary-satellite orbits. The RR incorporates the decisions of the World Radiocommunication Conferences (WRC) and includes ITU-R Recommendations incorporated by reference. The WRC is held every three to five years. The *Radio Regulations* (ITU-R, 2008) which are developed by the WRC and verified by signatures of the ITU Member States include:

- definitions of terms including allocation, allotment and assignment of frequencies
- the specific allocation of different frequency bands to different radio services;
- the mandatory technical parameters to be observed by radio stations, especially transmitters;

- frequency assignment coordination procedures
- radio frequency interference regulations
- other administrative and operational procedures and provisions.

ITU-R Recommendations constitute a set of international technical standards developed by the Radiocommunication Sector of the ITU. They are the result of studies undertaken by Radiocommunication Study Groups and approved by ITU Member States (countries). Topics covered by *ITU Recommendations* and *ITU Reports* include:

- The use of a vast range of wireless services including those that mandate new telecommunication technologies such as CRS and DSA
- Spectrum management and the efficient use of the spectrum
- Radiowave propagation
- Systems and networks for wireless communications.

The use of *ITU-R Recommendations* and *ITU-R Reports* is not mandatory. However, since these documents are developed by technical experts from administrations (countries) around the world, communications systems operators, the telecommunications industry (including manufacturers and national and international standards organizations) and other organizations from all over the world that deal with radiocommunication matters, these documents are highly regarded and are implemented worldwide. Free online access to all current ITU-R Recommendations & Reports is now provided to the general public: http://www.itu.int/pub/R-REC and http://www.itu.int/pub/R-rep

As will be discussed in the following sections, the ITU-R activities in regard to CRS and DSA include the following:

- WRC-12 Agenda Item 1.19 (ITU-R, 2007a).
- Resolution 956
- Agreement on definitions for cognitive radio systems and software defined radio.
- Development of ITU-R Reports and Recommendations on software defined radio, cognitive radio systems and related technologies.

4.1 World Radiocommunication Conference Agenda Item 1.19 and Resolution 956

The World Radiocommunication Conference (WRC) is organized by ITU to review and revise as necessary the ITU-R Radio Regulations. An important topic for the next WRC (January 2012 in Geneva) are issues related to software defined radio and CRS/DSA. One of the many Agenda Items of WRC-12 is Agenda Item 1.19:

> *"To consider regulatory measures and their relevance, in order to enable the introduction of software-defined radio and cognitive radio systems, based on the results of ITU-R studies, in accordance with Resolution 956 of WRC-07." (ITU-R, 2007a).*

ITU-R Resolution 956 states in part (ITU-R, 2007a):

"resolves to invite ITU-R

1. *to study whether there is a need for regulatory measures related to the application of cognitive radio system technologies;*

2. *to study whether there is a need for regulatory measures related to the application of software-defined radio.*

Resolves further that WRC-11 consider the results of these studies and take the appropriate actions.[1]

The ITU-R studies relative to the potential need for additional regulatory measures is a critical step in the evolution of CRS. As previously discussed in regard to Figure 1, regulatory policies are one of the critical piece-parts in the evolution of CRS/DSA. Technologies which are developed to satisfy the business cases for CRS/DSA must be compliant with existing regulatory measures. The starting point for these regulatory measures is the *ITU-R Radio Regulations*. The following section provides information on the work of ITU-R Study Groups and Working Parties that are underway to satisfy Resolution 956. These results will be reviewed during WRC-12 as part of discussions relative to WRC-12 Agenda Item 1.19. Because of the complexity of the regulatory issues, it can reasonably be expected that these studies will be continued beyond WRC-12, i.e., into the next ITU-R Study Period and probably even beyond that although there may not be a continuation of the agenda item.

4.2 Conference Preparatory Meetings

There are two Conference Preparatory Meetings between WRCs; one is immediately after the conclusion of a WRC and is known as CPM1; the second CPM typically is a few months before the following WRC and is known as CPM2. CPM1 develops the program of study for the ITU-R Study Groups. CPM2 prepares a consolidated report to be used in support of the work of World Radiocommunication Conferences, based on:

- contributions from administrations and the Radiocommunication Study Groups concerning the regulatory, technical, operational and procedural matters to be considered by the following WRC;
- the inclusion, to the extent possible, of reconciled differences in approaches as contained in the source material, or, in the case where the approaches cannot be reconciled, the inclusion of the differing views and their justification. (ITU-R, 2007b).

4.3 ITU-R study groups and working parties

One of the major activities of the ITU-R is that associated with the ITU-R Study Groups and Working Parties. (ITU-R, 2010) The topics of study for ITU-R Study Groups are known as *QUESTIONS*. (ITU-R, 2011) These are documents that describe the scope and nature of studies that are assigned to Working Parties for study. This work ultimately results in one or more *ITU-R Recommendations* and *ITU Reports*. The *QUESTIONS* are assigned to Study Groups which in turn designate a Working Party within the Study Group to lead the effort of developing ITU-R *Reports and Recommendations* in response to the *Question*.

The primary Study Groups working on CRS studies are ITU-R Study Group 1 and ITU-R Study Group 5. Each of these Study Groups have been assigned *QUESTIONS* and work programs which guide the work of these Study Groups on CRS, DSA and related technologies.

[1]Note: The WRC that was originally scheduled for 2011 has been rescheduled for 2012 and is now referred to as WRC12.

4.3.1 ITU-R Study Group 1 – Spectrum Management

The purpose of Study Group 1 (SG 1) is the development of "spectrum management principles and techniques, general principles of sharing, spectrum monitoring, long-term strategies for spectrum utilization, economic approaches to national spectrum management, automated techniques and assistance to developing countries in cooperation with the Telecommunication Development Sector." (ITU-R, 2010).

Working Party 1B (WP1B) is the Working Party under SG 1 that is responsible for the development of spectrum management methodologies and economic strategies. WP1B is responsible for Agenda Item 1.19 and is working on developing a response which will be sent from Study Group 1 to the CPM2 for WRC-12 in response to Resolution 956 from WRC07. This work within ITU-R is likely to continue well beyond WRC12 but is likely to be focused on additional studies that generate ITU-R Reports and ITU-R Recommendations rather than changes to the Radio Regulations.

4.3.2 ITU-R Study Group 5 – Terrestrial Services

ITU-R QUESTION 241-1/5 is entitled, "Cognitive Radio Systems in the Mobile Systems." And is assigned to ITU-R Working Party 5A (Land mobile service above 30 MHz*(excluding IMT); wireless access in the fixed service; amateur and amateur-satellite services). There are several parts of Question 241-1/5 (ITU-R, 2010). however, the key part of the Question is:

> *"How can cognitive radio systems promote the efficient use of radio resources?"*

Although, WP5A has been assigned this *Question,* other Working Parties within Study Group 5 are also working on this topic. For example, ITU-R WP5D (International Mobile Telecommunications Systems) is working on an ITU-R Report entitled, *"Cognitive radio systems specific for IMT systems."* In the ITU-R, the term International Mobile Telecommunications (IMT) means a family of radio interfaces for mobile telecommunication services. The term is applied to 3rd and 4th generation commercial mobile telecommunication services.

A primary difference between the work of WP5A and WP5D on CRS is in their focus; whereas WP5D is focused on the IMT application of CRS, WP5A is focused primarily on non-IMT applications of CRS. Although the current reports being developed within ITU-R Working Parties 5A and 5D are likely to be completed in the near future, it is likely that work in this arena by these two working parties is likely to continue for several years. It is expected that ITU-R *Recommendations* and additional ITU-R Reports will be produced in addition to the ITU-R *Reports* currently being developed.

4.4 Other ITU activities

Radiocommunication Assemblies (RA) are responsible for the structure, program and approval of Radiocommunication studies. They are normally convened every three or four years and may be associated in time and place with World Radiocommunication Conferences (WRCs). Amongst other responsibilities, the Assemblies are responsible for:

- Suggesting suitable topics for the agenda of future WRCs;
- Approving and issuing ITU-R Recommendations developed by the Study Groups;

- Assigning *Questions* to the Study Groups and setting the program of work for the Study Groups;
- Establishing and disbanding Study Groups according to need.

5. IEEE standards development related to CRS and DSA

A number of papers have been published in recent years that provide snapshots of international standards activities on CRS/DSA including those of the Institute of Electrical and Electronics Engineers (IEEE). (Sherman, et. al. 2008; Prasad, et. al., 2008; Granelli, et. al., 2010, and Stanislav, et. al. 2011) The branch of the IEEE that is responsible for standards development is the IEEE Standards Association (IEEE-SA). (http://standards.ieee.org/).

Under the IEEE-SA, there are standards boards associated with several IEEE Societies that have undertaken standards development relative to CRS/DSA. Examples are:

- IEEE Communications Society (ComSoc) Standards Board (CSSB)
- IEEE Electromagnetic Compatibility Society (EMC) Standards Development Committee (EMC SDCom).
- IEEE Computer Society Standards Activity Board.

The ComSoc Standards Board http://www.comsoc.org/about/standards/sponsors standards in communications and networking technologies and applications including the following technologies):

- Ad Hoc and Sensor Networks
- Cognitive Networks
- Communications and Information Security
- Communications Quality and Reliability
- Communications Software
- Communications Switching and Routing
- Computer Communications
- Communications Networks
- Internet Multimedia Communications
- Network Operations and Management
- Optical Communications and Networking
- Power Line Communications and Networking
- Radio, Satellite & Space, and Wireless Communications and Networks
- Signal Processing and Communications Electronics
- Transmission, Access, and Optical Systems

More information on ComSoc standards activities may be found at:
http://committees.comsoc.org /standards/

The IEEE Electromagnetic Compatibility Society (EMC) also has keen interest in CRS/DSA because of its mission to develop and facilitate the exchange of scientific and technological knowledge in the discipline of electromagnetic environmental effects and electromagnetic compatibility. The EMC Society mission includes the development of electromagnetic compatibility standards and naturally includes interest in CRS/DSA because electromagnetic compatibility is at the heart of issues related to the ultimate deployment of CRS/DSA. For more information: http://www.emcs.org/standards/sdcomindex.html.

Two major efforts within the above IEEE Societies on CRS/DSA standards development are:

- The DySPAN Standards Committee (DySPAN-SC) which is sponsored by the Communications Society (ComSoc), and
- The IEEE 802 Committee which is sponsored by the Computer Society.

The IEEE EMC Society Standards Development Committee and the Communications Society Standards Board were the original sponsors of the IEEE standards organization that is now known as the DySPAN Standards Committee (DySPAN-SC) which is described in the next section.

The organizational relationships of key elements of the IEEE Standards Association involved with CRS/DSA standards development are depicted in Figure 2.

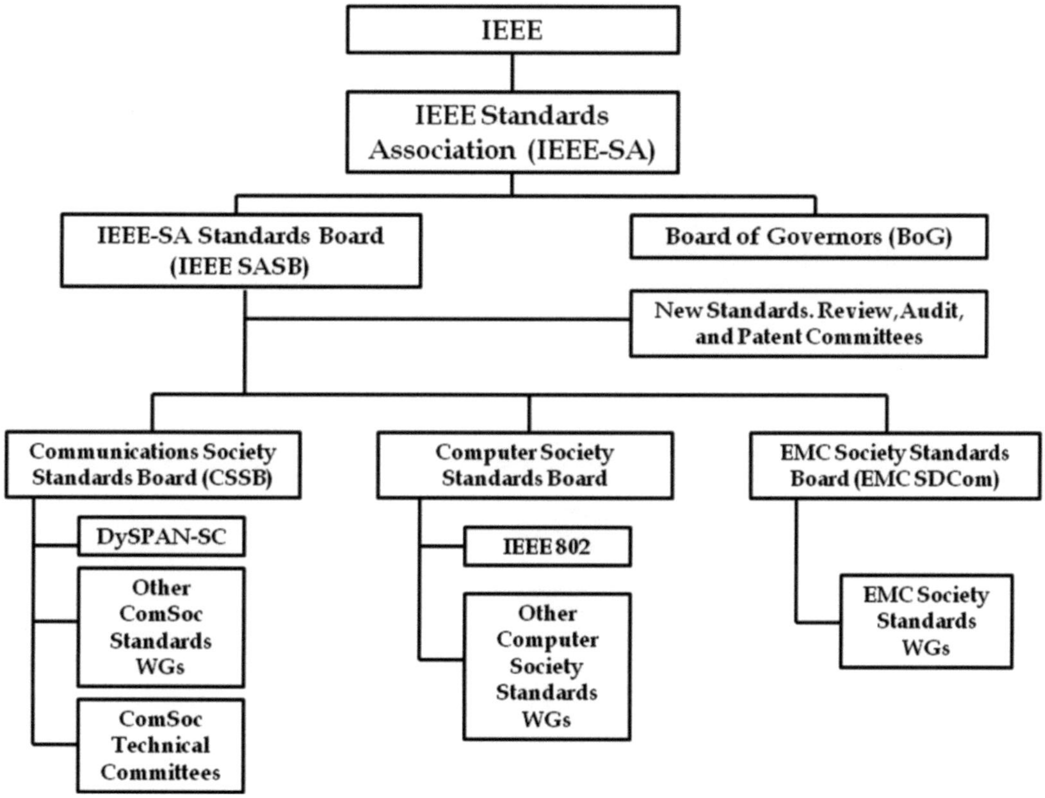

Fig. 2. IEEE Standards Organization.

5.1 IEEE DySPAN Standards Committee

The DySPAN (Dynamic Spectrum Access) standards committee had its beginning in early 2005 jointly under CSSB and EMC. It was initially known as the 1900 Committee and later as the Standards Coordinating Committee 41 (SCC41). The scope of the DySPAN-SC includes the following:

- dynamic spectrum access radio systems and networks with the focus on improved use of spectrum,
- new techniques and methods of dynamic spectrum access including the management of radio transmission interference, and
- coordination of wireless technologies including network management and information sharing amongst networks deploying different wireless technologies.

Additional information on DySPAN-SC may be found at www.dyspan-sc.org.

Figure 3 is the DySPAN-SC organization chart. The detailed technical work of the IEEE DySPAN-SC is performed by the IEEE 1900.x Working Groups (WGs) which are shown in the figure and which are described in the following subsections. Several of the 1900.x WGs have completed baseline standards and will be developing follow-on standards that are extensions of the baseline standards. Therefore, it is expected that each of these IEEE 1900.x WGs will be in existence for many years.

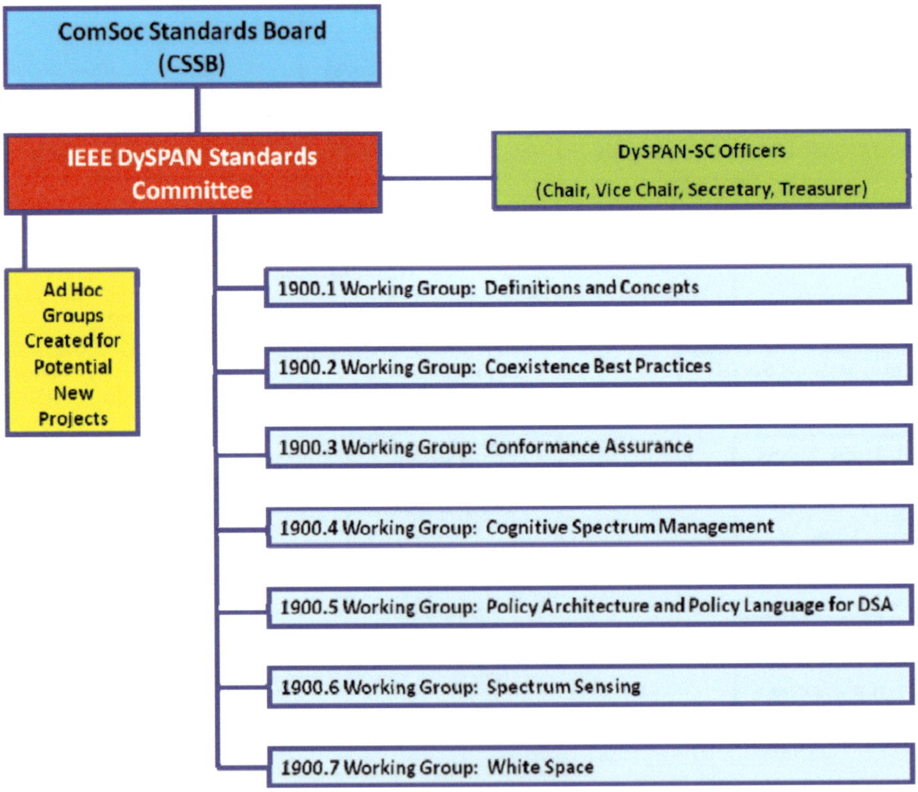

Fig. 3. Organizational chart of the DySPAN Standards Committee showing working groups

5.1.1 IEEE 1900.1 Working Group

The IEEE 1900.1 WG is responsible for DSA definitions and concepts and published its first standard in 2008. This standard provides definitions and explanations of key concepts in the fields of spectrum management, cognitive radio, policy-defined radio, adaptive radio, software-defined radio (SDR), and related technologies. The standard goes beyond simple, short definitions by providing amplifying text that explains these terms in the context of the technologies that use them. The document also describes how these technologies interrelate and create new capabilities while at the same time providing mechanisms supportive of new spectrum management paradigms such as CRS and DSA. (IEEE 2008a).

The body of this standard is "normative" meaning that the material is a required part of the standard. Because of the complexity of the technology, the 1900.1 Standard provides amplifying material in "informative" annexes. Although these "informative" annexes are not a required part of the standard, these annexes are helpful to users of the standard in fully understanding the meaning of the "normative" definitions. A key annex is the annex that describes the relationship of terms such cognitive radio, software defined radio, software-controlled radio, intelligent radio, policy-based radio, and reconfigurable radio.

The IEEE 1900.1 Standard ™ defines *"cognitive radio"* (CR) and *"cognitive radio network"* but does not define *"cognitive radio system"* which is a term defined by the ITU-R and the US National Telecommunications and Information Administration and provided earlier in this chapter. (ITU 2009 and NTIA 2011). The IEEE and ITU-R definitions for CR and for CRS respectively are consistent however. The note in the IEEE definition for CR states the cognitive functionality typically associated with a cognitive radio is beyond the functionality normally associated with a *"radio."* The key is having a common understanding of what is meant by "radio."

The IEEE 1900.1 Standard ™ defines "radio" as being a "technology for wirelessly transmitting or receiving electromagnetic radiation to facilitate transfer of information." In this context, it is important to understand and differentiate between "radio" and "radio systems." Radio systems include software control of the radio proper. This radio control may include cognitive and adaptive features. The IEEE 1900.1 Standard ™ provides figures and text useful in the understanding of these aspects of cognitive radio, cognitive radio systems and radio control mechanisms.

Figure 4 is an illustration of a basic concept derived from the concepts provided in the IEEE 1900.1 Standard ™. The figure illustrates what is meant by software control of a software defined radio. A software defined radio is defined by IEEE 1900.1 Standard ™ as:

> *A type of radio in which some or all of the physical layer functions are software controlled. Software control is the use of software processing within the radio system or device to select the parameters of operation.*

The parameters of operation include the radio frequency (rf), the modulation, the transmitted power level, etc., i.e., the parameters that one typically associations with radio operation. In a software defined radio, some or all of these radio operational parameters may be implemented in software and may be changed by a real-time software control mechanism. The software control mechanisms may be very complex and may include cognitive and adaptive functionality that responds in real-time to a changing rf environment, changing operational policies, changing location, etc. More will be said in Section 5.1.5 about the policy-based radio control mechanism.

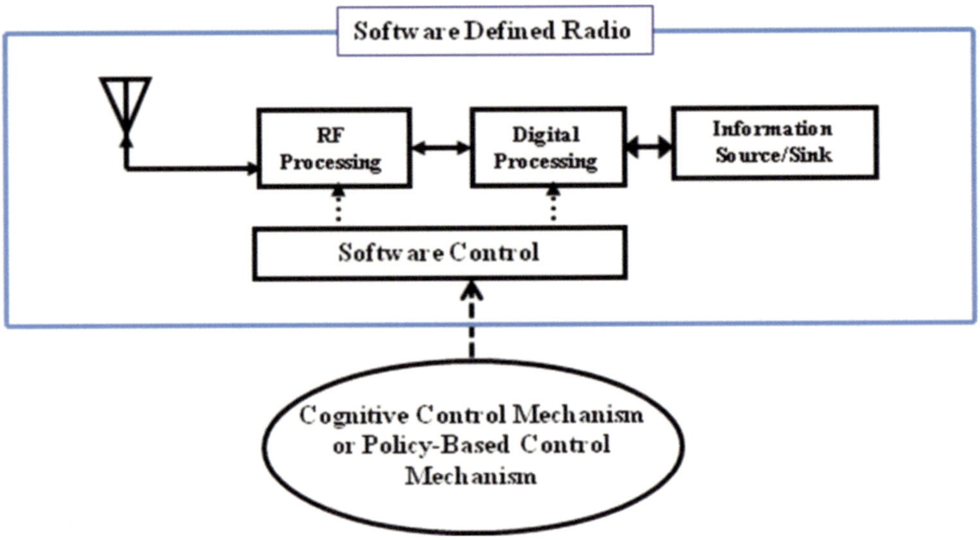

Fig. 4. Illustration of software control of software defined radio

5.1.2 IEEE 1900.2 Working Group

The IEEE 1900.2 WG has produced a standard which provides technical guidelines for analyzing the potential for coexistence or, by way of contrast, interference between radio systems operating in the same frequency band or between different frequency bands. (IEEE, 2008b). Additional work in this arena is needed and is likely to be conducted by one or more of the IEEE 1900.x WGs that operate under the DySPAN-SC.

5.1.3 IEEE 1900.3 Working Group

The IEEE 1900.3 WG was dissolved prior to the completion of any standard. It was chartered to produce a standard which would:

- specify techniques for testing and analysis to be used during compliance and evaluation of radio systems with dynamic spectrum access capability, and which would
- specify radio system design features that simplify the evaluation challenge.

The potential for deployment of CRS/DSA systems may be affected by the ability of regulatory agencies and industry stakeholders to verify that a system conforms to applicable technical and policy requirements. Thus, the future reactivation of the IEEE 1900.3 WG is needed. The initiation of the IEEE 1900.3 Project in 2006 was undoubtedly premature. The topic of what kind of new regulation, if any, is being vigorously debated within the ITU-R and within various national regulatory bodies. It can be reasonably expected that the IEEE 1900.3 WG activities will be resurrected as the ITU-R progresses its work on Agenda Item 1.19 to consider regulatory requirements for CRS and SDR and as national regulatory bodies take action on this critical issue.

5.1.4 IEEE 1900.4 Working Group

The IEEE 1900.4 WG has published a standard which defines the building blocks comprising

- network resource managers,
- device resource managers, and
- the information to be exchanged between the building blocks. (IEEE 2009)

This standard enables coordinated network-device distributed decision making which will aid in the optimization of radio resource usage, including spectrum access control, in heterogeneous wireless access networks. The standard is limited to the architectural and functional definitions at a first stage. The corresponding protocols definition related to the information exchange is now being addressed during a second stage of IEEE 1900.4 WG activities.

The purpose of the IEEE 1900.4 project is to improve overall composite capacity and quality of service of wireless systems in a multiple Radio Access Technologies (RATs) environment, by defining an appropriate system architecture and protocols which will facilitate the optimization of radio resource usage.

In addition to the baseline IEEE 1900.4 Standard ™, an amendment to this standard has recently been completed. (IEEE 2011a). This amendment modifies the IEEE 1900.4 standard to enable mobile wireless access service in white space frequency bands without any limitation on used radio interface (physical and media access control layers, carrier frequency, etc) by defining additional components of the IEEE 1900.4 system. This amendment is integrated into the baseline standard.

An additional project of the IEEE 1900.4 working group is the P1900.4.1 project which enhances the baseline standard by providing detailed specification of interfaces and service access points defined in the baseline standard, This enables distributed decision making in heterogeneous wireless networks and the use of context information for this decision making.

5.1.5 IEEE 1900.5 Working Group

The IEEE 1900.5 WG has produced a standard which defines a vendor-independent set of policy-based control architectures and corresponding policy language requirements for managing the functionality and behaviour of dynamic spectrum access networks. This standard also defines the relationship of policy language and architecture to the needs of at least the following constituencies: the regulator, the operator, the user, and the network equipment manufacturer. (IEEE 2012).

Figure 5 is a simplified user interpretation of the architecture provided in the IEEE 1900.5 Standard ™ for a policy-based DSA radio system (PBDRS) functional architecture. This figure may be viewed as an expansion of the more general architecture provided in Figure 4. The additional details are those involved in the software control mechanism and consist of the system strategy reasoner (SSR), the policy enforcer (PE) and the policy conformance reasoner (PCR). Essentially these components are designed to ensure compliance by the PBDRS to policies which may change depending on many parameters such as user requirements, rf environment, geographical location, etc. Some of these changes may be real-time; for example changes to the CRS operations due to changes in the rf environment. This type of capability is needed to reach the ultimate goal of more efficient use of the spectrum.

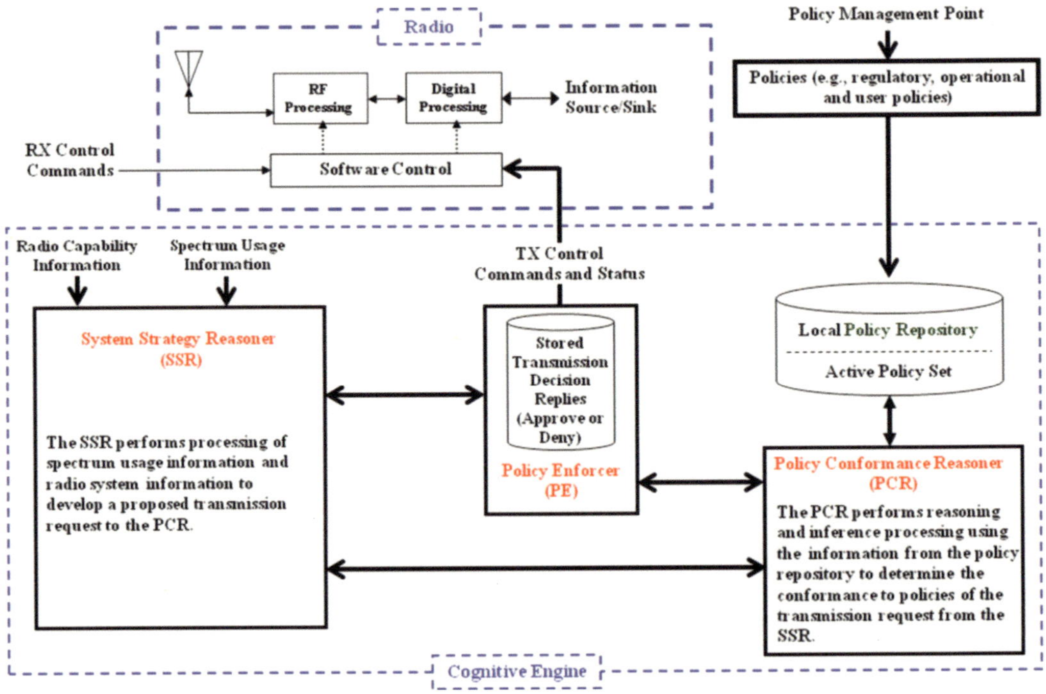

Fig. 5. Simplified interpretation of IEEE 1900.5 policy-based DSA radio system architecture.

The PCR, SSR and the PE represent functionalities that may be distributed, i.e., this functionality may not reside within a single radio device or even within a radio node. This is the reason for the emphasis on the system and network and not just the radio per se (meaning not just the receiver and transmitter aspect).

The policies that come from the policy management point (PMP) may be regulatory policies, operational policies or even user policies. The input to the PMP may be human readable policies that are converted into digital policies that are machine readable.

The IEEE 1900.5 Standard ™ (IEEE 2012) describes the communications between the SSR, PE and PCR which are shown in Figure 5; however the standard does not specify the details of these interfaces. One follow-on standard to the base standard will specify these interfaces in detail. The detailed specification of these interfaces may be helpful to regulatory agencies that will be addressing certification issues for these types of radios which can be viewed as a new paradigm. Information from the research community including ComSoc Technical Committees such as the Technical Committee on Cognitive Networks will be helpful in the development of this follow-on standard.

In addition to the functional architecture just described, the IEEE 1900.5 Standard ™ also describes requirements for a policy language to be used by the PCR. This language will be a declarative language. This type of language expresses what is to be accomplished rather than how it is to be accomplished. This is in contrast to languages such as C, C ++, and Java which are imperative or procedural languages that manipulate the state of the program through an executed sequence of commands specified by a control structure. A follow-on to the IEEE 1900.5 Standard ™ will provide a detailed specification of this policy language based on the requirements listed in the base standard.

5.1.6 IEEE 1900.6 Working Group

The IEEE 1900.6 WG has produced a standard which defines the information exchange between spectrum sensors and their clients in radio communication systems. The logical interface and supporting data structures used for information exchange are defined abstractly without constraining the sensing technology, client design, or data link between the sensor and client. The purpose of this standard is to define spectrum sensing interfaces and data structures for DSA systems and other advanced radio communications systems that will facilitate interoperability between independently developed devices and thus allow for separate evolution of spectrum sensors and other system functions. (IEEE 2011b).

The standard contains system models that allow for multiple sensors and for multiple cognitive engines or data archives. In other words, the sensor data acquisition can be distributed with multiple sensors, the sensor processing can be distributed with multiple cognitive engines, and the sensor data storage can be distributed with multiple data archives.

Sensing of the rf environment is a critical part of the original CRS/DSA concept. It is well recognized, however, that sensing of the rf environment is itself not always sufficient to have a realistic picture of the rf environment. For instance, sensing will not detect the usage of spectrum by some passive users of the spectrum such as radio astronomy. Therefore, sensing may need to be supplemented with other sources of information on the rf spectrum usage such as data bases. Chouinard (2010) discusses the practical limits to rf sensing and specifically notes that although it is a good approach to be considered in general for cognitive radio, it has been a challenge for us in the TV white space bands.

5.1.7 IEEE 1900.7 Working Group

In June 2011, the IEEE 1900.7 WG commenced work on a standard that specifies a radio interface including medium access control (MAC) sublayer(s) and physical (PHY) layer(s) of white space dynamic spectrum access radio systems supporting fixed and mobile operation in white space frequency bands, while avoiding causing harmful interference to incumbent users in these frequency bands. The standard provides supports the IEEE 1900.4 Standard ™, the IEEE 1900.4a Amendment (IEEE (2011a) for white space management, and the IEEE 1900.6 Standard ™ to obtain and exchange sensing related information (spectrum sensing and geolocation information).

The IEEE P1900.7 Standard will enable the development of cost-effective, multi-vendor white space dynamic spectrum access radio systems capable of interoperable operation in white space frequency bands on a non-interfering basis to incumbent users in these frequency bands. This standard facilitates a variety of applications, including the ones capable to support high mobility, both low-power and high-power, short-, medium, and long-range, and a variety of network topologies. This standard is a baseline standard for a family of other standards that are expected to be developed focusing on particular applications, regulatory domains, etc.

One result of TV broadcasting systems being upgraded from analog to digital technology is that it frees up channels under certain conditions (e.g., that the channel is not being used by licensed wireless microphones). Unlike the white space work in IEEE 802 which is discussed later in this chapter, the IEEE 1900.7 WG is not focused on the TV white space bands (TVWS). The intent of 1900.7 is the development of a family of standards that could be used for any portion of the spectrum having white space, i.e., spectrum that is not being utilized.

5.2 IEEE 802

The IEEE-802 Local Area Network/Metropolitan (LAN/MAN) Standards Committee (LMSC) develops LAN/MAN standards primarily for the lowest 2 layers of the ISO Reference Model for Open Systems Interconnection (OSI). The first meeting of the IEEE "Local Area Network Standards Committee", Project 802, was held in February of 1980. Like the DySPAN-SC, there is strong international participation in IEEE 802. Some IEEE 802 Standards are published as International Standards Organization (ISO) Standards. (IEEE 802, 2008). Additional general information about IEEE 802 may be found at: http://www.ieee802.org and http://standards.ieee.org/about/get/

Like the DYSPAN-SC, the primary work of IEEE 802 is performed in the IEEE 802 Working Groups. Only the IEEE 802 WGs that are involved in CRS/DSA standards are briefly discussed in the following subsections. Information on all of the IEEE 802 WGs may be found at: http://www.ieee802.org/dots.shtml

The work of the 802 WGs is a practical example of the concepts illustrated in Figure 1 regarding relationships between standards development, technology, and regulatory policies. The work of these WGs is a direct result of decisions made by regulatory agencies regarding TV white space. (FCC 2004, FCC 2008, FCC 2011). Clearly there are business case questions related to this work - new LAN business case potential for the IEEE 802.11af WG and new WRAN business cases for the IEEE 802.22 WG for example.

The CRS/DSA activity within IEEE 802 is the direct result of regulatory actions taken by the US Federal Communications Agency (FCC) and other national regulatory bodies. In the US, this started in 2002 by the FCC issuance of a Notice of Inquiry (NOI) to explore the possibility of allowing access to the TV broadcast bands for license-exempt devices on a non-interfereing basis for operation in VHF and UHF TV bands. (FCC, 2004). The IEEE 802 views this FCC action and the resulting decisions (FCC, 2008 and FCC 2011) as a potential business oportunity to:

- enhance IEEE 802.11 LAN functionality
- use of the TV white space bands for rural area broadband coverage using the IEEE 802.22 Wireless Regional Area Networks.

These IEEE 802 standards development efforts are discussed briefly in the following subsections.

5.2.1 IEEE 802.11 Wireless LAN Working Group

The IEEE 802.11 WG is well known for the family of IEEE 802.11 Standards. http://www.ieee802.org/11/. In December 2009, a Project Approval Request (PAR) was approved by the IEEE SASB to develop a new standard that will be part of the IEEE 802.11 family of standards. The title of this proposed standard is:

> *"Standard for Information Technology - Telecommunications and Information Exchange Between Systems - Local and Metropolitan Area Networks - Specific Requirements - Part 11: Wireless LAN Medium Access Control (MAC) and Physical Layer (PHY) Specifications; Amendment: TV White Spaces Operation."*

This standard will be known as IEEE 802.11af. The main difference between this standard and other members of the well known IEEE 802.11 family of standards is that IEEE802.11af will be a based on cognitive radio for operation in the TV White Spaces. The TV White Spaces is spectrum already allocated to the TV broadcasters but is not being used. The only additional prerequisite for IEEE 802.11af in the US, is the need for a database that maintains data about used frequencies in the TV band.

With the global transition to digital TV, sub-Gigahertz rf spectrum is becoming available. Much of this spectrum is for unlicensed, license exempt and/or lightly licensed use. The IEEE 802.11af standard will make the necessary MAC and PHY changes to enable 802.11 products to take advantage of this additional spectrum. The purpose of the IEEE 802.11af standard is to allow 802.11 wireless networks to be used in the TV white space. The scheduled completion date for the IEEE 802.11af Standard is during 2013. http://www.ieee802.org/11/Reports/802.11_Timelines.htm

5.2.2 IEEE 802.16 Broadband Wireless Access WG

The IEEE 802.16 Working Group on Broadband Wireless Access Standards develops standards and recommended practices to support the development and deployment of broadband Wireless Metropolitan Area Networks. This WG does not have projects that specifically develop CRS/DSA standards. However CRS/DSA technology may be incorporated into the standards developed by the IEEE 802.16 WG such as the wireless metropolitan area network (MAN) air interface which has been approved by ITU-R Working Party 5D as an IMT Advanced (Internal Mobile Telecommunications) air interface. http://ieee802.org/16/index.html

5.2.3 IEEE 802.18 – Radio Regulatory Technical Advisory Group

The IEEE 802.18 Radio Regulatory Technical Advisory Group (RR-TAG) supports the work of the IEEE 802 LMSC and the IEEE 802 WGs including IEEE 802.11 (WLAN), IEEE 802.15 (WPAN), IEEE 802.16 (WMAN), IEEE 802.20 (Wireless Mobility), IEEE 802.21 (Handoff/Interoperability Between Networks), and IEEE 802.22 (WRAN). This support includes actively monitoring and participating in radio regulatory matters worldwide as an advocate for IEEE 802. One example is representing IEEE in the ITU-R.

http://ieee802.org/16/index.html

5.2.4 IEEE 802.19 – Wireless Coexistence WG

The IEEE 1900.1 WG

- develops standards for coexistence between wireless standards of unlicensed devices, and
- reviews coexistence assurance (CA) documents produced by other IEEE 802 WGs developing new wireless standards for unlicensed devices.

The IEEE 802.19.1 subgroup is in the process of developing a standard entitled:

> *Standard for Information Technology - Telecommunications and Information Exchange Between Systems - Local and Metropolitan Area Networks – Specific Requirements - Part 19: TV White Space Coexistence Methods*

The purpose of the standard is to enable the family of IEEE 802 Wireless Standards to most effectively use TV White Space by providing standard coexistence methods among dissimilar or independently operated TV band devices (TVBD) networks and dissimilar TVBDs. This standard addresses coexistence for IEEE 802 networks and devices and will also be useful for non IEEE 802 networks and TVBDs. Completion in 2012.

http://ieee802.org/19/index.html

5.2.5 IEEE 802.22 – Wireless Regional Area Networks WG

The IEEE 802.22 Working Group on Wireless Regional Area Networks (WRANs) is responsible for developing standards for a cognitive radio-based PHY/MAC/air interface for use by license-exempt devices on a non-interfering basis in spectrum that is allocated to the TV Broadcast Service.

The IEEE 802.22 WRAN Standard uses cognitive radio techniques to allow sharing of geographically unused spectrum allocated to the television broadcast service, on a non-interfering basis. The goal is to bring broadband access to hard-to-reach, low-population-density areas which are typical of rural environments. A particularly difficult problem is that of ensuring that there is no interference to low-powered licensed devices such as wireless microphones.

IEEE 802.22 Provides Three Mechanisms for Incumbent Protection:

- Sensing
- Database Access
- Specially Designed Beacon

One or more protection mechanisms can be adopted based on the regulatory domain requirements.

Stevenson, et.al (2009) and Mody and Chouinard (2010) provide overviews of the IEEE 802.22 standards work.

Projects have been approved by IEEE SASB to develop the following standards:

- IEEE 802.22: Standard for Information Technology -Telecommunications and information exchange between systems - Wireless Regional Area Networks (WRAN) - Specific requirements - Part 22: Cognitive Wireless RAN Medium Access Control

(MAC) and Physical Layer (PHY) specifications: Policies and procedures for operation in the TV Bands

- IEEE 802.22.1: Standard to enhance harmful interference protection for low power licensed devices operating in TV Broadcast Bands
- IEEE 802.22.2: Information Technology - Telecommunications and information exchange between systems – Wireless Regional Area Networks (WRAN) - Specific requirements - Part 22.2: Recommended Practice for the Installation and Deployment of IEEE 802.22 Systems
- IEEE 802.22a: Amendment to IEEE Standard for Wireless Regional Area Networks - Part 22: Cognitive Wireless RAN Medium Access Control (MAC) and Physical Layer (PHY) specifications: Policies and procedures for operation in the TV Bands. - Management and Control Plane Interfaces and Procedures and Management Information Base Enhancements.

The IEEE 802.22 baseline standard and IEEE 802.22.1 standard have been published. (IEEE 802.22, 2010) (IEEE 802.22, 2011).

6. Other activities related to standards for CRS and DSA

There are numerous other organizations that are involved either directly or indirectly in the development of national and international standards for CRS, DSA and related technologies. It is not possible in this short chapter to discuss all of these organizations. One source that provides a list of standards setting organizations (SSOs) is ConsortiumInfo.org. For example, information on SSOs for telecommunications may be found at:

http://www.consortiuminfo.org/links/linkscats.php?ID=26

It is important to realize that some standards such as those produced by the IEEE and ITU-R are not implemented directly. Instead, numerous alliances have been created to develop more specific documents that are based on what comes out of the IEEE. Examples of this are:

- The WiFi Alliance creates specifications based on IEEE 802.11 Standard.
- The WiMAX Forum creates specifications based on the IEEE 802.16 Standard
- The WRAN Alliance creates specifications based on the IEEE 802.22 Standard and other standards.
- The 3GPP and other organizations create specifications that are responsive to ITU-R Recommendations. These specifications are then adopted by the ITU-R.

The following subsections briefly describe some other organizations involved in CRS/DSA and related technologies. The list is not considered to be all inclusive.

6.1 European Telecommunications Standards Institute

The European Telecommunications Standards Institute (ETSI) produces globally-applicable standards for information and communications technologies (ICT), including fixed and mobile telecommunications and internet technologies. ETSI is officially recognized by the European Union as a European Standards Organization. ETSI has more than 700 member organizations from 62 countries in 5 continents.

More information made be found at: http://www.etsi.org/WebSite/homepage.aspx

Mueck (2010a) provides a summary of ETSI work on reconfigurable radio systems (RRS), SDR and CR standards and the European Regulatory Framework. The cognitive radio efforts within EYSI are focused on

- a cognitive pilot radio channel, and
- a functional architecture for management and control of RRS.

The ETSI RRS Technical Committee (TC) develops studies and standards complementary to the IEEE DySPAN-SC and IEEE 802 standards development activities. The focus of the ETSI is on:

- SDR standards
- CR/SDR standards addressing the specific needs of the European Regulatory Framework,
- CR/SDR TV white space standards adapted to the digital TV signal characteristics in Europe.

The ETSI RRS TC comprises four WGs (Mueck, 2010a).:

- WG1 focuses on system aspects and develops proposals from a system aspects point of view for a common framework in RRS TC.
- WG2 focuses on SDR technology with a particular interest in radio equipment architecture and proposes common reference architectures for SDR/CR radio equipment.
- WG3 focuses on cognitive management and control including a functional architecture for radio resource management and a cognitive pilot channel.
- WG4 focuses on public safety and collects and defines the related RRS requirements from relevant stakeholders in the public safety and defense domain.

These ETCI RRS TCs have completed numerous studies regarding CRS, SDR, and CR as the result of several Europe Union funded programs. One focus now in ETSI RRS TC is on utilizing the results of these research studies in developing appropriate normative standards. Mueck (2010b) provides additional information.

6.2 Ecma International

Ecma International was originally founded as the European Computer Manufacturers Association; however the name is now officially Ecma International without reference to the original ECMA acronym. Since 1961 Ecma International and its predecessor organization have developed international standards in information technology and telecommunications. More than 370 Ecma Standards and numerous Technical Reports have been published, many of which have also been adopted as international standards and/or technical reports.

Ecma International has published a standard on medium access control layer (MAC) and physical layer (PHY) specifications for operation in TV white space. (Ecma, 2009). This standard specifies MAC and PHY for personal/portable cognitive wireless networks operating in TV bands. The standard also provides specifications for higher layer protocols. It specifies a number of incumbent protection mechanisms available to meet regulatory requirements.

More information may be found at:
http://www.ecma-international.org/activities/index.html

6.3 IEEE ComSoc technical committees

The IEEE Communications Society has more than 24 technical committees. These technical committees define and implement the technical directions of ComSoc. All ComSoc Members are encouraged to participate in one or more of these technical committees. These committees usually meet twice a year at major conferences. Throughout the year, these committees also play a major role in determining which events (conferences, workshops, etc.) are technically co-sponsored by ComSoc.

The ComSoc Technical Committees may be tasked by ComSoc Working Groups to provide assistance on specific technical issues that may arise during the development of a standard being developed under ComSoc Standards Board sponsorship. Of particular interest to standards activities responsible for standards on CRS/DSA are the Cognitive Networks, Radio, and Wireless Technical Committees.

More information may be found at: http://www.comsoc.org/about/committees/Technical

6.4 Wireless Innovation Forum

The Wireless Innovation Forum™ (WInF) is a non-profit corporation whose predecessor organization was started in 1996. Formerly known as the SDR Forum, the WInF is dedicated to driving technology innovation in commercial, civil, and defense communications worldwide. The focus of the WInF is on SDR, CR, and DSA. The Forum has produced many technical reports on these topics. The Forum is not an internationally accredited standards development organization. However, the technical reports developed by the Forum have been input to standards organizations such as IEEE DySPAN-SC and ITU-R. For example, the Forum contributed many inputs to the development of the IEEE 1900.1 Standard ™.

More information on the Wireless Information Forum may be found at:
http://www.wirelessinnovation.org/page/About_the_Forum

6.5 WRAN Alliance

The Wireless Regional Area Network (WRAN) Alliance promotes products and services based on wireless telecommunications standards for broadband services using TV band frequencies.

The WRAN Alliance has the objective of promoting standards that will permit the provision of broadband connectivity at reasonable cost in rural areas and in developing areas. This is viewed by the WRAN Alliance as a means bridging the digital divide.

The WRAN Alliance is initially focusing on promoting the use of the IEEE 802.22 Standard on Wireless Regional Area Networks, in variety of different regulatory domains and for a variety of different use cases. Use cases include but are not limited to the following:

- broadband access over large distances,
- broadband access for remote and rural areas,

- long-range backhaul,
- smart grid and critical infrastructure monitoring,
- defense,
- homeland security,
- healthcare, and
- small office.

More information on WRAN Alliance may be found at: www.wranalliance.org/

6.6 Wireless World Research Forum

The objective of the Wireless World Research Forum (WWRF) is to formulate visions on strategic future research directions in the wireless field, among industry and academia, and to generate, identify, and promote research areas and technical trends for mobile and wireless system technologies. The WWRF is not an internationally accredited standards development organization. However, the results of the WWRF research have significant impact on the direction of standards development in many international standards development organizations. The WWRF has one group focused on cognitive networks for wireless internet and another on spectrum issues.

More information may be found at: http://www.wireless-world-research.org/

7. Conclusion and future work

This chapter has provided a roadmap to international standardization activities for cognitive radio systems, DSA systems and related technologies. Numerous standards have already been published by several standards bodies. In addition, there are many technical reports related to standards that have developed and are continuing to be published by a variety of organizations such as the ITU-R, WWRF, ETSI, and the Wireless Innovation Forum. Although these technical reports are not standards per se, they nevertheless provide the foundation for future standards development.

To a certain extent this roadmap to CRS/DSA standards development could be considered to be a snapshot in time. However, the information on organizations and standards provided herein is essentially a description of the foundations of the CRS/DSA standardization effort. It is anticipated that all of the organizations mentioned herein will still be active in CRS/DSA standardization for at least the next decade. Furthermore, it is anticipated that the specification standards mentioned herein are in most cases baseline standards that are the foundation for work that will be ongoing for the next decade or more.

The CRS/DSA standardization effort is very complex because ultimate success of this technology is dependent on the efforts of the regulatory, technical research, standards development, and business communities. Although these communities are all part of the effort for ultimately successfully fielding these technologies, these communities work independently.

A driving force is the need to more efficiently utilize the rf spectrum. Thus, additional regulatory action may be needed to free up more spectrum for CRS and DSA -- the FCC rulings related to TV white space provide an example of what may be forthcoming. But clearly, these types of rulings can be controversial. The research community can come up with technology proposals for better utilization of the spectrum, but it must be within regulatory boundaries. Standards development organizations can develop new standards that incorporate results from the research community, but these standards also must stay within the regulatory boundaries. But the regulatory boundaries can be changed depending on the influence of industry on regulators - this influence is the result of business cases for the use of spectrum.

Finally, it can be expected that the ITU-R, IEEE DySPAN-SC, IEEE 802 and many other organizations identified in this chapter will continue their standardization efforts for at least another decade. Regulatory boundaries, technology, business requirements will all evolve during this period of time and will be driven by the need to more efficiently utilize the spectrum.

8. References

Chouinard, G. (2010). RF Sensing in the TV White Space, 8th Conference on Communications Networks and Services Research (CNSR2010) held at McGill University, 12-14 May 2010.

Ecma (2009). MAC and PHY for Operation in TV White Space, ECMA-392, available at: http://www.ecma-international.org/publications/standards/Ecma-392.htm

FCC (2004). Notice of Proposed Rule Making, In the Matter of Unlicensied Operation in the TV Broadcast Bands, US FCC Docket 04-186, May 25, 2004.

FCC (2010). Second Report and Order and Memorandum or Opinion and Order , in the Matter of Unlicensed Operation in the TV Broadcast Bands Additional Spectrum for Unlicensed Devices Below 900 MHz and in the 3 GHz Band, Oct. 18, 2006.

FCC (2011). Unlicensed Operation in the TV Boradcast Bands, Additional Spectrum for Unlicnsed Devices Below 900 MHz and in the 3 Ghz Band, FCC DA 11-131, January 26, 2011.

Granelli, F., Pawelczak, P., Prasad, V., Subbalakshmi, K., Chandramouli, R., Hoffmeyer, J., and Berger, H., (2010). Standardization and Research in Cognitive and Dynamic Spectrum Access Networks: IEEE SCC41 Efforts and Other Activities, *IEEE Comm. Mag.*, January 2010.

IEEE (2008a). IEEE Standard Definitions and Concepts for Dynamic Spectrum Access: Terminology Relating to Emerging Wireless Networks, System Functionality, and Spectrum Management.

IEEE (2008b). IEEE Recommended Practice for the Analysis of In-Band and Adjacent Band Interference and Coexistence Between Radio Systems.

IEEE (2009). IEEE Standard for Architectural Building Blocks Enabling Network-Device Distributed Decision Making for Optimized Radio Resource Usage in Heterogeneous Wireless Access Networks.

IEEE (2011a). 1900.4a, Standard for Architectural Building Blocks Enabling Network-Device Distributed Decision Making for Optimized Radio Resource Usage in Heterogeneous Wireless Access Networks - Amendment: Architecture and Interfaces for Dynamic Spectrum Access Networks in White Space Frequency Bands, 16 September 2011.

IEEE (2011b). IEEE Standard for Spectrum Sensing Interfaces and Data Structures for Dynamic Spectrum Access and Other Advanced Radio Systems.

IEEE (2012). IEEE Standard Policy Language Requirements and System Architectures for Dynamic Spectrum Access Systems. (Approved for final publication in January 2012).

IEEE 802 (2008) Overview and Guide to the IEEE 802 LMSC http://www.ieee802.org/IEEE-802-LMSC-Overview-and-Guide-01.pdf

IEEE 802.22 (2010), IEEE Standard for Informatin Technology - Telecommunications and information exchange between systems – Local and metropolitan area networks – Specific requirements: Part 22.1 Standard to Enhance Harmful Interference Protection for Low-Power Licensed Device Operation in the TV Broadcast Bands, *IEEE Std 802.22.1TM-2010*, 1 November 2010

IEEE 802.22 (2011). Standard for Information Technology -Telecommunications and information exchange between systems - Wireless Regional Area Networks (WRAN) - Specific requirements - Part 22: Cognitive Wireless RAN Medium Access Control (MAC) and Physical Layer (PHY) specifications: Policies and procedures for operation in the TV Bands, July 2011

ITU-R (2007a). WRC-12 Agenda Item 1.19: Software Defined Radio (SDR) and Cognitive Radio Systems (CRS) http://www.itu.int/ITU-R/information/promotion/e-flash/4/article5.html

ITU-R (2007b) RESOLUTION ITU-R 2-5, Conference Preparatory Meeting, (1993-1995-1997-2000-2003-2007)

ITU-R (2008). International Telecommunication Union Radio Regulations, http://www.itu.int/pub/R-REG-RR/en

ITU-R (2009). Definitions of Software Defined Radio (SDR) and Cognitive Radio System (CRS), ITU-R Recommendation SM 2152

ITU-R (2010). ITU-R Study Groups, http://www.itu.int/dms_pub/itu-r/oth/0A/0E/R0A0E0000010001PDFE.pdf

ITU-R (2011). List of QUESTIONS assigned to ITU-R Study Group 5, http://www.itu.int/pub/R-QUE-SG05/en

Mitola, J. (20000), An Integrated Agent Architecture for Software Defined Radio, Dissertation Royal Institue of Technology, Sweden, ISSN 1403-5286, 8 May 2000.

Mody, A. And Chouinard, M. (2010). Enabling Rural Broadband Wireless Acess Using Cognitve Radio Technology, IEEE 802 Doc. 802.22-10/0073r03, available at: http://www.ieee802.org/22/Technology/22-10-0073-03-0000-802-22-overview-and-core-technologies.pdf

Mueck, M. (2010a). ETSI Reconfigurable Radio Systems: Status and Future Directions on Software Defined Radio and Cognitive Radio Standards. IEEE Comm. Mag., September 2010.

Mueck, M. (2010b). TC RRS Activity Report 2010. http://portal.etsi.org/rrs/activityreport2010.asp

NTIA (2011). Manual of Regulations and Procedures for Federal Radio Frequency Management (Redbook), US Government Printing Office Bookstore, Stock Number 903-008-00000-8 ISBN 0-16-016464-8

Prasad, V., Pawelczak, P., and Hoffmeyer, J., (2008). Cognitive Functionality in Next Generation Wireless Networks: Standardization Efforts, *IEEE Comm. Mag.*, April 2008

Sherman, M., Mody, A., Martinez, R., Rodriguez, C., Reddy, R. (2008). IEEE Standards Supporting Cognitive Radioi and Dyanmic Spectrum Access, *IEEE Comm. Mag.*, July 2008..

Stanislav, R., Harada, H., Murakami, H. And Ishizu, K. (2011), International Standardization of Cognitive Radio Systems, *IEEE Comm. Mag.*, March 2011.

Stevenson, C. , Chouinard, G., Lei, Z,, Hu, W, Shellhammer, S., Caldwell, W., (2009). IEEE 802.22: The First Cognitve Radio Wireless Regional Area Network Standard, *IEEE Comm. Mag.*, January 2009. http://www.inf.ufrgs.br/~cbboth/802_22commag.pdf